Ecosystem Dynamics

Ecosystem Dynamics

From the Past to the Future

Richard H.W. Bradshaw
University of Liverpool

Martin T. Sykes
Lund University

WILEY Blackwell

Contents

Acknowledgements

This book has developed from the academic journeys that we have both enjoyed from the 1970s until today. Between us we have held various posts in academia in Denmark (R.B.), Ireland (R.B.), Sweden (R.B., M.T.S.), New Zealand (M.T.S.), the United Kingdom (R.B., M.T.S.) and the USA (R.B.), and both colleagues and experiences from all these countries have influenced the content of the book. R.B. proffers special thanks to John Birks, Colin Prentice and the other occupants of Room 28, Tom Webb III and Herb Wright, who all helped awaken my interest in palaeoecology. Many friends from the Faculty of Forest Sciences, SLU, Sweden combined to teach me both Swedish and Forest Science. Tack Olle Zackrisson och Pelle Gemmel for creating this opportunity. I also thanks special friends at the short-lived but exciting Department of Environmental History and Climate Change, GEUS, Denmark, where Bent Odgaard, the two Peters (Rasmussen, Friis Møller) and Anne Birgitte Nielsen, among others, showed me how Denmark was a true cradle of paleoecology and worthy of respect, despite being a small, flat country with little natural vegetation! In Sweden and Denmark, I led or participated in seven EU-funded projects, which generated important parts of the material covered here. A special thanks to Thomas Giesecke and other European Pollen Database colleagues for many of the good ideas in Chapter 4 and for introducing Teutonic rigour to my thinking.

M.T.S. would like to especially thank three people who provided significant opportunities: Bastow Wilson from Otago, New Zealand, who gave an anxious mature student the opportunity to do a PhD; Eddy van der Maarel from Uppsala, who provided the first postdoctoral position in 1989, albeit initially for 5 months, which provided a vital stepping stone to more than 24 years in Sweden; and Colin Prentice, then in Uppsala, who provided the opportunity to integrate my earlier computing background and ecology training into forest gap modelling and much more. I would also like to acknowledge the valuable discussions and informal contributions made by many colleagues, particularly those involved in various European-scale projects over a number of years, including ALARM, ATEAM, CLIMIT, CLIMSAVE, DECVEG, ECOCHANGE, EPIDEMIE, FIREMAN, MACIS, RUBICODE and others. I also thank various Lund scientists, including Almut Arneth, Dörte Lehsten, Veiko Lehsten, Paul Miller, Honor Prentice, Jonathan Seaquist, Ben Smith and others from the Departments of Physical Geography and Ecosystem Science and Ecology, as well as past PhD students, especially Thomas Hickler and Marie Vandewalle, who all gave support in myriad ways.

R.B. also acknowledges support from M.T.S.'s Lund colleagues while writing the book and thanks his old Lundian friends on the other side of the door in Kvartärgeologi, who were always ready for coffee and conversation. A special thanks to the DYNAMITE team for providing distraction from the book and good field discussions, plus some exceptional food and drink.

We both acknowledge the significant contribution Margaret P. Sykes made as a facilitator in a number of meetings between the authors, where much of the structure of the book was developed. She also read all chapters, more than once, and provided valuable editing, comments and corrections.

We both thank Sandra Mather for her cheery temperament and skilled help with figure production, even when all of the figures arrived at the last minute. Thanks also to readers Anne Birgitte Nielsen, Tom Webb III, Gina Hannon, Abigail G. Sykes and Julian M. Sykes-Persson for many useful comments and to Louise Bradshaw for help with the references. Krister Larsson, Thomas Giesecke, Gina Hannon, Peter Rasmussen and Jennifer Clear all provided photos or figures. A special thanks to Jean Clottes for the wonderful lions from the Chauvet Cave.

M.T.S. thanks the staff at the Geolibrary in the Geocentrum, Lund, in particular Rolf Hall and Robin Gullstrand, who responded with enthusiasm to his requests for books and articles.

We acknowledge with thanks financial support from:

- The EU FP6 ERA BiodivERsA – FIREMAN (Fire Management to Maintain Biodiversity and Mitigate Economic Loss) 2009–2013, funded in Sweden by FORMAS and in the UK by NERC.
- LUCCI (Lund University Centre for Carbon Cycle and Climate Studies), financed by a Linnaeus grant from the Swedish Research Council VR, who financed R.B. for a visiting professorship in Lund.
- LUCID (Lund University Centre of Excellence for Integration of Social and Natural Dimensions of Sustainability), financed by a Linnaeus grant from the Swedish Research Council VR, who provided some funding to M.T.S.
- STINT (The Swedish Foundation for International Cooperation in Research and Higher Education), who supported Lund–Liverpool exchange, including several book activities through project DYNAMITE.
- The Hasselblad Foundation, who awarded M.T.S. the 2012 Natural Scientist writing stipend, which allowed a 2 month visit to the Hôtel Chevillon, Grez-sur-Loing, France in November and December 2012 in the final phase of the writing.
- Dr Joel Guiot, Directeur de la fédération de recherché, CEREGE, Europole Mediterranéen de l'Arbois, Aix-en-Provence, France, who provided working space and fruitful discussions for 3 months in Spring 2011 to M.T.S. as visiting professor in the critical start-up phase of the book.

Finally, both authors would like to acknowledge the love and support of their families, without whom it would have been a whole lot tougher.

About the Companion Website

This book is accompanied by a companion website:

www.wiley.com/go/bradshaw/sykes/ecosystem

The website includes:

- Powerpoints of all figures from the book for downloading
- PDFs of tables from the book

1

Where Are We and How Did We Arrive Here?

'I could calculate your chance of survival, but you won't like it.'
The Hitchhiker's Guide to the Galaxy, *Douglas Adams, 1979*

1.1 Why this book?

In January 2013 the Australian Bureau of Meteorology had to increase the temperature range of its standard weather forecasting chart by 4 °C to a maximum of 54 °C, adding deep purple and pink to its colour palette. The new colours were put into immediate action as old temperature records toppled in the latest heatwave, following what the government's Climate Commission called an 'angry summer'. This is just one symptom of change that will influence global ecosystems, and it is important for all whose livelihoods and food are linked to the land to know what these impacts will be. Providing projections of the impacts is a challenge that we can meet either by studying records of past warmer periods or by using models to forecast the future. These two approaches are linked, as forecasting models have to build on knowledge and experience from the past. This link between models and data is one of the motivations and central themes of our book.

This book is about the long-term dynamics of the terrestrial ecosystems of the Earth. These ecosystems only cover 29% of the surface of the planet, but that is where we live, produce much of our food and gather most of our raw materials. The oceans and freshwaters of the world are also of vital importance for civilisations past and present, but in this book we concentrate on the dry land systems. We need to understand the changes taking place around us in order to be able to manage and exploit ecosystems in appropriate ways in the future. To do this we must have adequate descriptions, both of the system dynamics and of the forces that cause or shape these dynamics. The current state of many ecosystems is a consequence of dynamics, forces and events that have operated over very long periods; the timespan we cover stretches back over 20 000 years, to the coldest part of the last ice age, and reaches forward 100 years into the future. There are several factors that influence ecosystem dynamics but the most important are climate change, human impact and the physiological constraints of individual species. Specialised geological techniques are needed to explore the past, and modelling carries our analyses into the future. Our

Ecosystem Dynamics: From the Past to the Future, First Edition.
Richard H.W. Bradshaw and Martin T. Sykes.
© 2014 John Wiley & Sons, Ltd. Published 2014 by John Wiley & Sons, Ltd.
Companion Website: www.wiley.com/go/bradshaw/sykes/ecosystem

combination of data and modelling helps us understand how we arrived at the present state of the world and where we might be headed.

We two authors have expertise in long-term ecosystem dynamics that result in population and range changes of individual species. We both have a botanical background, so we use more plant than animal examples, although humans are the species most often mentioned. For long-lived trees, we examine both rapid events like forest fires and slow events like the range changes that occurred as Europe became revegetated after the last ice age. The ecosystem concept comprises species interactions with soils, water and the atmosphere, but inevitably our treatment reflects our own interests and experience and may be uneven. As biologists we are more conversant with the biotic components and processes than the abiotic, and there is more coverage of large plants and animals than of microorganisms. Humans are at the centre of this book. We analyse the development through time of the way people interact with the ecosystems that they have come to dominate. We write little about the present day as that is well covered in numerous other sources.

Models are another central topic in the book. They are sophisticated tools for integrating our knowledge of the Earth system and exploring the future. Models have a symbiotic relationship with data, which we examine in this book. Models draw on data during their construction and must be tested against yet more data. Models are not real life but can be used as experimental tools to explore the nature and relationships of the systems under study to generate future scenarios and even forecasts. They give outputs that may or may not be correct but can be assessed for validity against available data, for example through hindcasting: the comparison of known past events with model output. Hindcasting can increase understanding of past ecosystems and boost confidence in the explanatory power of models, and it is one important focus of this book. The overall aim of ecosystem modelling is to improve insight into and understanding of the complex interactions within an ecosystem, such as the responses to past or future variations in climate. Models can generate insight but do not necessarily provide definite answers, because they are, after all, only models. Once the model output has been validated against data, the model becomes a more effective and convincing tool with which to explore possible futures.

1.2   Ecosystems in crisis

The human race has now moved into the driving seat of all terrestrial ecosystems and the control panel is complex. There is no owner's manual and several systems are already careering out of control. There is an urgent need to understand these controls and to use our power wisely. This book provides the background information needed to ensure a long, sustainable relationship between planet Earth and its new managers and prepares the ground for writing that owner's manual. The control panel has warning lights and touch controls that alter land cover, emission of gases, hydrology, soil properties, genetic diversity and several other factors. Guidelines are needed for the appropriate settings and there is some urgency as several warning lights are flashing, including those labelled 'greenhouse gas emissions', 'rapid climate change', 'biodiversity loss', 'food security' and 'water supply'. These warning lights show that the resilience of several ecosystems is being put to the test.

There are enormous issues at stake, including the future of ecosystem goods and services such as agriculture and silviculture, water and soil resources and the carbon and nitrogen cycles. There is active debate about how many people can survive on the planet in the longer term. Can the Earth support 12 000 000 000 people or are we threatened by severe and painful population reductions, such as occurred in the distant past? What are the prospects for the long-term survival of the human race? The term Anthropocene has been introduced to describe the last 200 years, in which 'our societies have become a global geophysical force', a process that has accelerated during the last 35 years (Steffen *et al.*, 2007).

It is not easy to put a precise date on when humans took over control of the planet from natural forces such as competition between species, natural selection, fire, weathering of rock and hydrological cycles, but the 200 years of the Anthropocene is one convenient estimate as good data exist from that period, although others have argued for the first millennium AD. Humans today move an order of magnitude more rock, soil and sediment during construction and agriculture than the sum of all other natural processes that operate on the surface of the planet put together. If the erosion of rock and soil caused by construction and agriculture were evenly distributed over ice-free continental surfaces, these human activities would now lower land surface by a few hundred metres per million years, as compared with an earlier estimated natural rate of a few tens of metres per million years (Wilkinson, 2005). We are also exploiting many of the same ecosystem processes that were operating in the past, but our exploitation is now so intensive that we have considerably amplified or modified their rates, properties and effects. Fire is one ecosystem process that has become totally altered through its deliberate use in agriculture and its suppression to protect forest resources, as well as the manipulation of fuels. The management of grazing animals and the selection of genotypes of plants and animals that are favourable for us are further examples of the ways in which we have modified ecosystem processes and properties. Our owner's manual draws on past experience to propose appropriate uses of the ecosystem controls.

Human society faces several developing crises in ecosystem services that are making warning lights flash on our control panel. The global economy is almost five times the size it was half a century ago and such a rapid increase has no historical precedent (Jackson, 2009). The associated increase in use of finite natural resources and management of increased land areas has led to rapid conversion of terrestrial biomes into agricultural land, plantations, wasteland and cities, with consequent loss of species and modification of ecosystem services. The Millennium Ecosystem Assessment (MEA) has identified those services that are rapidly degrading, which include freshwater, wild foods, wood fuel, soil volume and quality, genetic resources and natural hazard regulation by wetlands and mangroves (www.maweb.org). Terrestrial ecosystems provide key components of the natural capital and services that fuel much current economic growth. They probably always fulfilled this function for hunter-gatherer societies and subsequent civilisations, but never on the scale posed by current demands. Agriculture lies behind much ecosystem transformation and is a contributory factor to major environmental concerns, including loss of biodiversity, overexploitation of freshwater, soil degradation and even climate change. Yet about one in seven people are chronically malnourished (Foley et al., 2011). The amount of land dedicated to cereal production per person has been reduced from 0.23 ha in 1950 to 0.1 ha in 2007, increasing the challenge involved in feeding the growing global population. There is also growing competition between individual services as terrestrial ecosystems become more heavily exploited. Most of the Earth's surface that is suitable for arable agriculture is now utilised, and maize, rice and wheat provide over 30% of their essential daily food to more than 4.5 billion people (Shiferaw et al., 2011). However, competing demands for the use of maize, both as animal feed and for its conversion into bioethanol for fuel, have driven up prices in an alarming manner, with significant social consequences.

It is sobering to consider the global development of population size since AD 1750 and its consequences , which include demands on some major ecosystem services such as water use and fisheries (Figure 1.1; Steffen et al., 2004, cited in Dearing et al., 2010). No complex analysis is needed to understand that as the population curve climbs, reduction of tropical forest area, number of species extinctions and loss of sustainable fish stocks are likely to follow, along with global gross domestic product (GDP) and other economic indicators, such as the number of vehicles on the road (Figure 1.1).

All around us we can see the ecosystem consequences of the rapidly increasing human population, which is coupled to the demands we make on our environment and fuelled by

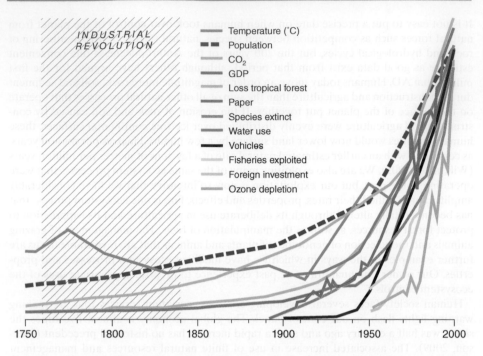

Figure 1.1 Changes in global states and processes since AD 1750 – including ecosystem services, climate variables and economic data – all show acceleration in rates from the mid-twentieth century (after Steffen *et al.*, 2004, Dearing *et al.*, 2010 and Ehrlich *et al.*, 2012)

technologies and energy sources developed during the industrial revolution. Recent and forecasted future global population dynamics are the herd of elephants in the room that underlie all that is written about current and future global change. While it is proving possible in several regions of the world to influence family size, which is usually closely linked to social equality and levels of education, the associated increasing consumption of natural resources is proving to be far harder to control (Ehrlich *et al.*, 2012). The inexorable spread of consumer culture from developed to developing economies is driven by too powerful forces to be managed by normal governmental regulatory tools. The usually cautious United Nations Secretary-General has called for revolutionary action in the developed world to replace the prevailing model of economic growth, which is driven by extravagant use of natural resources. He has described this model as 'a global suicide pact'.

Every previous period of human society has had its concerns, worries and prophets of doom. Here we are with more people than ever before, who are living longer and have access to resources and knowledge on an entirely different scale from previous generations, yet many researchers share the foreboding of the Secretary-General and feel that the increasing pressure on global ecosystems is precipitating a crisis that will be impossible to resolve by technological means alone and will result in social dislocation and suffering. This book provides a background to the consequences for terrestrial ecosystems of the current state of affairs and reviews the tools that can help explore possible future scenarios.

Rockström *et al.* (2009) introduced the concept of a safe operating space for the Earth, in which they feel their way towards the planetary boundaries of the Earth system (Figure 1.2). They identify nine critical processes for which thresholds of control variables such as atmospheric carbon dioxide concentration should be defined – although this is easier said than done. If these notional thresholds are crossed, the consequence could be 'unacceptable environmental change'. They suggest that three processes have already exceeded these safe operating limits, namely climate change, the rate of biodiversity loss and interference with the global nitrogen cycle. A particular concern they raise relates to 'tipping points', which describe the tendency of complex Earth systems not to respond smoothly to changing pressures but rather 'to shift into a new state, often with deleterious or potentially even disastrous consequences for humans' (Rockström *et al.*, 2009). There is an urgent need to identify potential thresholds within Earth systems, which once crossed will alter their states in ways that could have alarming consequences for civilisation. Our agriculture is highly dependent on the regularity of monsoon systems and the timing of spring, for example, so we need to understand any nonlinear responses of these climatic features to global warming. Could a reduction in the cover of arctic sea ice elicit a nonlinear response in the northern hemisphere growing season? This feels like an immediate concern as farmers bemoan one problem after another and survey the poor condition of their winter cereals during late, cold springs. Many of the Earth's subsystems do appear to react in a nonlinear, often abrupt, way and are particularly sensitive around threshold levels of certain key variables. If these thresholds are crossed then important subsystems, such as a monsoon system, could shift into a new state, posing problems for sustained agricultural production. The concepts of critical tipping points and nonlinear responses to pressure on Earth systems have become a significant research issue in global ecology and will be explored further in this book.

1.3 Relevance of the past

A case has been made that the Anthropocene begun about AD 1800 as humans entered into a new phase of their relationship with the Earth and became a 'global geophysical force' that threatens important life support systems (Steffen *et al.*, 2007). Does this mean that the earlier history of our relationship has little relevance for current issues? In some ways this would be convenient as there is a good deal of relatively well organised information about the changing state of the world during the last 200 years. To delve further back into the past (by several millennia) requires specialised research techniques, relying for example on pre-instrumental records for the reconstruction of past climate. Even philosophers are divided about the value of the past in planning for the future. 'History is bunk' is a view that is widespread. Its advocates argue that continually changing circumstances mean that little of value can be learnt from history, as shown by military commanders who use outdated strategies and fail to exploit opportunities presented by new technologies (Munro, 2001). Others argue for the value of analysis of past trends in forecasting the future, even if it is just to pick the winning horse in a race. A strong case has been made for the existence of uniform processes of change in physical Earth systems, such as mountain building or rock weathering; this idea was first proposed by Charles Lyell (1830–33) as the principle of uniformitarianism. Darwin was strongly influenced by Lyell's ideas and transferred this physical concept across to biological systems during the development of his theory expounding the slow but continuous evolution of species through natural selection. Whichever view one adopts concerning the value of the past, few can dispute that history is an inescapable part of our culture and helps define us as human beings. We all enjoy stories around the campfire and this book includes many such stories, as well as examples of how our long-term cultural heritage influences current management issues.

I (R.B.) have found from personal experience that understanding the history of a region can help resolve management conflicts. I saw this in the establishment of a National Park in Sweden, where landowners were upset by compulsory purchase orders for beech forests that they had managed for two generations. History showed that many of the features that the park was designed to protect, including the beech stands themselves, had largely

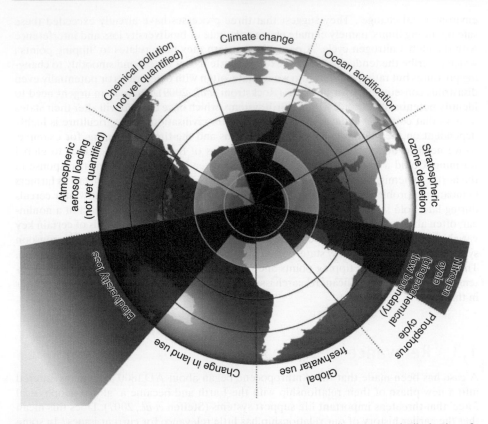

Figure 1.2 Proposed safe operating space for nine planetary systems (inner green shading). The safe boundaries for three systems (rate of biodiversity loss, climate change and human interference with the nitrogen cycle) have already been exceeded (red wedges). Source: Rockström *et al.* 2009. Reproduced by permission of Nature Publications

developed as a consequence of earlier human activities. This knowledge surprised both the state authorities and the landowners and placed the negotiations on a new common footing, based around the idea that recent management would have to be explicitly acknowledged in the planning of the park and not simply ignored or eradicated. There are several other lines of research and policy that draw on evidence from the past. We use the models that form an integral part of this book to explore and understand past climate–vegetation interactions through hindcasting, adopting uniformitarian principles; if the models succeed in reproducing the important properties of past observations then our confidence in using them as forecasting tools is increased. It can be harder to model human impact on ecosystems, as human behaviour tends to be less predictable than that of plants. The Ruddiman hypothesis, which is discussed in Chapter 7, is a good example of how introducing the long-term history of anthropogenic deforestation into analyses of the biogeochemical cycling of CO_2 and CH_4 has stimulated debate and increased our understanding of atmosphere–biosphere interactions (Ruddiman *et al.*, 2011a). The modern relevance of what we can learn from the archaeological record about the collapse of past civilisations is more controversial.

A debatable question for our time is whether the archaeological record provides evidence for over-exploitation of an irreplaceable natural resource and a consequent collapse of a civilization (Tainter, 2006). Tainter argues that there are probably no known examples from the past of pure ecological collapse brought about by overpopulation leading to resource depletion. The case of the Third Dynasty of Ur (c. 2100–2000 BC) in southern Mesopotamia comes close, but the final collapse can be attributed to poor management of agriculture – which may well have relevance for today. This civilisation built a canal from the River Tigris in modern Iraq into a fertile but arid region to generate high yields of wheat and barley, which facilitated population increase and city expansion (Jacobsen & Adams, 1958). A side effect of the irrigation was the raising of saline groundwater to the soil surface, which gradually reduced cereal yields and particularly affected the less salt-tolerant wheat. Written records and archaeological data show a slow but cumulatively large reduction in food supply and the consequent collapse of several cities (Jacobsen & Adams, 1958). There is evidence that methods for avoiding salinisation, including the use of frequent fallow periods, were known at the time but were not employed by those in charge. So, poor leadership decisions contributed to the collapse of this civilisation, rather than the direct over-exploitation of resources.

Soil salinisation is an example of a tipping point, where an ecosystem changes to another state, with consequences for the services that people have come to depend upon; such tipping points are obviously of concern to civilisation. Study of the past provides other examples of heavily exploited ecosystems or systems with low resilience that passed critical tipping points, leading to long-lasting and sometimes irreversible change, bringing undesirable social consequences. While these examples are unlikely to provide direct analogues for modern times, they illustrate certain principles that could influence current management. Prehistoric deforestation and the cultivation of soil on sloping ground or porous bedrock in the Mediterranean region and in the Burren of western Ireland caused soil erosion and loss of fertility (Figure 1.3). Early agricultural activities have interacted with climate change, leaving abandoned civilisations and cities that could no longer be sustained by the surrounding landscapes. Widely discussed examples include the collapse of the Central American Mayan civilisation c. AD 900 and of the Anasazi of New Mexico c. AD 1200, where severe drought sharply reduced the population size, resource consumption and political complexity that could be maintained in these societies (Tainter, 2006). Modern societies have developed more resilience to many of the types of localised environmental stress that caused serious problems to past isolated communities, but there are still lessons to be learnt that are surprisingly relevant today.

1.4   Forecasting the future

Increasing numbers of scientists and commentators are pointing out the mostly negative outcomes that result from our current wholesale exploitation of the planet. Some of these are likely to occur in the immediate future or at least within the lifetimes of our children and grandchildren.

The European heatwave of the summer of 2003 was probably the hottest summer for at least the last 500 years. It was focused on France, Germany and Italy, where there were thousands of heat-related deaths, especially among the elderly. Agricultural crops were also badly affected. Was this just natural variability or can the effects of anthropogenically increased CO_2 concentrations on climate be blamed, at least in part? As with anything to do with climate change, there are both believers and nonbelievers. Stott *et al.* (2004) tried to address the issue using a method involving both the decadal-mean seasonal mean changes in summer temperatures and the change in risk of exceeding a threshold to estimate the contribution of increased greenhouse gases on risk of high mean summer temperatures in continental Europe. They concluded that it was very likely that anthropogenic effects on climate at least doubled the risk of a heatwave of the magnitude seen in 2003. In addition, it is likely that such heatwaves will be occurring frequently by 2050 and that unless atmospheric

Figure 1.3   Limestone pavement showing through the thin soils of the Burren, Co. Clare, Ireland. The Burren was covered by forest earlier in the Holocene and the soils were deeper (photo Richard Bradshaw)

CO_2 concentrations are controlled, the 2003 heatwave will be considered relatively mild in comparison to those that will occur. Current climate change has no known endpoint and the prospect of temperatures rising beyond human endurance in some parts of the world, in addition to rising sea levels, will present a serious challenge to future generations.

In reality, forecasting the future is impossible, but it is possible to build scenarios that can allow explorations of 'what if?' questions. In this book we use three different types of scenario to explore possible futures over the next 50–100 years. Of course, we must bear in mind that political timescales are usually short and that there are few recent examples of successful implementation of policy on longer ones. In practice, much of the focus on forecasting of the future has to do with relevance to the short-term political cycles that exist in most countries. A typical example is the European commitment to halt biodiversity loss by 2010, which has clearly failed. There may have been some improvements as a result, but the goal was always unrealistic given the lack of will or even interest among many politicians, as well as the general public in many places. However, around 2050 is when model projections indicate that significant climatic changes will be showing their effects. It is also quite clear that in order to reduce these possible effects, serious action needs to be taken now or in the immediate future, and even as we write windows of opportunity are closing.

Of course, many initiatives have value and contribute at least to our understanding of the present situation and do give warnings about the future, whether they are ignored or not. For example, the MEA was initiated by the United Nations in 2000 'to assess the consequences of ecosystem change for human well-being and the scientific basis for action needed to enhance the conservation and sustainable use of those systems'. The MEA places human welfare in focus and emphasises the importance of ecosystem integrity for sustained welfare. It also strives to forge new links between science and policy, which is an area in need of support in several nations.

The MEA, in common with most ecosystem research, has a limited time perspective, chiefly combining observations from recent decades with forecasts for the future. This book accesses specialised data from the past and adopts a much longer timeframe, designed to place the current state of terrestrial ecosystem affairs into a broader perspective, as Michael Mann did with his famous 'hockey stick' portrayal of northern hemisphere mean temperatures over the last 600 years (Mann *et al.*, 1998). A deeper background outlining how the current state of affairs developed does influence interpretation and helps choose between the various ways ahead. This book contests the belief that the present is so different from the deep past that there is little to be learnt from ecosystem states thousands of years ago. Many important ecosystem processes still operate and there is much to be learnt from past experience.

1.5 Chapter details and logic

The book comprises 10 chapters. Following this introduction, two introduce the data and the models that respectively provide and explore the evidence discussed in the subsequent chapters. The models of Chapter 2 are a central feature of the book. They are tools for integrating our knowledge of the Earth system and exploring the future. Models come in many forms and the focus here is on both equilibrium and dynamic models of climate, ecosystems and species, as well as models of complex socioecological systems. Earth system models integrate general circulation models with dynamic ecosystem models and can, for example, incorporate the important feedbacks from ecosystems to climate. The vegetation models are presented in the order they were developed and linked to a brief history of global plant geography. The chapter covers the important linking of biogeochemistry to biogeography and reviews dynamic vegetation modelling from the landscape to the global scale, and introduces the concept of plant functional types. Finally, it discusses agent-based modelling and how human land use decisions are or can be incorporated into model frameworks.

Chapter 3 presents the data used to describe and interpret past ecosystem dynamics. Models are built from generalisations that are largely consistent with data, they are parameterised using data and they are validated by comparison with data. Modellers and their models would like to have ecosystem data of high spatial and temporal resolution, neatly organised into accessible databases. Unfortunately, data are rarely available in this form. Some recent climatic data and remotely sensed data on land cover are fairly well organised, but they cover rather short periods compared with the timespan of this book. A myriad of data types, including directly observed phenological data, 'proxies' such as pollen data from which past vegetation dynamics can be reconstructed and even cave paintings, must be evaluated to piece together a fragmentary story of past ecosystem dynamics that can then be explored with models. Chapter 3 introduces the major data sources from the past and outlines the assets and limitations of each data type.

The past relationships between climatic change and ecosystem dynamics are examined in Chapter 4. We wrestle with some complexity here because past ecosystem dynamics are a useful source of information about past climates, but they also provide important feedbacks to the climate system. Our main interest, however, is in discovering how past climatic change influenced ecosystems, because this provides a good basis for forecasting the effects of future climatic change on terrestrial ecosystems and their services. Models do help to tease apart these complex interactions and develop our understanding of climate-forced ecosystem change. Climatic changes during the last ice age resulted in global-scale vegetation dynamics that appear to be synchronous on different continents, allowing for some limitations of dating control. The climatic changes that define the Late Glacial period (21 000–11 500 years ago) were synchronous throughout Europe and resulted in rapid ecosystem responses that have been analysed in some detail. The reforestation of Europe during the Holocene (11 500 years ago to the present day) was a seemingly more complex process, influenced by climate but also by the location of species refugia, dispersal biology, competition, atmospheric CO_2 concentrations and human activities. There is more and better-dated ecosystem evidence from the Holocene than from earlier periods and the apparent increase in complexity probably arises because we can examine the records with greater spatial and temporal resolution than before. Examples comparing the model results with data from Europe and Africa illustrate a range of climate–ecosystem relationships. The chapter concludes with a look at megafaunal extinctions, taking genetic evidence into account, and a review of recent, directly observed species range shifts and plant productivity changes.

Short-lived, irregular episodic events can be difficult to incorporate into models, yet they prove to have made a significant impact on ecosystem dynamics during the Holocene. Chapter 5 examines the role of fire, pathogens and wind on ecosystems. People and their domestic animals have been an important source of these hard-to-model disturbances, and their ecosystem impacts are covered in more detail in Chapter 6, just as longer-term climatic events, such as drought, were examined in Chapter 4. Increasing knowledge of the profound ecosystem effects of short-lived, episodic events in the past has rendered the classical concept of successional climax of limited practical value for many ecosystems. Past disturbance of natural and human origin has caused long-term changes of state for major ecosystems in Australia, New Zealand and elsewhere, pointing to the existence of alternative stable states. Fire is a disturbance agency with a very long history of human influence that is forecast to increase its ecosystem impact under future climates in many regions. Interactions between disturbance agencies are increasing and are likely to pose management issues for the maintenance of several ecosystem services in the future.

We begin to unravel the very long relationship between human activities and ecosystem dynamics in Chapter 6. The use of fire is the oldest management tool employed by early peoples, stretching back for quite possibly millions of years, so there are few convenient baselines from which we can assess the extent of current human impact on ecosystems. The transition from hunter-fisher-gatherer societies to settled agriculture began ecosystem conversions that still continue today. The timing and scale of these conversions are under continual revaluation and are important because of their influence on the global carbon cycle and feedbacks to the climate system. A 300 year period of frenzied construction of megalithic monuments accompanied the spread of agriculture into the northwest margins of Europe, with at least 30 000 monuments built in Denmark alone. A subsequent socio-economic collapse with the return of forest on to recently established fields was an early example of the boom-and-bust dynamics so familiar to Western societies today. Island systems are sensitive recorders of the biological impact of human settlement, and mass extinction of birds in particular has been a typical outcome on islands in the South Pacific during recent millennia. Island floras also change rapidly following settlement, through the introduction of exotic species, the increased abundance of previously rare species and the consequent loss of dominance of other plants. Total extinction of plant species has not been recorded as often as has extinction among large animals. Modelling human activities can be difficult, but one attempt suggests that considerable modification of the global land surface had already taken place prior to industrialisation, which supports a general conclusion from this chapter that past human impact on ecosystems has often acted over longer periods of time and to greater effect than many researchers previously believed to be the case. The last 200 years of human impact have been more intensive, but it is important not to underestimate the significance and scale of earlier human ecosystem impact.

Chapter 7 examines the changing demands made by the global community and the expectations that societies have become accustomed to in the form of goods and services provided by ecosystems. Rapid increases in human population are placing pressures on these services – pressures that need to be understood and reacted to through appropriate management. Here we explore the long-term development of these pressures and show how different services have been valued through time. The multiple origins of agriculture and abandonment of hunter-gathering lifestyles are examples of major past transitions in the development of provisioning services.

The first two chapters of Charles Darwin's *Origin of Species* are titled 'Variation Under Domestication' and 'Variation Under Nature' and emphasise how easily human intervention in natural processes can result in large effects. In this case, breeding programmes exploit natural variation in species for the benefit of society, which is an example of the use of genetic resources as an ecosystem service. It can be argued that we exploit several ecosystem properties in the same way to maximise the benefits we derive from such ecosystem services as food, fibre and water. We explore future scenarios for ecosystem services, which show that food security is likely to remain beyond reach for many people. Some ecosystem services are already being degraded through the increased demand for provisioning services. Biodiversity and its services suffer to some degree under all scenarios.

It can be hard to evaluate the commercial importance of cultural ecosystem services for society, but we argue in Chapter 8 that these are no luxury. They have been central to social behaviour through time and have influenced mythology and religion. Cultural services contribute to the humanity of our species and have inspired artistic creation for millennia. Cultural services retain importance through our relationship with and love of both the 'natural' world and the diverse cultural landscapes we have created through time. Hiking, camping, hunting and visits to the zoo pay homage to cultural ecosystem services. Support for traditional, nonmechanised methods of food production link to our fond memory of earlier landscapes and rural life. All these cultural attributes contribute to conservation policies, which are complex expressions of how we value many ecosystem properties, but particularly their cultural services.

Chapter 9 describes how current conservation policy is beset by potential paradoxes when viewed with a long-term perspective. We fight the spread of recently introduced species while protecting others that were introduced as a result of human activities long ago. We prize rarity for its own sake. We sometimes try to mimic 'natural' conditions that that we know little about. Often we fight change, and conservation can be a case of 'keeping the landscape the way it was when I was young', as Sir Arthur Tansley once wisely said. We can enter into costly interventions and fight losing battles against very successful invaders to atone for our earlier misdeeds. In my garden, I (R.B.) continually intervene to protect losers from successful species with rapidly increasing populations such as magpies, crows, pigeons, the whining collared dove and the neighbours' cats. It is estimated that Swedish cats kill over 16 million small garden birds each year, and with a mere 70 million nesting pairs in the country it feels like the neighbours have introduced a new and unnatural threat to the nation's wildlife. Another neighbour traps and kills the numerous and wily magpies that rob the nests of the smaller birds we all like to see. Magpies were first recorded in Ireland in 1676 and bred in Dublin in 1852. Their numbers expanded in urban Dublin during the late 1900s and an elderly colleague told me that he was amazed at the increase in numbers of magpies in Dublin during his lifetime as we watched a group of five squabbling outside his window. These beautiful and intelligent birds do eat eggs and young birds but are often unfairly blamed for the similar work of cats, squirrels and rats. A non-specialist predator like the magpie is unlikely to have any lasting effect on small bird populations. With so many changes to nature over such long periods of time and so many shifting demands on ecosystem services, it feels appropriate to assess how we have arrived at the current situation and our conservation policies. Given likely future scenarios it feels timely to explore the current priorities for human–ecosystem interactions, management and conservation, which is what we cover in Chapter 9.

Finally, in Chapter 10 we pull all the threads together and return to some of the issues raised in this chapter with fresh insight gained from the book as a whole.

1.6   For whom is the book intended?

Long-term ecosystem dynamics influence everyone on the planet, whether they are aware of it or not. This book has developed from joint research and teaching undertaken by both authors and the primary targets are graduate students and graduate courses or seminars that cover ecosystem dynamics. We also believe it should be of interest to a wider audience that is not engaged in academic research but wants some background to the issue of global change and wishes to understand the importance of a time perspective in ecosystem dynamics. We hope that the book has much to offer to students from the undergraduate level onwards and to researchers looking for background in subjects complementary and relevant to their own special field of knowledge. We intend to use it as the basis for a Master's course at Lund University, Sweden, which attracts students from the fields of ecology, geography, geology, archaeology and other disciplines within the biological and Earth sciences. While many readers will select groups of chapters, a few may choose to read the book from cover to cover following the logic of our chapter structure. The topics covered are not of interest only to researchers, teachers and students. Anyone with a garden or smallholding is aware of the seasonal and annual changes in vegetation and will have opinions about longer-term changes, whether they be the arrival of Spanish killer slugs, wetter summers or the ability to sit out in the garden later in the year than usual. Ecosystem dynamics also brush up against policy in many nations, because 'degradation and loss of habitats and species are compromising the ecosystem services that sustain the quality of life for billions of people worldwide' (Bradshaw *et al.*, 2010a). As well as containing valuable biodiversity, forests help regulate the global carbon cycle and lock up considerable quantities of carbon. Deforestation has been estimated to account for 20% of global emissions of CO_2 during the last 50 years (Global Carbon Project, 2012).

This book is not light reading, but its analysis of part of the science underlying current concerns about the effects of global change is of relevance to everyone. We have tried to make it accessible to the interested but nonacademic reader. The book lies between the subject of biology and that part of Earth sciences that deals with the recent ice ages and interglacials, where the records lie in the roughly 2.6 million year 'soft rock' deposits of the Quaternary period. Most of the text focuses on the Holocene and uses models to explore future scenarios, touching on a range of disciplines, including biological and environmental sciences, Earth sciences, archaeology, ethnology, climatology and genetics. The book will present the biological basis for ecosystem models and will introduce readers with backgrounds in Earth sciences, physics or modelling to the fascinations of long-term vegetational and ecosystem dynamics. The background, rationale and use of ecosystem models will introduce palaeoecologists and biologists to these powerful research tools. We have tried to make the text authoritative but readable. We hope that we have also made the contents accessible to interested members of a broader public outside the immediate research community. The book is intended to summarise evidence for the current state of global ecosystems and explore potential future developments, which are topics that affect us all.

1.7   Four key questions and the links to policy

The book aims to show how and why ecosystems change, on a day-to-day basis, season-ally, annually and on millennial timescales. There has been a widespread tendency to take ecosystems and the goods and services they provide for granted. Many can see the conse-quences when we alter ecosystems by our own actions but are surprised when they alter as a result of other influences, such as climate change or their own internal dynamics. It is important to know more about the variability of individual ecosystems and to investigate how they respond to and recover from disturbance. Nature can be portrayed in the media as delicate and sensitive, but many ecosystems have long histories so are probably quite resilient. We need to know how much punishment they can really withstand.

Human exploitation and manipulation of ecosystems and climate change are two of the major forces affecting ecosystems today. It is important to know what their effects are, how they interact and what the consequences will be, both for the ecosystems themselves and for the human race. If an important ecosystem service such as the pollination of food crops falters, we need to know the reasons why, if we are to fix the problem. We have identified four key questions that have guided the planning of this book. (1) How have ecosystems changed in the past? (2) How much of this change is attributable to human activities? (3) How much change is anticipated for the future? (4) What are the appropriate ecosystem management measures by which to prepare for the future?

The fourth question provides a link to policy issues, which are largely beyond the scope of this book but should nevertheless retain close links to the scientific evidence. Many scientists separate themselves from decision-making processes, arguing that it is their job to provide the background while expert decision makers generate the policy. However, evidence-based environmental policy can be disappointingly rare, with economic or legal arguments often carrying more weight. Scientists in many Western countries keep themselves (or are kept) further from policy decisions than those in China or Germany, for example, where qualified research scientists often hold the highest positions in government. The economic and political costs of environmental policy need to be factored in with the social benefits, but where these benefits are weakly defined or cannot be quickly realised there has been a tendency toward more talk than action over important environmental issues.

Scientists have begun to engage more fully with policy than in the past and have had to face the personal attacks and public scrutiny that seem to be an integral part of Western politics. Professor Ray Bradley of the University of Massachusetts has published exten-sively on climate change and writes of the 'the urgency of now' and the need for decisive political leadership. He argues that the science of climate change is now broadly under-stood but that there remains a worrying reluctance within many governments to act in an effective way over environmental issues, partly because of pressure from influential sec-tors of society whose wealth rests on exploitation of fossil fuels and other natural resources (Bradley, 2011). In his latest book he describes global warming as 'the number one environ-mental issue of our time', yet he rues the fact that 'some prominent politicians have refused to accept scientific evidence of human responsibility and have opposed any legislation or international agreement that would limit greenhouse gas emissions' (Bradley, 2011). Prof. Bradley has also had personal experience of cynical attempts to 'destroy the reputations of scientists researching climate change by deliberately undermining the credibility of their research' (Bradley, 2011). It is now important for all citizens of democracies to understand some basic scientific facts, as environmental issues are topical politics. It needs to be widely understood that absolute proof is unattainable in the environmental sciences and that there will always be some level of uncertainty associated with scientific conclusions, so that 'when someone says that society should wait until scientists are absolutely certain before taking any action, it is the same as saying society should never take action' (Gleick *et al.*, 2010). Lack of action over the issues of ecosystem degradation and climate change poses a dangerous risk for the planet and responsible citizens must engage with environmental policy and share the heavy responsibility over what we leave for future generations.

The journalist George Monbiot has raised particular concern over ecosystems becoming incorporated into mainstream economics, which is necessary but shows up the inflexibilities and inadequacies of traditional economic practice (Monbiot, 2012). He is critical of the 'commodification' of 'natural capital' and 'green infrastructure'. While an acknowledgement of the value of nature by the market is an important development, 'appointing the landlord as the owner and instigator of the wildlife, the water flow, the carbon cycle, the natural processes that were previously deemed to belong to everyone and no one' is a privatisation and transfer of natural values into a ruthless market place (Monbiot, 2012). Financial markets have poor credentials for environmental management and indeed have already precipitated the current global environmental crisis. There is a desperate need for a new approach to a democratic economic valuation of nature that incorporates a long-term view and includes the interests of the wider public and not simply those of major financial institutions.

The world faces several troubling environmental challenges that are closely linked to society, politics and economics. This book examines ecosystem dynamics from the deep past into an uncertain future. We do not pretend to suggest ways of protecting the spendthrift way of life to which more and more of us aspire, but rather give the long-term background to the current state of our terrestrial ecosystems and help with the development of the modelling tools needed to support the policy development necessary for the future.

2
Modelling

2.1 Introduction

Ecosystems are dynamic; that is, they change through both time and space. Understanding these changes involves exploring how species, ecosystems and biomes respond at different scales to changing environments, past, present and future. This exploration usually involves data from a variety of sources and of varying quality and extent, often in conjunction with some sort of simulation modelling. Computers have developed hugely in speed and complexity over the last 50 years, and the increasing and extensive use of them to model at least some of the highly complex and interlinked systems of the Earth reflects this. In this chapter, a selected number of different but related models relevant to the themes of the book are introduced. They include models of climate, ecosystems and species, as well as models that simulate interactions with society, for example through land use. As with the rest of the book's historical approach, Chapter 2 is structured around the historical development of models and is based in part on a similar approach in Prentice *et al.* (2007).

Models are not real life, but they can be used as experimental tools to explore the nature and relationships of the systems under study. They give outputs that may or may not be correct but can be assessed for validity in some way against available data. The aim is to give improved insights into and understanding of the complex interactions within an ecosystem. Models can generate insight but cannot necessarily provide definite answers, since they are, after all, only models. Once some sort of validation of the model output against data has been done, they can be used as tools to explore possible futures. There are many models, operating at a range of spatial and temporal scales. Initially, models were relatively simple, but recent developments are leading to highly complex integrated models of Earth systems, including important new features in a climate-change context such as feedbacks from the ecosystem to the climate. In addition, the critical role of modern society in ecosystems is being recognised through the integration of more direct human influences on ecosystems and feedbacks, for example by the inclusion of societal ('market') pressures and drivers.

This section provides some background to ecosystem models. It is not an exhaustive introduction but aims to give enough background and sufficient examples to allow the general reader to be able to follow the major themes of this book. Examples of structures and of some of the typical outputs are intended to show development. Specific algorithms for any particular model are not included, as this level of detail would be incompatible with the general broad approach of the book. However, as most models reference the original papers in which they were developed, the interested reader can explore them in more

Ecosystem Dynamics: From the Past to the Future, First Edition.
Richard H.W. Bradshaw and Martin T. Sykes.
© 2014 John Wiley & Sons, Ltd. Published 2014 by John Wiley & Sons, Ltd.
Companion Website: www.wiley.com/go/bradshaw/sykes/ecosystem

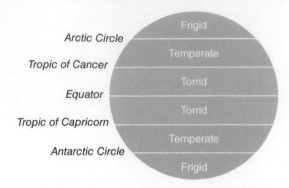

Figure 2.1 Simple model of the Earth divided into five different climate zones, based on the work of Aristotle (credit Matt Rosenberg). Reproduced by permission of Matt Rosenberg, geography.about.com

detail as they wish. Models that are discussed include equilibrium and dynamic models of species, vegetation (biomes) and ecosystems, statistical and mechanistic models, broad- and fine-scale models, climate (general circulation models-GCMs), Earth system models (ESMs) and agent-based models (ABMs). All models have a range of assumptions, caveats and uncertainties that users must be aware of before attempting to interpret their outputs, and these will also be discussed.

2.1.1 How did these models develop?

Looking for some explanation of the world we live in is not a new human occupation. In particular, the role of the environment and its interactions with humans and society has long been a preoccupation, and frequently a necessity for survival. Models of these relationships are not new either, as the ancient Greeks were devising them hundreds of years ago. In around 350 BC, Aristotle, reviewing the work of Pythagoras and his disciple Parmenides, from the sixth century BC, agreed with them that a simple model of five zones could describe the climate of the Earth (Sanderson, 1999). This model is based on the geometry of the Earth and Sun and classifies geographical regions across the globe from the two 'frigid zones' at the poles, through the two 'temperate zones' to the 'torrid zone' at the equator (Figure 2.1). Both the frigid zones and the torrid zone were considered to have environments that were too inhospitable for humans to live in. In this latter point the model was evidentially wrong, but one of the reasons for constructing a model is to explore ideas about our place in the scheme of things. It is interesting to note that this simple division remains the basis for some global maps still published today.

Providing definitive or even 'correct' answers is not always possible or even desirable. Forecasting the weather using models is not an exact science, for example, but it still provides important information that many parts of society rely upon. Models act as summaries of our knowledge of a system and help identify gaps within this knowledge.

2.1.2 Climate data, climate and Earth system models

Ecosystems and species respond in various ways to their environment. The environment is a complex of climate and soil variables. Climate comprises a number of variables, including different aspects of temperature and precipitation, and ecosystem models need some data describing these aspects. These data might be instrumental recordings from measurement stations, historical data of some sort, proxy data (e.g. using tree ring widths as a proxy for summer temperature) or data derived from models of climate. They can be so-called 'equilibrium' or 'snapshot' descriptions of various climate variables, such as mean monthly temperature, monthly precipitation or cloudiness, or they can be time series of various lengths (e.g. daily, monthly, yearly). Various datasets of different climate variables extrapolated on to a global grid for the recent historical period (the twentieth century) are available, such as the Climate Research Unit dataset from the University of East Anglia (University of East Anglia).

However, exploration of the future or the past when data are patchy or incomplete requires some sort of modelling of the climate system. Climate models and GCMs are used to model global and regional climates both for the past and for the future. They are a simplification of reality and yet highly complex in their relatively detailed descriptions of the climate system, even though not all aspects of this system are as yet understood. The theoretical background, methodologies and application of these types of model are described elsewhere (see further reading suggestions at the end of this chapter).

Climate, however, should not be viewed in isolation from the rest of the Earth system and in fact is intimately linked to what is happening on the land surface, among other factors. Until recently, GCMs tended either to ignore the processes occurring in terrestrial ecosystems or to view the land surface as highly simplified green slime. It is clear, however, that while both species and ecosystems are influenced by the climate system they themselves also influence it, through what are known as feedbacks. Such feedbacks can typically be related to solar radiation and atmospheric water content. There are substantial differences in the albedo or reflectance of solar radiation from a forest-covered landscape as opposed to an open agricultural landscape, especially when covered by snow. Moisture is transferred back to the atmosphere through evaporation from any surface. Plants contribute to this further through transpiration from their leaves; as they grow, they take up atmospheric CO_2 through their stomata so that they can carry out photosynthesis, and at the same time water is transferred back to the atmosphere through evapotranspiration via the open stomata of leaf surfaces. Such feedbacks to the atmosphere affect the climate. Different types of ecosystem and even species have different feedbacks and can therefore influence the climate in different ways. In recent years this has been recognised and significant advances have been made in linking different types of model together into what has become known as an ESM. Typically, such a model integrates both responses to climate and feedbacks to the atmosphere; for example, an ESM might link an atmospheric general circulation model, a model of atmospheric chemistry (including aerosols), models of land, models of ecosystem dynamics (including hydrology) and models of ocean circulation and marine processes. Including more accurate representation of processes and feedback should help reduce the uncertainties involved in current projections with regard to future climate change. ESMs are discussed in more detail in Section 2.4.

Summary: Introduction

- Ecosystems are dynamic in space and time and can be simulated using computer models.
- Both equilibrium and dynamic models of climate, ecosystems, species and socioecological systems will be described.
- Modelling is not new; the ancient Greeks devised a global model of the Earth's environment based on five different zones: a division that can still be seen today in published maps.
- General circulation models (GCMs) are used to model the changing climate in both the past and the future.
- Feedbacks from ecosystems to the climate system are important. Earth system models (ESMs) integrate a range of different models to account for this.

2.2   Background ecosystem, vegetation and species models

2.2.1   Vegetation models

The science of ecosystems, vegetation and species and their interactions has developed with increasing rapidity over the last couple of hundred years, and these developments form the basis for the current state of environmental and biological modelling. Important contributors to this development include von Humboldt, who, in an essay on the geography of plants (von Humboldt and Bonpland, 2008), argued that plant science is not only about describing new plant species and their characteristics but also concerns their geography or their associations and distributions in different climates. He integrated a 'tableau' through equatorial South America of various physical attributes, such as air temperature, geology, barometric pressure, the intensity of the colour of the sky, the limit of perpetual snow, agricultural cultivation with vegetation composition and distribution, to show how distributions were related to altitude and how they varied in systematic ways. Such synthetic approaches involving plant geography have contributed to the focused and increasingly rapid integration of various other disciplines, such as plant physiology and biogeochemistry, vegetation dynamics and biophysics, stimulated in part by concerns over the impacts of climate change (Prentice et al., 2007).

Plant geography, or phytogeography, is concerned with the distribution of species or vegetation types through space and with relating this distribution to some sort of environmental response or control. Plant sociology, or phytosociology, is a related field that explores the associations formed among species, for example as a result of competition between individuals. Groupings that are identified can then be classified and quantified into the concept of plant communities. There is a whole science concerned with phytosociology, which is not relevant in any detail here (but for further information, see for example the work of Braun Blanquet, Viktor Westhoff and others). Classifications into plant communities and associated habitat descriptions are useful ways of describing vegetation and classifying habitats for conservation (e.g. the Natura 2000 sites), as discussed in Chapter 9. Modelling of plant geography can be done at the species level or expressed through vegetation groupings variously described as vegetation types, plant formations and biomes. The historical backgrounds to these different levels of complexity are similar.

One of the major historical influences was plant physiologist Wladmir Köppen. Building on the work of the ancient Greeks, he produced the first quantitative classifications of global climate zones based on phytogeography in 1900 (Kottek et al., 2006). He made the important link between the physiological responses of plants and the influence of climate, using five vegetation groups devised by the Swiss/French botanist De Candolle and the original climate classifications of Aristotle (further subdivided to include moisture). With the additional later input of Rudolf Geiger, Köppen produced global maps of climate zones based on native vegetation types or biomes as indicators or expressions of climate in a given region. Climate variables such as annual and monthly temperatures and precipitation were used, as well as the seasonality of precipitation. This climate classification remains widely used today and has been updated by Kottek et al. (2006) with new digital global temperature and precipitation data (Figure 2.2).

Holdridge (1947), exploring the relationships between mountain and other vegetation in Haiti, devised a system of 'plant formations' or life zones based on simple climate variables, which he then developed for the global scale. These life zones were originally placed within a 3D figure but are most often shown as a triangle of an array of hexagrams: one axis describes annual precipitation, a second the ratio of potential evaporation to mean annual precipitation and a third humidity classes. Mean annual temperature (represented

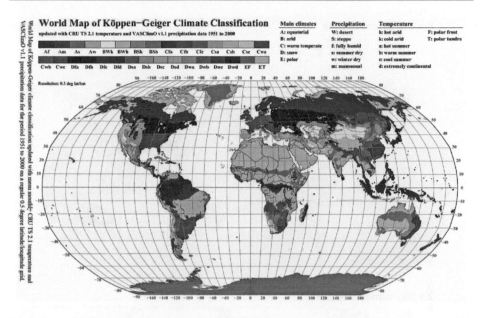

Figure 2.2   Mapped Köppen–Geiger climate classification, using mean monthly temperature and precipitation datasets for the period 1951–2000 (Kottek *et al.*, 2006). Source: Franz, 2006. Courtesy Schweizerbart Science Publishers, www.schweizerbart.de

as biotemperature based on growing season length and temperature) is located perpendicular to the base. This model was used to produce the first projections of potential natural vegetation in a climate-change context by Emanuel *et al.* (1985), who took climate data of varying quality from 8000 meteorological stations worldwide and interpolated them on to a 0.5 × 0.5° global grid. The Holdridge classification was applied using interpolated biotemperature and annual precipitation values to produce a world map of the life zones under 'current' climate (Figure 2.3). A dataset of global average temperature increases corresponding to a doubling of CO_2 concentration, produced by a GCM experiment by Manabe and Stouffer (1980), was used in a further simulation. The results showed large-scale patterns of potential natural vegetation that are sensitive to changes in average temperature, particularly in high latitudes. One of the first firm forecasts for the future with real impact was the almost complete extinction of the boreal forest as the climate becomes more suitable for temperate forest (Figure 2.3).

Building on basic theory, Elgene Box (1981) pointed out that vegetation does not vary in space as a fixed plant formation or vegetation type or biome, but rather that individual plants within formations respond to the environment individualistically, which is a cornerstone of modern ecological theory (see Gleason, 1926). Box argued that ideally it is better to work at the species level within a vegetation type. This is of course impossible at many scales, and certainly globally, not least because of the number of species involved and the severe lack of relevant data. Some grouping is thus required based on a similarity in ecological requirements; in other words, on life forms that have similar attributes or 'traits' and can therefore be expected to respond in similar ways to the environment. For example, trees grow by producing carbohydrates for energy through photosynthesis in their leaves. Some trees are deciduous – that is, they lose their leaves in the winter or in the dry season – while some are evergreen, maintaining leaves all year. These two approaches are clearly different and the question in any given climate could be: Is it 'better' to drop leaves and thus reduce energy costs by limiting the loss of carbon through maintenance respiration in the

Recent current climate

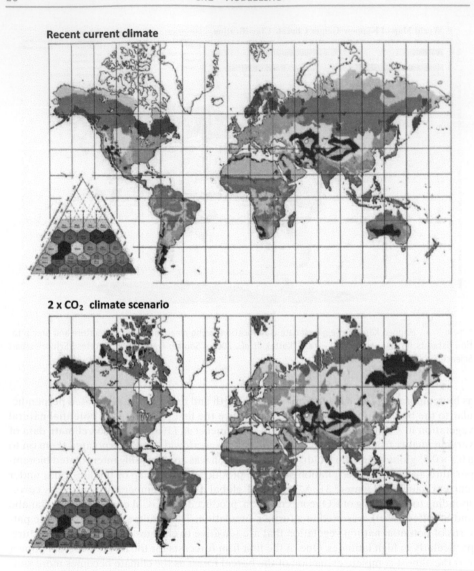

2 x CO_2 climate scenario

→ Poleward shifts of boreal and temperate forests
→ Expansion of steppe, dry woodland and desert

Figure 2.3 World map of potential vegetation based on the Holdridge classification. Top: current state, using climate data from meteorological stations. Bottom: projection based on global average temperature increases corresponding to a doubling of CO_2 concentration. Source: Emanuel *et al*, 1985. Reproduced by permission of Springer

cold or dry season, but incur significant costs associated with producing new leaves in the following season, or to keep leaves all year despite the associated maintenance cost? The outcome of competition for resources between plants in any given climate can depend on such leaf attributes.

There is a substantial literature concerned with traits, which has a long history in both genetics and ecology. In ecology, traits form the basis for plant classifications based on function. The trait concept is not new and perhaps the most well known early classification, and the basis for the majority of later work, is Raunkiaer's (1934). He defined plant life forms according to the traits that allowed them to survive the unfavourable season, be it dry or cold. Such a trait might be expressed as a seed in the case of the annual life form or as a bulb, which survives underground, in the case of monocotyledonous species such as lilies. Similarly, woody plants survive the unfavourable season through the use of a protective bud that protects against local extremes in climate. Raunkiaer's approach, although simple, shows a functional response to climate, which has formed the basis for the development of the trait concept for a whole range of functions, based on a more mechanistic or physiological approach. It forms the basis for the concept of Plant Functional Types (PFTs) which can be used to model vegetation and biomes at broader scales and thus ignore species as a concept.

Box was the first to combine climate responses, traits and PFTs or life form distributions in a numerical global terrestrial model of vegetation. He classified global vegetation types by using combinations of all dominant and codominant ecophysiognomic life forms, such as tropical rainforest trees and mediterranean evergreen shrubs. These life forms were related to macro- or broad-scale climate variables by means of enclosing envelopes that expressed distributions with upper and lower boundaries for each climatic variable. The limits for these climatic variables were determined by the correlations between observed vegetation and local climatic conditions. This concept of climate, or of 'bioclimatic' envelopes, remains a significant element of various current modelling approaches, including both statistical and physiological species and ecosystem models, as we shall see later in this chapter. Box defined each life form by a combination of six physiognomic characteristics or traits: general structural form (tree, shrub, graminoid), plant size (tall, normal, short), leaf type (broad, narrow/needle, absent), relative leaf size, structure of photosynthetic surface (e.g. sclerophyllous) and seasonal photosynthetic habit (e.g. summer green). The model included eight climatic variables – including mean temperatures of the warmest and coldest months, annual and monthly information on precipitation and a moisture index reflecting potential evapotranspiration (PET) – which were thought to represent those factors influencing the energy and water budgets of the plant, including the seasonal variation. The model used a world climate database of 1125 sites and the results were validated at independent sites from around the globe. The experiment showed that a simple and highly generalised model was able to predict in a general way the main features of the present-day vegetation distribution at the global scale. In addition, the basic vegetation structure was determined in the main by general and seasonal levels of temperature and available moisture.

It was Ian Woodward (1987), however, who explicitly related the physiological responses of vegetation to its environmental conditions, in a global model of biome distributions. Physiognomy was used to classify vegetation into five types: evergreen broadleaf, deciduous broadleaf, needleleaf, shrub and herb. Vegetation responses were related to cold tolerance and water availability through the response of the leaf area index (LAI), which expresses the balance between precipitation and evapotranspiration. He produced maps of LAI for the meteorological stations of the globe and compared them to the distribution of the observed physiognomic types, so that for example high LAI corresponded to tropical rain forests. Modelled LAI compared well to observed measurements of LAI, though these were rather few, and modelled maps of the distribution of the different physiognomic types compared well to observations at the global scale. Ron Neilson et al. (1992), working in the 48 contiguous US states, developed a model based on a set of rules that related the seasonality of temperature, precipitation and runoff to the physiology of plants during different stages of their life cycles. His model showed that the boundaries of the major biomes in the USA coincided with regional patterns of climate and runoff seasonality.

These approaches to explicitly physiology-based modelling were in turn important for the relatively more detailed global biome model of Prentice et al. (1992).This approach, along with others built on similar principles, marked a significant step towards the development

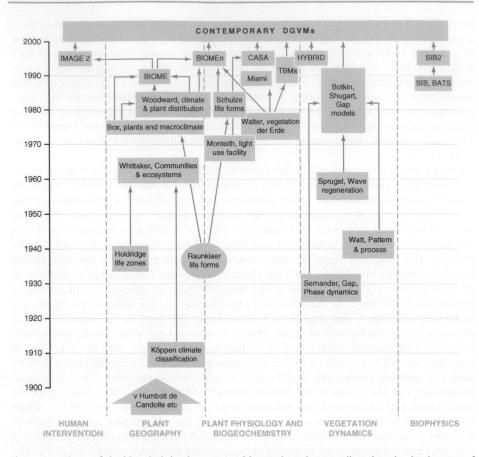

Figure 2.4   Some of the historical developments and innovations that contributed to the development of DGVMs. Source: Prentice *et al*, 2007. Reproduced by permission of Springer

of dynamic global vegetation models (DGVMs) (Figure 2.4) and, more recently, towards an integrated ESM. BIOME (designated BIOME1 as later versions were developed) was designed to be as simple as possible to use to make global-scale simulations, requiring as few PFTs (14) as possible. The choice and number of PFTs was also limited by the need to provide environmental constraints based on known physiological responses. This type of information is lacking for many species, and even where species are grouped by function into PFTs, data remain sparse. In addition, the driving variables had to be derived from climate data at an appropriate scale for a global model. The description of the PFTs and their constraints tells the story of the physiological response-based thinking behind the model (Table 2.1; the columns in the table are mostly empty because values were assigned only if there was a known or hypothesised mechanism that could be identified).

This was different from the Box model, where observed (current) upper and lower limits for various environmental variables are imposed. This difference in approach is important; if models are used to explore the impact of different climate elements on vegetation responses then it is possible that the boundaries and interactions between variables seen under observed current climate conditions may not have been the same in the past and may not be the same in the future.

Table 2.1 List of PFTs used in the BIOME model and their associated environmental constraints. T_c, mean temperature of the coldest month; GDD_5, number of growing degree days over and including 5 °C; GDD_0, number of growing degree days over and including 0 °C; α, Priestley–Taylor coefficient (available soil moisture); D, dominance class for each PFT (Prentice *et al.*, 1992)

	T_c		GDD_5	GDD_0	T_M	α		D
	Min	Max	Min	Min	Min	Max	Max	
Trees								
Tropical evergreen	15.5					0.80		1
Tropical raingreen	15.5					0.45	0.95	1
Warm-temperate evergreen	5.0					0.65		2
Temperate summergreen	−15.0	−15.0	1200			0.65		3
Cool-temperate conifer	−19.0	5.0	900			0.65		3
Boreal evergreen conifer	−35.0	−2.0	350			0.75		3
Boreal summergreen		5.0	350			0.65		3
Non trees								
Sclerophyll/succulent	5.0					0.28		4
Warm grass/shrub					22.0	0.18		5
Cool grass/shrub			500			0.33		6
Cool grass/shrub				100		0.33		6
Hot desert/shrub					22.0			7
Cold desert/shrub				100				8

An additional 'dummy type' is defined for computational consistency; this type has dominance class 9 and no environment limits, representing the 'plant type' that would occur under conditions unfavourable for any other type (e.g. ice caps).

Typically in these sorts of simulations, the vegetation responds to a few important bioclimatic variables, such as the minimum temperature of the coldest month. Functionally this is because low temperatures can slow metabolic processes in cells or even kill plants as the cell contents freeze and then expand to burst the cell wall. The role of data in model development and validation is important. Observed data for woody species' responses to minimum temperature were summarised by Woodward (1987) but, given the lack of global data on minimum temperatures, Colin Prentice *et al.* (1992) devised a regression between the minimum temperature tolerated by various woody plant types and the mean temperatures of the coldest month. This relationship has been used widely in the modelling literature as a surrogate for the absolute temperature of the coldest month. Plants also need a certain level of warmth for growth and for the completion of various annual cycles, such as flowering and fruiting. This indicator of growing season warmth is often expressed in models as growing degree days (GDDs); that is, the annual total of number of degrees per day above a specified limit, usually 5 °C. The GDD limits for various types were found by comparison between global maps of GDD and estimated distributions from actual maps of vegetation produced by Olson *et al.* (1983). Temperature is important, but in many ecosystems, especially drier ones, so too is the amount of moisture in the soil. Different variables were used in the past, but here the ratio of actual to equilibrium evapotranspiration (α, the Priestley–Taylor coefficient) spread through the year as a measure of annual growth-limiting drought stress was used. A dataset of monthly mean values for temperature (°C) extrapolated to the mean elevation of the grid cell, precipitation (mm) and sunshine or cloudiness (%) on a $0.5 \times 0.5°$ global grid provided the basic data from which to produce the bioclimatic limits within each grid cell. The basic climate data were produced by interpolating from c. 6300 globally distributed weather stations (Leemans and Cramer, 1991) and were used together with a global dataset on the same grid of simple soil textures (Zobler, 1986). In general, temperature and GDD data are more reliable than precipitation data, as precipitation varies tremendously at a range of scales in time and space, partly due to surrounding landscape features such as mountains and lakes. The bioclimatic values for each grid cell were then compared to those limits required by each of the PFTs. Those PFTs that could survive within the specified bioclimate of a grid cell were amalgamated using a set of rules, including one based on a dominance hierarchy which said that only the most dominant types were to be used to describe the vegetation type of the cell. Comparisons were made with actual maps of vegetation produced by Olson *et al.* (1983) and were considered to be generally in good agreement, except where intensive agriculture obscured the natural patterns.

The BIOME model was used to project vegetation distributions both in the future and in the past. For example, Prentice *et al.* (1993b) used this model to explore changes in biome distributions and terrestrial carbon storage since the last glacial maximum (LGM). They used modern climate data, GCM output for the LGM, digitised ice sheets and sea levels 120 m below current levels to compare changes in vegetation patterns and terrestrial or land-based carbon with current potential values and patterns. Data on vegetation patterns were not available globally but comparisons were made using pollen-based regional-scale constructions. Simple comparisons with available evidence showed that BIOME modelled the vegetation at the LGM reasonably well, at least at the places where data were available. In temperate regions, pollen-based reconstructions implied much less forest than is found in current climates, in agreement with the model, although data also suggested changes towards the equator. Calculations of changes in carbon storage were made, but these were fairly primitive, using published biomass and soil carbon densities for each biome multiplied by the area of the biome at each time period. The results showed that terrestrial carbon storage had increased by 300–700 Pg C since the LGM (Figure 2.5).

This approach was of course full of assumptions and uncertainties, not least in how carbon storage was measured. What was really needed was a mechanistic way of modelling carbon stored in or moving through an ecosystem. Such methodological improvements involve the integration of plant geography modelling with techniques relevant to ecosystem processes, such as the flows of water and CO_2; that is, the biogeochemistry of the system. These developments involve integration with other modelling paradigms, as will be discussed later in the chapter.

Summary: Vegetation models history

- In 1900, Köppen produced the first quantitative classifications of global climate zones based on plant geography, using five vegetation groups and climate variables, including annual and monthly temperatures, precipitation and the seasonality of precipitation (Kottek *et al.*, 2006).
- Holdridge (1947) devised a 3D figure of plant formations based on simple climate variables, which was used by Emanuel *et al.* (1985) to make the first projections of the effect of climate change on future global vegetation distributions.
- Box (1981) highlighted the fact that vegetation does not vary as fixed formations; rather, individual species respond to their environment individually. It is impossible to model all species, so groups or types based on similar traits, which express similar physiological functions and responses to the environment, need to be used, e.g. Raunkiaer's trait classification.
- Woodward (1987) explicitly modelled the global distribution of vegetation types or biomes using their LAI response to the environment.
- Prentice *et al.* (1992) developed the BIOME global model with a limited number of PFTs and their physiological responses to known environmental constraints (bioclimatic variables).

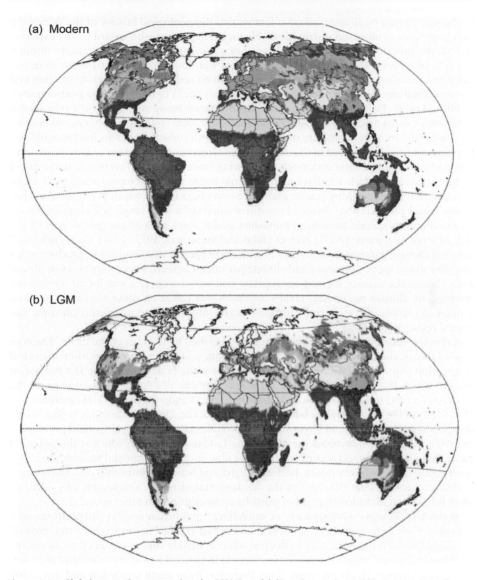

(a) Modern

(b) LGM

Figure 2.5 Global vegetation maps using the BIOME model (Prentice *et al.*, 1993b). Top: modern climate data (9000 meterological stations on a 0.5° grid; Leemans and Cramer, 1991). Bottom: GCM output (ECMWF T21 MODEL) (Saetersdal *et al.*, 1998)

2.2.2 Species-level modelling

Plant geography as a discipline is concerned not only with the geography of vegetation but also with the distribution of individual species and their interactions with the environment. Individual species are of course often very important in their own right, acting as natural keystones for the maintenance of a particular ecosystem or for example having conservation or exploitation value for society. The distributional ranges of individual species have been shown to be related to large-scale or macroclimate variables, sometimes interacting together (Huntley *et al.*, 1995). Eilif Dahl, a Norwegian ecologist, showed that in addition to the role played by climate, edaphic factors and the geological history of the flora were also important to observed distributions. History is important with regard to any explanation of the present-day presence or absence of a particular species, for example through the role of refugia. The type of soil and level of nutrients also play a part; for example, calcium-tolerating species are normally only found on soils with high levels of calcium and are restricted elsewhere even when the climate may seem to be right. In a posthumously published book, Dahl (1998) mapped various isotherms based on biologically meaningful parameters, using macroclimatic data from meteorological stations. The isotherms considered were the mean temperature of the coldest month (which relates to frost sensitivity), the amount of heat received in the summer, the respiration equivalent (which is an accumulated sum of temperatures through the growing season, weighted according to the effect of temperature on dark respiration), mean maximum annual temperatures, total mean precipitation in the growing season and a drought stress index (which is potential evaporation minus precipitation). Maps of the distributions of a wide range of European woody and herbaceous species have been published as dot maps on a 50 km grid as part of the *Atlas Florae Europaeae* (AFE) project (Jalas and Suominen, 1972–1994). Dahl correlated selected climate parameters individually with these species distributions to produce maps showing where species presence and climate parameter agreed, where a species was absent even though the climate seemed appropriate and where a species was found outside the appropriate climate parameter. Dahl's approach was based on plant responses to those aspects of climate that seem to be physiologically important and not purely on some statistical relationship.

A species' relationship to its environment is described by the idea of the niche. The concept of the species niche has developed through time and a number of important historical descriptions must be considered. The Grinnellian niche is determined by the habitat or area in which the species lives, while the Eltonian niche is a species' actual place in the environment and its relationships to other organisms (Vandermeer, 1972). However, it was Hutchinson (1957) who really clarified the concept. The Hutchinsonian niche is a multidimensional hypervolume of environmental variables in which a species could potentially exist. This description includes the idea of the fundamental niche, which is the potential niche that a species should occupy based on the abiotic or environmental variables to which it responds (such as temperature, humidity, light, soil water or nutrients) plus the species' realised or actual niche (the part of the fundamental niche the species actually occupies after it has been excluded from other parts by competition with other species).

In the last 30 years, various ways of modelling equilibrium species distributions with regard to their niche have been proposed, and in fact this has become one of the 'growth' industries in environmental and ecological science, mainly due to the need to assess the effects of climate change on species, biodiversity in general and ecosystems and their services (see Chapter 7). At the same time, there has been much confusion and discussion about the value of some of these methodologies. For example, most of these approaches assume that they are modelling the actual or realised niche of the species, based on the current observed distributions delimited by current climate and after local competition from other species. In other words, distributions are in equilibrium with climate and how a species responds to climate affects its establishment, growth and competition with other species. However, given that rapid climate change is likely to lead to disequilibrium in climate, it is perhaps incongruous to use an equilibrium model to make projections for any different climate where there will be transient changes in its various aspects. This alone would suggest the need for an approach that can express the dynamics of change.

Other issues include the question of how to choose from the multitude of ever more complicated and obscure statistical methods, the lack of important variables, the relevance of broad scales for some of the policy and conservation requirements at local scales, the general validity and usefulness of results that seem to work for current climate conditions but may not under different climates and a wide range of assumptions and uncertainties.

We intend to highlight some of the different approaches to equilibrium modelling of species ranges but not to discuss the various attributes too deeply. In fact, to explain and explore the current plethora of statistical methods alone would require another book. We will describe briefly both purely statistical equilibrium models and those with a more physiological or mechanistic logic as a background for later chapters.

Statistical methodologies do have a great advantage over other approaches in that the basic method is to compare 'current' plant or animal distributions with as many available 'current' climate variables as possible in order to find some statistical significance in these relationships. The advantage is that this can now be done for hundreds of species (e.g. plants, animals, insects) since the advent of digitised data on species distributions and climate on similar spatial grids (e.g. Thuiller *et al.*, 2005; Araújo *et al.*, 2006). More physiological methods that aim to reflect some of the actual mechanisms that interact with the environment are likely to involve a much more restricted species dataset, as the data on responses for many species – both plant and especially animal – are just not available, or even known. Thus this restriction tends to limit this type of modelling to relatively few species or groups of species, classified by similar traits (PFTs).

2.2.3 Equilibrium physiologically-based modelling of species

More physiologically-based equilibrium approaches follow the trends described earlier with regard to historical progression towards biome and vegetation modelling and in fact build on them. Results from a number of papers, produced around the same period from different biomes, sometimes using different methods, highlight the value in using climatic or bioclimatic variables that can be directly related to a plant's physiology. Some of the more significant conceptual advances include a method using nonlinear functions or 'ecological response surfaces' (Bartlein and Webb, 1986) that describe the way in which the abundance of species is related to two or more environmental variables with a physiological basis. This approach was developed for a number of major tree species in North America, using surface pollen samples transformed to vegetation composition and then related to climate data. Eight pollen types were used (*Picea, Pinus, Betula, Quercus, Tsuga, Fagus, Carya* and prairie forb pollen) with two climate variables: mean July temperature and annual precipitation, used as proxies for moisture and growing season warmth, both of which are required for plant growth. These relationships showed that the optimum for *Picea* pollen in eastern North America was where annual precipitation was around 1000–1200 mm and July temperatures around 12–15 °C. Prentice *et al.* (1991) used response surfaces to describe the relationships between three climatic variables (mean July and January temperatures and mean annual precipitation) and surface pollen samples in eastern North America to infer past climates over the last 18 000 years. A later paper by Huntley *et al.* (1995) applied them directly to selected European tree, shrub and herb species using the AFE distribution maps and three bioclimatic variables (mean temperature of the coldest month, GDDs above 5 °C and AET/PET), as originally devised by Prentice *et al.* (1992). Response surfaces of the relationship between distributions and current climate were calculated using locally weighted regressions of species presence or absence to the bioclimatic variables in the AFE grid cells. Future equilibrium distributions were then projected using outputs from GCMs to show the potential range of all modelled species, all of which showed changes in range – some significant – but said little about transient responses. Such responses require a dynamic approach and will be discussed later in this chapter, initially with regard to forest modelling.

Similar approaches were developed for other biomes elsewhere. Lenihan (1993) derived response surfaces for tree species in the boreal zone of Canada using five bioclimatic parameters: annual snowfall, degree days, absolute minimum temperature, annual soil moisture deficit and actual evapotranspiration over the summer months. He argued that these values influenced plant responses more directly than the annual or monthly values of precipitation or temperature used earlier, as any correlations with the latter might not exist under changed climates. The response surfaces were related to species dominance, considered more appropriate than presence/absence at the large grid scale of 0.5°, using the logistic regression approach of Bartlein and Webb (1986). Lenihan concluded that most of the variation in probability of dominance could be explained by species' individualistic responses to the climate controls in the different regions and that the results were more interpretable in terms of actual mechanisms than earlier studies.

Leathwick and McLeod (1996) used a bioclimatic modelling approach to predict changes in New Zealand indigenous forests under global warming scenarios. Distributions of 41 tree species taken from historical forest inventory and catchment survey data were related to climate and environmental variables considered to have some physiological basis, including solar radiation, water balance, lithology and drainage. Distributions of species were then projected across a New Zealand-wide grid, this time with an increase in temperature of 2 °C. Results suggested that the model was successful in describing the existing pattern of New Zealand's native forest and that in the future there was likely to be a significant disequilibrium between the current pattern and the changed climate.

Another equilibrium species model inspired by the work of Dahl and Woodward and based on the physiological constraints on plant growth was developed in part as a spinoff from the forest community dynamic model FORSKA2 (Prentice *et al.*, 1993a; see Section 2.3.1). STASH (STAtic SHell) (Sykes *et al.*, 1996) is a bioclimatic model used mainly in Europe and North America to describe current, future and past species distributional ranges that has the added value of providing a calculation of potential growth in any grid cell under a particular climate based on potential net assimilation. Bioclimatic variables include mean coldest-month temperature as a surrogate for the importance of minimum temperatures on a species' range and a drought index based on the Priestley–Taylor coefficient evaluated over the assimilation period for evergreen trees and over the growing period for deciduous trees. This approach is considered relevant for northern Europe in allowing for finer discrimination among environments with relatively slight summer drought stress. It also uses the number of growing degree days over 5 °C (GDD5) as applied in the BIOME model, with the added aspect of a response to chilling. The requirement for a number of chill days (below 5 °C in the winter) has been suggested to be important for some species. European beech (*Fagus sylvatica*) has been shown experimentally by Murray *et al.* (1989) to respond to the length of the chilling period. This chilling response delays budburst until after any damaging late spring frosts that could harm the newly emerging leaves. In this model, species parameter values come either from actual silvics data (Prentice and Helmisaari, 1991) or from a comparison of isoline maps with distributional maps. Simulations using current climate data show a good agreement with maps of species distributions, though with some discrepancies. They also show that growth declines (though not to zero) towards northern limits and to a lesser extent from west to east, due to reduced growing season length. In addition, the maximum temperature of the coldest month experienced can influence species presence and these limits can be abrupt. Spruce (*Picea abies*), for example, shows greater growth southwards until a limit related to becoming too warm in winter acts on regeneration, excluding the species from a grid cell (Figure 2.6). It is not clear what the physiological reason may be but, given that for spruce the limit is close to zero (−1.5 °C), it can be surmised to be related to the presence or absence of snow and the protection or otherwise that this may give to seedlings and saplings in the early growing period of late winter and early spring.

Holocene explorations with the STASH model tried to pinpoint the reasons for the distributional history of one of the most common European trees, the beech

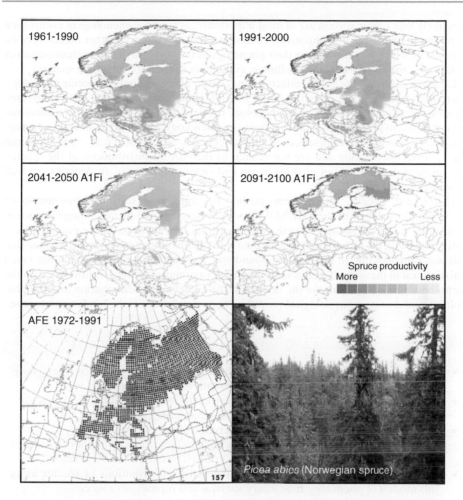

Figure 2.6 Series of *Picea abies* (spruce) distribution maps modelled using STASH and climate data from the recent past, present and future (using an A1Fi scenario (see chapter 7)). The darker the green, the more productive the species is simulated to be in that grid cell. Bottom: map of the actual natural distribution of spruce from the recent past from AFE (Jalas and Suominen, 1972–1999)

(Giesecke *et al.*, 2007). Climate data simulated for 6000 years ago from PMIP (Paleoclimate Modelling Intercomparison Project, http://pmip.lsce.ipsl.fr) were used to drive the model and the resultant modelled beech distribution at 6000 BP compared to pollen data. Major discrepancies in the distribution between modelled output and pollen-described distributions were shown, but no one factor could explain the patterns. Many uncertainties and assumptions in the species and climate modelling outputs, as well as in the pollen data, were highlighted (see Chapter 4). It seems likely that a dynamic model including competition, population dynamics and migration, perhaps with added aspects such as the role of humans, may better describe beech's distribution. Some of the bioclimatic parameter values used by STASH for individual tree species were later incorporated into dynamic vegetation models (DVMs).

Climate change affects the developmental timing or phenology of species. Longer growing seasons involve warmer spring temperatures and extended warmer days in the autumn.

It is likely that earlier springs are of most importance, at least for phenological events such as budburst and flowering. This may well have a significant effect on other members of an ecosystem, which may not change their phenology in the same way (e.g. synchronisation of pollinators with the flowers they pollinate, availability of food for migrating species).

Changes in phenology and its effects are not easily assessed, though specific phenological models have been developed (e.g. PHENOFIT; Chuine and Beaubien, 2001) and some dynamic ecosystem models include phenology routines. PHENOFIT is in fact a combination of different process-based models, based on phenology, frost injury, survival and reproductive success. Most parameters, except those related to drought survival, are derived from the observation of traits and are not related to the current distributions of the species. The present distributions of quaking aspen and sugar maple in North America are well described by the model.

Morin and Thuiller (2009) went further by comparing statistical bioclimatic or niche-based models with the PHENOFIT process-based model by describing the current and future distributions of 15 North American trees. They showed more potential local extinctions with the niche-based models than with the phenology model and suggested that this was probably due to the fact that phenotypic plasticity and adaptation are not taken into account in the niche-based models. The model assumptions are also different, with the niche-based models relying on the realised niche to define the distribution and the phenology model using a physiological modelling approach based on the fundamental niche.

Summary: Equilibrium physiologically-based models

- Equilibrium physiologically-based models tend to be applied to relatively few plants and only where known physiological information is available.
- Ecological response surface models relate vegetation composition via pollen samples to two or more environmental variables that have a physiological basis.
- STASH gives tree species distributions derived from a few physiologically important climate variables, including response to winter chilling, and can give an estimate of how much of a species is to be found in a grid cell based on net assimilation.
- PHENOFIT is a phenology model that describes distributions of species using sub-models of phenology, frost injury, survival and reproductive success.

2.2.4 Statistical equilibrium modelling of species

Purely statistical approaches to the description of species ranges in terms of their environment have been extensively explored and developed over the last 30 years. Guisan and Zimmermann (2000) effectively review the background to and theories behind these models and we will therefore not discuss them to any great extent. Such models are described under different names, including 'bioclimatic envelope', 'niche-based', 'habitat' and 'species distribution models' (SDMs), but they all have a basically similar methodology.

The objective of these models is to find a species' 'climate or environmental envelope', which best describes its current range. This is in principle the same as in the more physiological models, except that often all possible environmental variables are included in the hunt for those that are statistically significant. These significant variables are then used to predict species distributions under different climates. The most common statistical methods used in the past include general linear models (GLMs), general additive models (GAMs), logistic regressions and artificial neural networks (ANNs). In a review of assessments of the accuracy of different methods, Heikkinen *et al.* (2006) found that most had problems, although these were often different. Segurado and Araujo (2004) evaluated seven methods using lizard and reptile species and concluded that there was a strong relationship between model performance and the kind of spatial distributions being modelled. Models generally did best with restricted-range species and worst with widespread species, which might be expected.

Of more importance for climate-change studies, however, there is evidence of increased variability across bioclimatic models when projecting into the future. Thuiller (2003), within the BIOMOD bioclimatic envelope modelling framework, assessed a number of statistical approaches (including GLMs, GAMs, classification and regression trees (CARTs) and ANNs) and concluded that on average all methods gave very good agreement between observed and modelled distributions. However, the accuracy of each model was species-dependent and differences between models describing current distributions were made worse in future scenarios. The variability across models in their projections of future ranges was large and could override variability arising from the use of a range of climate-change scenarios (Thuiller, 2004). Thuiller concluded that relying on just one technique was flawed and that a framework giving a range of models and then selecting the most accurate was likely to give stronger predictive capabilities. Thuiller (2004) and Araújo *et al.* (2005) used consensus analysis based on ordination to produce composite variables that summarised the greatest amount of information from four different statistical methods and produced forecasts that significantly reduced uncertainties. However, they concluded that the averaging of projections only really works when the models themselves have good overall performance and good data are available; in other words, the usefulness of the consensus approach depends on model and data quality. Statistical modelling uncertainties and challenges have been extensively discussed in the literature and the reader is referred to Zimmermann *et al.* (2010) for more information.

Summary: Statistical equilibrium modelling of species

- Statistical equilibrium models have as their objective the discovery of the species climate or environmental envelope that best describes its current range, often using all available environmental variables. The statistically important variables are then used to project future ranges.
- They can be used to describe the ranges of thousands of species, both plant and animal, so long as their distributions are digitised.
- There are many possible methods, some better than others. Their accuracy is often species-dependent. Variability between models increases when they are used to project into the future. Consensus approaches significantly reduce uncertainties.

2.2.5 Some uncertainties and assumptions that apply generally to bioclimatic models

There are a range of other assumptions and uncertainties that are relevant to the themes of this book, which must be borne in mind when considering the results from equilibrium bioclimatic species models, whether statistical or physiologically based, if for example the approach relies on species being in equilibrium with current climate. However, this is merely an assumption. The approach is likely to be species-dependent, not least because of differences in dispersal rates among species and hence the ability to track climate. Svenning and Skov (2004) demonstrated departures from equilibrium for European tree distributions, showing that many species have a potential range well beyond their native range, probably reflecting dispersal limitations and constraints on postglacial expansion. Araújo and Pearson (2005) suggest that the assumption of equilibrium with climate may be valid for some species groups but varies substantially across groups of organisms; reptiles and amphibians, for instance, are poor dispersers and are therefore less likely to be in equilibrium with current climate. In addition, genetic differences in populations at different range margins in current distribution maps may in fact lead to false presumptions about how the range might change. Northern populations may respond to climate change in a different way to southern populations. If the climate change is mostly related to temperature change and thus has a strong latitudinal element to it then boundaries may well shift together, as might be predicted. If the climate change is more complicated, however, involving for example a west–east component related to continentality or drought, then genetic differences in extremes of the range may have unexpected consequences, including increasing the difficulty of fitting models.

Dispersal and movement across the fragmented landscapes of Europe and North America are often mentioned when assessing the reliability of outputs from bioclimatic models and the lack of realistic dispersal in these models is usually used to discredit the approach. Dispersal and species abundance affect competition and thus current distributions and their climate relationships are no real indicators of future distributions.

Accurate outcomes from modelling require that samples be taken across the complete environmental gradient in which a species is found, including sites defining the boundaries of species' environmental ranges. Incomplete sampling can strongly influence species response curves, especially at the lower and upper ends of the environmental range, restricting predictive ability and giving spurious projections into the future. In modelling done in Europe, for example, the 'full' range or climate space of a species is often presumed to be Europe or even the European Union. This is a moot point given the success of many invasive European species in places like New Zealand and elsewhere, where climates may be subtly or not so subtly different.

In summary, modelling bioclimatic envelopes can give reasonable results at the broad scale (continental) and such models can be used at that scale to make assessments about possible futures if various caveats and uncertainties are taken into account. Downscaling species' bioclimatic envelopes from broad to fine or very fine scales expands the range of problems and unknowns related to methodologies, data, drivers and output that cascade down through the scales, compounding uncertainty. Further assumptions must be made at finer scales, related for example to landscape heterogeneity (slope, aspect), local climate variation, soil variability in type and nutrients. In addition, they do not address competition or stochastic events, either ecological (e.g. pests or diseases) or climatic (e.g. extremes: wind, frost, water), which are particularly important at the finer scale, where local events and disturbances can have devastating effects on populations locally. Nor do they really take into account the fact that climates, CO_2 and land use are changing at a rate unknown in the Holocene and are interacting in ways known and unknown, leading to new and original landscapes. The future is likely to see a mixture of landscapes that have no analogue now or in the past, landscapes that have some similarity to existing ones and landscapes that remain the same, while some current landscapes will be lost forever.

2.2.6   Models of intermediate complexity

So where to go with this approach? It is clearly useful in a broad-sweep way to show the interactions between a few or many species and climate, especially in a climate-change context. Many of the issues are, however, addressed to varying degrees through dynamic ecosystem modelling, as will be discussed later in this chapter. But dynamic modelling, as will also be shown, is reliant on relevant data, much of which are absent, especially at the species level. Huntley *et al.* (2010) suggest the use of 'models of intermediate complexity', an approach proposed to address the problems of modelling Earth systems. ESMs of intermediate complexity (EMICs) (Claussen *et al.*, 2002) aim to describe the natural Earth system in which humans are viewed as an external driver by simulating some of the important processes in a more simplified way than is seen in complex models but in a more complex way than is seen in simple conceptual models. Huntley *et al.* (2010) suggest a fruitful approach along these lines might be the development of an integrated model of submodules in which each simulates different elements, such as the suitability of the climate and habitat, population dynamics and dispersal. An example of development towards such an integrated model is the work of Keith *et al.* (2008), who linked dynamic habitat suitability statistical models with stochastic population models and simulated plant populations in the South African fynbos under different climate scenarios. Their results suggested that the survival or extinction of any species under different climates was dependent on interactions between life history, disturbance regime and pattern of distribution. Anderson *et al.* (2009) built on this approach to explore the range limits of species under different climates, as well as extinction risk, metapopulation dynamics and dispersal. The sensitivity of range margins and the different responses between the leading and trailing edges were shown to be important. Given the lack of relevant data for the more complex dynamics models, especially with regard to answering questions relevant to conservation of species, it is likely that the development of such intermediate-complexity models could be a fruitful avenue for further research.

2.2.7 Biogeochemistry integrated into equilibrium biome models

A major step in the progression to dynamic ecosystem models was the integration of biogeochemistry models and plant geography models of the BIOME type. Biogeochemistry models were initially developed to simulate global net primary production (NPP). Various different approaches were explored. The Miami model, so called as it was first presented at a symposium in Miami in 1971, predicted NPP from annual average precipitation and temperature to produce a global map of terrestrial NPP. Similarly, the Montreal Models predicted primary productivity from evapotranspiration (Lieth, 1975). However, it was Melillo *et al.* (1993) who developed the first process-based equilibrium model of global patterns of NPP and soil nitrogen cycling using current climate and CO_2 concentrations. The terrestrial ecosystem model (TEM) was driven by data on climate, elevation, vegetation, soils and water availability on a $0.5 \times 0.5°$ global grid. These data were either from published information or, in the case of some of the vegetation, from calibration of the model to fluxes and pool sizes at intensively studied field sites. The model estimated the carbon and nitrogen fluxes and the size of the pools based on the potential vegetation for the globe. The results were similar to the mean of other published estimates. The same authors also estimated future fluxes from climate data simulated by four different GCMs with a doubling of atmospheric CO_2 concentrations. Their simulations showed that responses in tropical and dry temperate regions were most influenced by changing CO_2, while in northern and moist temperate regions the effects of temperature on nitrogen availability were the most important (Figure 2.7).

There are a number of other models of this type, called generally terrestrial biogeochemistry models (TBMs), and many are still widely used, such as Century (Parton *et al.*, 1993). Integrating biogeochemistry into equilibrium plant geography models is an important next step in the history of ecosystem models.

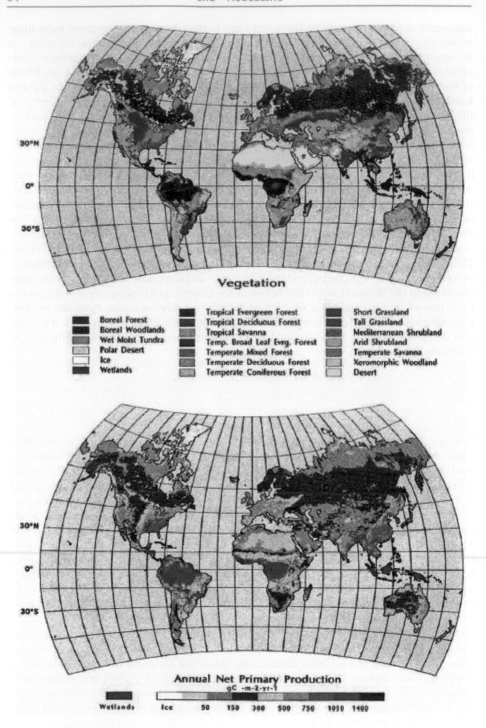

Figure 2.7 Output from the TEM model. Top: potential natural vegetation. Bottom: modelled NPP. Source: Melillo *et al*, 1993. Reproduced by permission of Nature Publications

2.2.8 Integrating biome and NPP models

An understanding of the complexity of the Earth system, especially in the context of the mechanistic modelling of climate-change effects, requires the integration of more and more processes that more closely represent ecosystems and their processes. This integration was significantly advanced through the Vegetation/Ecosystem Modeling and Analysis Project (VEMAP), an international project involving leading ecosystem modelling groups from the USA, the UK, Sweden and Germany. The programme supported the development of different models that could address the responses of biogeography and biogeochemistry in the conterminous USA to changing climate through time and space. It was an intercomparison of the different models and their responses to changing climate and CO_2 concentrations, as well as an extensive sensitivity analysis of models and results. In the first phase, equilibrium models were compared, and in the second the aim was to model transient ecosystem dynamics through development of more dynamic approaches in both biogeochemical models and DGVMs.

In the first phase a number of different models linked biogeography with biogeochemistry. DOLY (Woodward *et al.*, 1995), for example, is a model that simulates both locally and at global scales the distribution of LAI and rates of primary productivity based on soil nutrients and water holding capacity. The MAPPS (Mapped Atmosphere–Plant–Soil System) model (Neilson, 1995) had the basic idea that the distribution of vegetation is controlled either by the availability of water for transpiration, especially in temperate latitudes (Figure 2.8), or by energy for growth, especially in high latitudes.

The LAI of both woody and grass life forms is modelled through their competition for light and water. The model simulates the distribution of a few life forms (rather than biomes), dominant leaf type, leaf phenology, thermal tolerances and LAI.

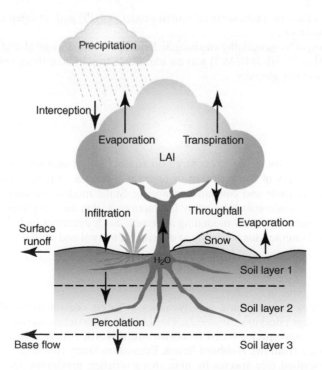

Figure 2.8 Vegetation–water balance model included in the MAPSS model (Neilson, 1995)

These characteristics are then combined into a classification not unlike the idea of biomes. The model was implemented at 10 km resolution in the USA and at 0.5° grid globally, as well as at watershed scale in the USA. Simulations with the model driven by GCM scenario data show changes such as a northward shift in vegetation and drought-induced forest dieback in eastern North America and eastern Europe (Neilson and Marks, 1994). Regional water patterns show greater spatial variation in their response than do LAI changes, even though they are linked together in the model. Haxeltine *et al.* (1996) built on this and earlier approaches to develop a new model (BIOME2), initially focused on Australia, which could simulate the structure of potential natural vegetation and phenology using biogeochemical calculations of the carbon and water fluxes. This model was based on the calculation of leaf area using foliar projective cover (FPC) as a function of available moisture to provide NPP of the different plant types. The same approach is involved in the well-known BIOME3 equilibrium model (Haxeltine and Prentice, 1996), which is a coupled carbon and water flux model that calculates for each PFT the LAI that maximises NPP. NPP calculations use a photosynthesis model based on the Farquhar model (Farquhar *et al.*, 1980), with maximum photosynthesis rates derived using a theory based on optimal nitrogen allocation. Competition between PFTs is simulated using the optimal NPP of each PFT as an index of competitiveness. A soil model in two layers allows competition between grass and woody vegetation based on the ratio of roots in the upper and lower soil layers.

Summary: Vegetation and biogeochemistry modelling

- Biogeochemistry models were initially developed to simulate global NPP; e.g. the Miami model.
- TEM was the first process-based global model of NPP and nitrogen cycling using climate and CO_2.
- The linking of biogeography and biogeochemistry in equilibrium global models (e.g. DOLY, MAPPS BIOME2&3) was an important step in describing vegetation and biogeochemistry globally.

2.3 Dynamic modelling

Up to this point the discussion has centred on equilibrium or snapshot models and their development in terms of the inclusion of relevant processes. But the environment and ecosystems are dynamic and constantly changing. Static models can only go so far and if we are to better understand the many interactions within an ecosystem, dynamic processes must be included in any modelling approach. Such dynamics include a large range of processes and interactions occurring at different temporal and spatial scales.

2.3.1 Local to landscape scales: forest gap modelling

Modelling of mixed species and age dynamics in forest ecosystems really began with the work of Botkin (Botkin *et al.*, 1972). The JABOWA model simulated the population dynamics of a mixed-species forest in northeastern North America and its output was compared to data from the Hubbard Brook Ecosystem Study (www.hubbardbrook.org). The model described tree species by nine characteristics: maximum age, diameter and height, height–diameter and leaf weight–diameter relationships, photosynthesis rate and available light, relative growth and climate, as well as relationship to soil moisture. The environmental information included elevation, soil depth, soil water-holding capacity, percentage rock in the soil and climate data including monthly temperature and precipitation. Competition between individuals was for light, and some species were more shade-tolerant than others. The model had a growth rate for each species that represented the growth on an optimum site with no competition. In each year the growth rate was reduced according to various species-specific factors such as shade tolerance, as well as site soil quality and climate. Saplings were planted in a plot each year according to their shade-tolerance and a suitable environment. Tree mortality could be age-dependent or related to low growth rates; there was also a close-to-random probability of death that represented such events as wind or lightning damage. Species interactions were simulated over 2000 years. The model was able to produce the general behaviour of the Hubbard Brook ecosystem by reproducing competition, succession and vegetation changes related to elevation in the ecosystem. Of course, as with all models there were various assumptions and uncertainties about parameters and generalisations. However, Botkin did conclude that a nonobvious result from the whole exercise, at least at that time, was that it was possible to represent the general behaviour of a complex forest using only a few characteristics – a conclusion that is perhaps nowadays often taken as a given in ecosystem modelling.

Further developments in the approach to gap modelling, for example, occurred in a variety of later models, such as SORTIE (Pacala et al., 1993) and FORSKA (Prentice and Leemans, 1990), among others. In the case of FORSKA, various improvements to the original gap modelling concepts were made, including a more realistic representation of the shape of a tree, from a flat plate (in which all the leaf area is located at the top of the tree) to a pyramidal shape (which distributes the light more evenly through the canopy). This was an important improvement in modelling the response to light, especially in higher latitudes. FORSKA2 (Prentice et al., 1993a) improved on the basic model by providing better landscape dynamics, including climate-driven simulations over hundreds of years to represent the history of the landscape, for example through the effects of changing climate on the dynamics of the forest.

Disturbances at the forest landscape scale as a whole can be both natural, such as wind or fire, and human-induced, such as logging. Such disturbances are important not only for a more realistic assessment of forest biomass and the successional processes but also in reflecting the role of changing climate. This might be through the provision of space or gaps for newly immigrating species, allowing them to establish themselves and to compete for resources with species already in the area, for example. In FORSKA2 the landscape was in effect an array of replicate modelled patches in which forest gap phase dynamics occurred. A stochastic generalised disturbance regime similar to that found in a forest was simulated over the whole landscape, with an increasing probability of disturbance with patch age. If a patch was disturbed, all vegetation was removed and the patch was replanted with species that could survive the current climate. The model was typically run for landscapes in the boreal and temperate regions of the northern hemisphere, where climate change and global warming are expected to have a strong effect. Typically, silvics data (e.g. from Prentice and Helmisaari, 1991) were used to define the growth parameters for each tree species used. In addition, species-specific environmental drivers developed from the equilibrium bioclimatic model STASH were used in test landscapes in Sweden (Sykes and Prentice, 1996). Simulations included different future climate-change scenarios, with and without disturbances and with and without the two extremes of migration: nonlimiting and no migration; that is, only species present in the current climate were allowed to establish. Results were complex, especially with regard to transient responses coupled with interactions with disturbance history and potential migration under a rapidly changing climate. Outcomes where no migration was allowed led to forests composed of early successional species and in general fewer tree species.

Some typical results are shown in Figure 2.9. Two different experiments from the same site in southern Sweden (Boa Berg) show the different effects of different levels of climate warming on the forest biomass and dynamics (Figure 2.9, upper two graphs). A warming over 100 years of 1–2 °C (upper graph) leads to the demise of *Picea* (spruce) and an increase in *Pinus* (pine), and at the same time an increased biomass in the deciduous species that are already there, giving as a result a much more mixed landscape. Under a greater warming (4–5 °C), *Picea* dies out and *Pinus* declines significantly (middle graph). There is a change in the nature of the forest landscape to a more open deciduous forest of much less biomass, dominated by *Betula* (birch) and *Quercus* (oak) species. Note that *Fagus* (beech) initially increases under the new climate, before declining due to increasing winter warmth, leading to a lack of the winter chilling required to avoid damaging late frosts during budburst. Further north, however, at Reivo, a similar increase in temperatures promotes the growth of *Picea*, leading to a landscape even more dominated by *Picea* (bottom graph).

Summary: Forest gap dynamic modelling

- Forest gap modelling is based on the ideas behind succession and the cycles that can be observed in plant communities, especially in forests.
- Old trees are blown down, forming a space or gap for colonisation by early successional species. This leads to a mosaic of differently aged populations of tree species in the forest.
- Later models are based on JABOWA, a model of population dynamics in a mixed-species forest.
- Models such as FORSKA2, which have improved representations of tree shape, for example, were developed for use at the landscape scale and for application in climate-change studies.

2.3.2 Regional to global scales: dynamic global vegetation modelling

Up to this point, broad- or global-scale modelling of vegetation distributions and NPP has been restricted to equilibrium snapshots in time. However, in order to better understand the effect of changing levels of atmospheric CO_2 concentrations on the dynamics of global terrestrial ecosystems, for example, dynamic modelling must be carried out at broad or global scales. There are clearly scale issues in doing this, however. Parameterisations of the various variables used in gap models, for example, are often site-specific, and to generalise in any sensible way to the degree required for the global scale is difficult. Measuring, through experiments or through collection of data on the physiological and ecological interactions between species or individuals, normally occurs at much finer scales, and the question is how to generalise in a way that is relevant for the modelling of ecosystems at the global scale. A number of attempts to move from the gap model local-scale modelling of dynamics to broader global scales have been published. Moorcroft *et al.* (2001) proposed a method using the ecosystem demography (ED) model that takes account of fine-scale spatial heterogeneity caused by stochastic processes normally modelled in gap models. From this, a size and age structure approximation is made, which allows broader-scale projections to be made without the need to model each individual as in a normal gap model approach. The model was tested over parts of South America and seemed to be valid for a range of environmental conditions and ecosystem types. Output was also compared at the regional

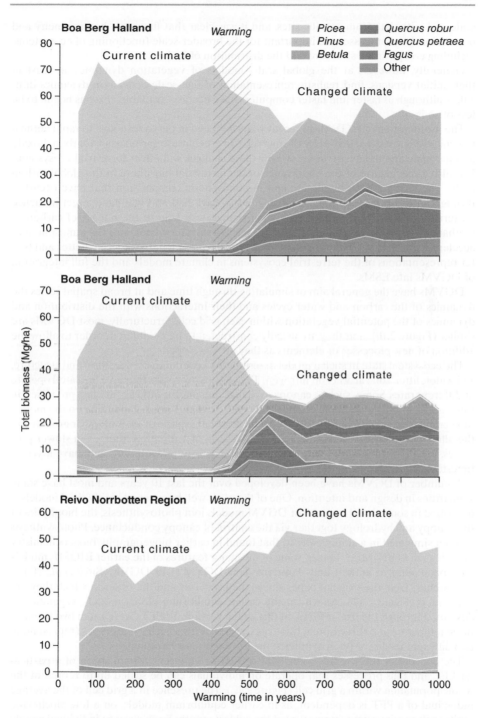

Figure 2.9 Simulations using FORSKA2 for two sites from southern (Boa Berg with two different climate changes) and northern (Reivo) Sweden. At each site, forest dynamics are simulated over 1000 years in 200 forest patches that also include a stochastic disturbance rate. Up to 19 tree species might establish if the climate is suitable. The current climate is used for the first 400 years and then the climate is allowed to warm over 100 years by 1–2 °C (Boa Berg top) or by 4–5 °C (Boa Berg middle and Reivo). After year 500, the climate is held steady at the new level. Source: Sykes and Prentice, 1995. Reproduced by permission of Springer

scale and with data from specific sites, and it was clear that fine-scale heterogeneity and population-level processes are important for the broader-scale functioning of ecosystems, including composition, structure and the distribution of carbon.

Generally, however, at the global scale, models of vegetation dynamics, at least in their initial versions, find realistic representation of fine-scale ecosystem dynamics difficult – although as faster and faster computers have become available, this has begun to be less of an issue.

The proliferation of DGVMs in recent years has been in part a response to rapid anthropogenic climate change driven by the increases in greenhouse gases such as carbon dioxide and methane and the desire to know how these changes will affect terrestrial ecosystems. It is also a recognition of the importance of the terrestrial biosphere in the global carbon cycle. The changes both in climate and in atmospheric composition that have occurred throughout the Holocene and continue to occur have had and will have direct influence on terrestrial ecosystems. These ecosystem changes can lead to the release of carbon or methane from the terrestrial ecosystem, which can then feed back into the atmosphere to accelerate the changes. They are also driving the development to integrate better and better representations of the terrestrial ecosystems in climate models and the full integration of DGVMs into ESMs.

DGVMs have the general aim of simulating through time and at broad spatial scales the dynamics of the carbon and water cycles and their interactions with the distribution and dynamics of the potential vegetation within each grid cell. Structurally, most DGVMs are similar (Figure 2.10), and they are usually constructed to be modular, in order to allow the addition of new processes or elements as they are developed.

The ecosystem state in such a model describes the vegetation composition and structure, soil water, litter, soil carbon and nitrogen. The various processes that are simulated operate at different rates. Some, such as canopy exchange and photosynthesis, are fast processes; in this case 'fast' can mean in the order of 1 day or less than 1 hour, depending on the model. Slower processes such as those involved with seasonal aspects of phenology or growth and the allocation of carbon may operate at a timescale of 1 month. Other, even slower processes, usually involving the dynamics and disturbance of vegetation, may have a yearly timestep.

A number of DGVMs have been developed over the last 10 years and most have some similarities in design and intention. One of the more well known and highly cited models is described in some detail here. Most DGVM models join photosynthesis, the biophysics of the canopy and hydrology together via the control of canopy conductance. Photosynthesis is often simulated in a similar way to that found in earlier biogeography/biogeochemistry models such as BIOME3. In fact, some of the major features of the earlier BIOME models have been adopted in the Lund–Potsdam–Jena DGVM (LPJ-DGVM; Sitch et al., 2003). A flowchart from the original paper shows the processes and the order in which they are performed (Figure 2.11). As with many earlier equilibrium global models, vegetation is described through the use of PFTs. In this global version, 10 PFTs were used: two tropical, three temperate and three boreal tree types, as well as two herbaceous types (representing two different modes of photosynthesis: the C_3 and C_4 mechanisms).

The concept of the average individual is used to represent a population of a particular PFT and thus processes that operate on individuals can be scaled up to represent the whole population within a grid cell (Figure 2.12). The presence in a grid cell of the average individual of a PFT is dependent, as in earlier equilibrium models, on a few bioclimatic limits such as minimum temperature of the coldest month. Each average individual woody PFT has various variables or attributes that describe it, such as crown area, leaf, sapwood, heartwood, root mass, the fraction of fine roots in the different soil layers, the minimum canopy conductance, fire resistance and leaf longevity, which is related to whether the PFT is evergreen, summergreen or raingreen. Herbaceous PFTs are treated much more simply.

Climate, PAR, (CO$_2$), N deposition, soil physical properties

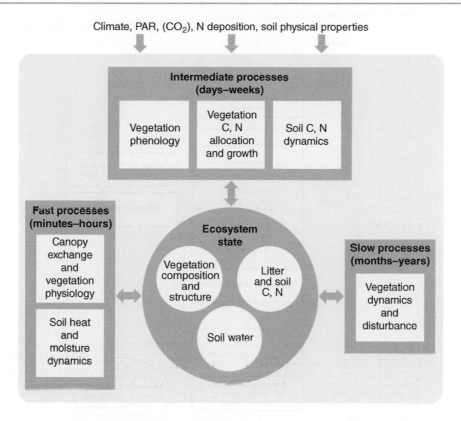

Figure 2.10 Typical structure of a DGVM. Source: Prentice *et al*, 2007. Reproduced by permission of Springer

The vegetation within a grid cell is described by the overall fractional cover of any PFT in the cell, which is found by defining the area of ground covered by foliage directly above an average individual (i.e. the FPC) and multiplying by crown area and population density. In any grid cell there can therefore be a mosaic of different populations of PFTs, which are assumed to be competing for resources. Establishment of individual PFTs occurs each year, depending on available space, but it can be reduced by the response to heat and water stress. Soil hydrology is modelled in a similar way to the BIOME3 model, using two soil layers of different but fixed thicknesses. The water content of these layers is updated daily based on precipitation, percolation, evapotranspiration, runoff and snow melt. Tree roots and herbaceous PFTs are distributed differentially between the layers, with fewer tree roots in the upper layer. Growth and allocation to different tissues, as well as reproduction and mortality, are also calculated for each average individual PFT. Maintenance or autotrophic respiration is calculated daily based on carbon : nitrogen ratios (C : N), the temperature for both above- and below-ground plant parts, phenology and tissue biomass. Mortality can occur for a number of reasons, including light competition, poor growth and heat stress. The biomass of dead individuals is moved to various litter and soil organic matter (SOM) pools. These pools can be fast, where the litter is returned quickly (around 2–3 years) as CO$_2$ to the atmosphere, or intermediate or slow, where decomposition occurs over approximately 33 and 1000 years, respectively, but this is dependent on soil temperature and moisture.

In a typical simulation, the relevant information that describes each PFT, as well as grid cell-based environmental data such as mean temperature, precipitation and cloud cover or solar radiation values for each month of each year, is required along with simple soil

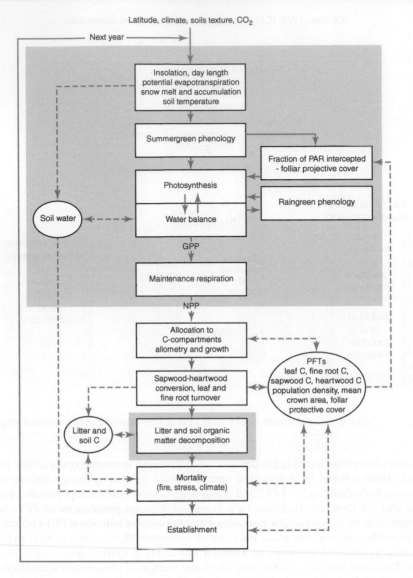

Latitude, climate, soils texture, CO_2

Next year

Insolation, day length
potential evapotranspiration
snow melt and accumulation
soil temperature

Summergreen phenology

Fraction of PAR intercepted
- folliar projective cover

Photosynthesis

Raingreen phenology

Soil water

Water balance

GPP

Maintenance respiration

NPP

Allocation to
C-compartments
allometry and growth

PFTs
leaf C, fine root C,
sapwood C, heartwood C
population density, mean
crown area, follar
protective cover

Sapwood-heartwood
conversion, leaf and
fine root turnover

Litter and
soil C

Litter and soil organic
matter decomposition

Mortality
(fire, stress, climate)

Establishment

Figure 2.11 Flowchart describing the order of various processes that occur during a simulation using the LPJ-DGVM (Sitch *et al.*, 2003)

data such as soil texture and percolation rate. Global CO_2 concentrations for each year of the simulation are also required. The climate data are further downscaled to daily values for use within the model. In a standard modelling experiment, the model needs a preliminary or 'spin-up' period before the simulation proper begins, as it starts with no vegetation and empty pools in each grid cell. This period, usually around 1000 years, is necessary for reaching some sort of equilibrium in the vegetation and in the litter and SOM pools. Often the first 30 years of historical data are cycled around and around for the spin-up period. Global historical time series data (from approximately 1900 to the early years of 2000) provided by the Climate Research Unit at the University of East Anglia on a global grid of 0.5° have become the standard dataset for use by many models. Projections into the future

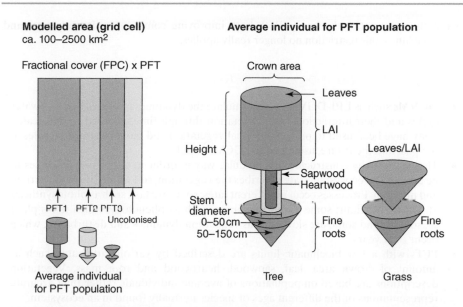

Modelled area (grid cell)
ca. 100–2500 km^2

Fractional cover (FPC) x PFT

PFT1   PFT2   PFT3
Uncolonised

Average individual
for PFT population

Average individual for PFT population

Crown area

Leaves

LAI

Height

Sapwood
Heartwood

Stem
diameter
0–50 cm
50–150 cm

Fine
roots

Tree

Leaves/LAI

Fine
roots

Grass

Figure 2.12 Simplification of vegetation dynamics in the LPJ-DGVM, where each PFT population is represented by the concept of an average individual (Sitch *et al.*, 2003)

are usually driven by climate data that is in part modelled by GCMs and might also include interpolations between the end of the historical dataset and the start of the projection, for example at 2071. Projections also often use different possible future scenarios based on the IPCC SRES projections (see Chapter 7).

Typical outputs from the LPJ-DGVM can be at site, regional, continental or latitudinal scales and include representations of the seasonal cycle of CO_2; that is, the flows between ecosystem and atmosphere (NEE/NEP) and the changes through the seasons. For example, carbon in the northern hemisphere is taken up in the summer and released to the atmosphere during the winter. Globally outputs can include various global values such as runoff for comparison with data, as well as global maps of potential vegetation, vegetation carbon, soil carbon, NPP and so on.

At around the same time, an additional development incorporated some of the dynamics described in gap models of the FORSKA2 type within a regional-to-landscape version called LPJ-GUESS (Smith *et al.*, 2001). This significantly improved the modelling of vegetation by explicitly including age classes in woody vegetation, which produced much better competition and thus more realistic dynamics and vegetation descriptions. LPJ-GUESS and LPJ-DGVM have the same modelling of plant physiology and biogeochemistry, but while LPJ-DGVM models populations through the idea of the average individual, LPJ-GUESS models individuals of different ages. Smith *et al.* (2001) compared both approaches over a range of sites and climates in Europe. Both models generally successfully predicted PFT composition and succession in modern natural vegetation, but in certain areas such as mixed deciduous–evergreen forests and areas of seasonal water deficit, LPJ-GUESS performed significantly better. This is related to the individual-level modelling of processes concerned with light competition and stress-induced mortality. This level of detail clearly improves the description of vegetation and of the dynamics of ecosystems and thus the projections at regional to continental scale. However, at the global scale this level of detail may or may not be required, and until recently it was extremely computationally heavy. With both improving computer code and faster and faster machines, this restriction no longer really applies.

Summary: dynamic global vegetation models

- DGVMs such as LPJ-DGVM aim to simulate the dynamics of the carbon and water cycles and their interactions with vegetation through time at broad spatial scales. They have been used to project potential vegetation and biogeochemistry cycles in the past and the future, using gridded GCM output.
- They are usually constructed in a modular way in order to allow new processes to be added. The ecosystem state describes the vegetation, soil water, litter, soil carbon and nitrogen. Processes occur at different rates: some are fast, such as photosynthesis at less than 1 hour; some are intermediate, such as phenology, which takes place seasonally; and some are slow, such as vegetation dynamics and disturbance, which occur over years.
- PFTs with a few bioclimatic limits are described by various attributes, such as amount of crown area, leaf, sapwood, heartwood and root mass. Vegetation descriptions are based on populations of average individuals rather than accurate representations of the different ages of species normally found in an ecosystem.
- Typical outputs include a representation of the seasonal cycle of CO_2, runoff, global distribution maps of vegetation or soil and plant carbon.
- The ideas from gap models have been used to further improve DGVMs for landscape simulations in order to produce better representations of vegetation, by including individuals and age classes (LPJ-GUESS).

2.4   Integrating models

The likely problems associated with rapid climate change have promoted the development of a range of different models, which explore in integrated ways the responses of different types of Earth system, both natural and human.

2.4.1   Earth system models

ESMs integrate GCMs into ecosystem models such as DGVMs; the former are concerned with physical climate dynamics and the latter with the dynamics of vegetation and biogeochemical cycles. An overview of the approach is given by Flato (2011). They tend – as yet, anyway – not to include much in the way of integration with economic or societal aspects, which are usually found within integrated assessment modelling (see Section 2.4.2). In the early days, climate models tended to include a fixed, very simple green slime approach to vegetation descriptions, along with a simple carbon cycle. However, with the realisation that feedbacks from the vegetation to the climate system could be important for the climate system, the integration of a more realistic representation of vegetation dynamics and the associated biogeochemistry was carried out (Figure 2.13).

Cox *et al.* (2000) published the first coupled climate–carbon model using the Hadley Centre coupled ocean–atmosphere HADCM3 GCM, an ocean carbon cycle (HadOCC) and the TRIFFID DGVM. They showed that CO_2 concentrations are higher in the coupled models than in the uncoupled by 2100, due to additional feedbacks, and that as a result global mean temperatures are 1.5 °C warmer. This is mostly the result of increased

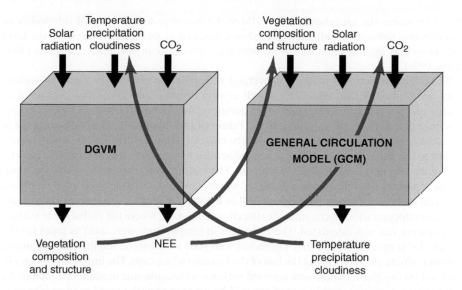

Figure 2.13   ESM structure accounting for ecosystem–climate feedbacks. Reproduced by permission of Ben Smith

CO_2 release due to increased soil respiration resulting from the warming. This approach, although full of uncertainties, shows the potential importance of feedbacks and the necessity of including more realistic ecosystem processes in climate simulations. In a later paper, the same authors (Cox *et al.*, 2004) raised the issue of increased forest dieback in the Amazon caused by raised CO_2 levels, leading to a warming circulation pattern that reduces rainfall in parts of the Amazon and causes further forest dieback and an increased feedback, with an increase of CO_2 to the atmosphere and a changing land surface. Huntingford *et al.* (2008) explored the uncertainties in the earlier projection and replaced TRIFFID with the more complex individual-based ED model, which includes size and age structure approximations (Moorcroft *et al.*, 2001; see Section 2.4). The outcome was a reduced but still significant forest dieback with soil moisture depletion and loss of Amazon forest. The importance of modelling feedbacks through coupled ESMs has also been shown using RCA-GUESS, which involves the coupling of LPJ-GUESS ecosystem model to the Swedish Rossby Centre regional atmospheric model, RCA3. Wramneby *et al.* (2010) used the coupled model over Europe to identify potential hotspots of vegetation where feedbacks are important. For example, when the forest expands upwards in altitude in the Scandinavian mountains there is a reduced albedo effect as the continuous snow cover is lost, leading to a positive feedback on climate and increased warming. In central Europe, increasing CO_2 stimulates vegetation growth through the fertilisation effect, giving a higher LAI and increased evapotranspiration, producing a negative feedback on the climate and thus mitigating the effects of climate change. In contrast, in southern Europe, where the climate becomes drier, especially in the summer, plant growth and cover decline, leading to reduced evapotranspiration and a positive feedback on climate.

2.4.2   Integrated assessment models

Another approach was initiated in the Netherlands, where an integrated global model was developed to explore the effect of climate change on various Earth systems. The integrated model to assess the greenhouse effect (IMAGE) is composed of coupled submodels or modules that simulate the effects of greenhouse gases on global temperatures and sea-level rise, as well as their impacts on the ecology and socioeconomy of selected regions (Rotmans, 1990).

IMAGE1 incorporated modules concerned with past and present estimates of various trace gases, such as CO_2 and CH_4, as well as with future scenarios. It also contained an energy module that considered economic, demographic, technical and policy issues, which was intended to make projections into the future of the supply and demand for energy in different regions. CO_2 emissions could then be calculated based on different scenarios with regard to the use of fossil fuels, which was influenced by population size, economic growth, fuel taxes and non-fossil fuel developments. The output from this module was the input to the atmospheric concentration modules, which included a carbon-cycle module linked to ocean and deforestation modules, as well as a trace gas (not CO_2) module. The various trace gas concentrations were input into the climate module, where the global mean surface temperature rise was calculated. These changes in temperature were used as input for the sea level-rise module, where they interacted with such elements as the thermal expansion of water, alpine glacial melt and the loss of the Greenland ice caps. The impact of changes in sea level on the Netherlands were assessed within a socioeconomic module that included a risk analysis. Various possible scenarios could be developed with regard to socioeconomic options for the future.

IMAGE1 was integrated into IMAGE2 as the first global integrated model that was explicitly geographical (Alcamo, 1994). It also included some of the ideas developed in the BIOME1 model (Prentice *et al.*, 1992), as well as those from the RAINS model of acidification (Alcamo and Hordijk, 1990). IMAGE2 had various goals, both scientific and policy-related, and importantly it could be tested against data. Scientifically, it aimed to explore the relationships between society and the climate–biosphere system, including (importantly) feedbacks, uncertainties and knowledge gaps (Alcamo *et al.*, 1994). In terms of policy, its goals included aids to decision makers that linked science and policy together in a geographically explicit way within a long-term perspective with respect to climate change, which explored links within the system and the side effects of policy actions. It also aimed to explore the role of different economic trends as well as of advances in technology and to provide a cost–benefit analysis of various measures that might mitigate climate change or provide adaptive approaches to it. The model consisted of three linked submodels (Figure 2.14): the energy–industry system, the terrestrial environment system and the atmosphere–ocean system. Emissions of greenhouse gases related to energy consumption and industrial production were calculated in the energy–industry system. Changes in global land cover based on economic and climatic factors were modelled in the terrestrial environment models. These were used to calculate the fluxes of greenhouse gases from the biosphere to the atmosphere. The atmosphere–ocean model simulated average temperature and precipitation patterns based on the increase of greenhouse gases in the atmosphere.

A significantly updated version, IMAGE2.1 (Alcamo *et al.*, 1998), made a number of changes, including to various methodologies modelling future regional energy use and improving land cover allocation. It could be used to assess the costs of controlling emissions and the impacts of biofuel usage, to link global forestry with climate change and to carry out climate-change assessments. It was used to develop baseline scenarios, based on global change indicators such as energy and food demand, changes in land cover, greenhouse gas emissions, climate change and impacts on crop productivity and natural vegetation (Alcamo *et al.*, 1996). It was also used to examine the long-term global consequences of reducing greenhouse gas emissions and the effects on climate change and land cover, as well as the cost.

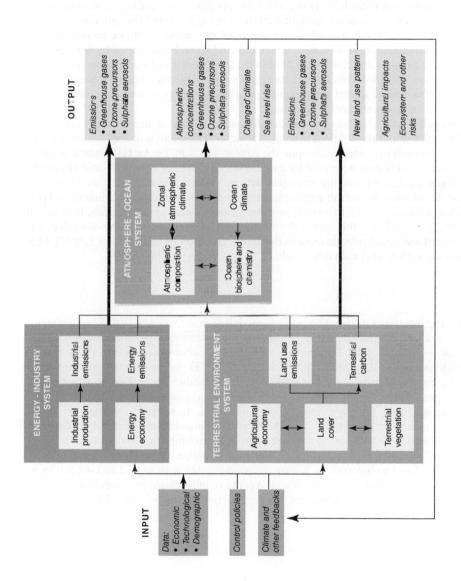

Figure 2.14   Diagram of IMAGE2, a global integrated model. Source: Alcamo, 1994. Reproduced by permission of Springer

2.4.3   Agent-based models

ABM is a bottom-up approach to modelling of particular relevance when discussing socioeconomic interactions within ecosystems and with their services. It has risen to prominence because it incorporates human decision-making with regard to land use in a mechanistically and spatially explicit way; this is done by taking into account social interactions and decision-making (Matthews *et al.*, 2007). In this type of modelling, several agents or individuals within a system are modelled interacting together and the outcomes of these interactions for the system as a whole are observed. An 'agent' might be an individual, a household, an organisation, a plant or an animal, and it can be described by various characteristics, behaviours and constraints. Within the context of a socio-ecological system (see Chapter 7), an ABM will contain various agents interacting in a landscape. Such a model can be used to explore spatial and temporal patterns of behaviour and their influence on, for example, the properties of an ecosystem.

Rounsevell *et al.* (2012) suggest the development and use of human functional types (HFTs), as this would allow ABMs to be used at larger scales. This is of course a similar generalising approach to – and builds upon the concepts of – PFTs. An HFT would in effect have a niche that is best described by its function (e.g. farmer), its traits and attributes, its preferences, decision-making strategies, actions and responses, as well as its physical location. In addition, in such an approach alternatives with regard to government and policy interactions would need to be built into the socioecological system models. Rounsevell *et al.* (2012) propose that this approach could be of considerable use to the ecosystem service concept and could provide links to dynamic vegetation models such as LPJ-GUESS. It would seem likely that such links could be integrated further into ESMs.

Summary: Integrated modelling

- ESMs: Integrate GCMs into ecosystem models such as DGVMs. The former are concerned with physical climate dynamics and the latter with vegetation dynamics and biogeochemical cycles. Feedbacks from the vegetation and ecosystem processes can be important to the climate system. Feedback of CO_2 to the atmosphere can occur as a result of increased soil respiration due to forest dieback and loss of vegetation.
- Integrated assessment models: IMAGE2.1 is a global model that explicitly geographically integrates the energy–industry, terrestrial environment and atmosphere–ocean systems. It has been used to model future regional energy use and land cover allocation, to assess the cost of controlling emissions and the impacts of biofuels and to link global forest and climate change.
- ABMs: Used to incorporate human decision-making with regard to land use in a mechanistic and spatially explicit way. Agents, or individuals, are modelled interacting together and the outcomes of these interactions as a whole are observed.

2.5   Further reading

Basic texts about climate science and models are available in *A Climate Modelling Primer* (McGuffie and Henderson-Sellers, 2005), while a detailed background to modelling can be found in *A Vast Machine, Computer Models, Climate Data and the Politics of Global Warming* (Edwards, (2010). The current IPCC (2007a) has a chapter on applications and evaluations.

3

Data

'It is a capital mistake to theorise before one has data.'

A Study in Scarlet, *Arthur Conan Doyle, 1897*

3.1 Introduction

Theophrastus (381–287 BC) was a student of Plato and Aristotle and was among the first in the Western world to write about ecology and collect data. In his *De Causis Plantarum*, he discussed the relationships between trees and their environment, distinguishing between the effects of climate, soils and species interactions on tree survival. He noted regional-scale deforestation, soil erosion and drainage in Crete and on the Greek mainland and speculated on their feedbacks to local climate and their probable effects on soil fertility. Observations like these are useful to the science of plant ecology, but need to be continued for long time periods or pooled together with data from other regions if they are to contribute effectively to the scientific study of long-term change in ecosystem composition, structure and function that we call 'millennial ecosystem dynamics'.

Data are important to this book for a number of reasons. First, the models described in the last chapter did not develop in a vacuum: their structure was obtained from field observations, and several model components are assigned values or 'parameterised' using data. Second, the models are tested against other sets of data to ensure that they reproduce the important features of those data and are therefore likely to become reliable forecasting tools. Third, models can also be used to look into the past, or for 'hindcasting'; this is a further type of data–model comparison, which can help us to understand why ecosystems have changed through time and which external agencies or pressures have had the greatest effects on them. So models need data for their construction, and hopefully models can help with data interpretation.

Beginning with rock and cave art and continuing with Theophrastus, there is a long, diverse history of useful observations and comments about ecosystem change, but even today there are rather few directly observed datasets that document decadal or longer changes in terrestrial ecosystems in a consistent manner. We have to make use of literature and archives, as did Reale and Dirmeyer (2000) when they collated written observations from 13 classical authors and selected archaeological evidence spanning a 1000 year period from c. 500 BC to AD 500 in order to draw a bioclimatic map of the Mediterranean region,

Ecosystem Dynamics: From the Past to the Future, First Edition.
Richard H.W. Bradshaw and Martin T. Sykes.
© 2014 John Wiley & Sons, Ltd. Published 2014 by John Wiley & Sons, Ltd.
Companion Website: www.wiley.com/go/bradshaw/sykes/ecosystem

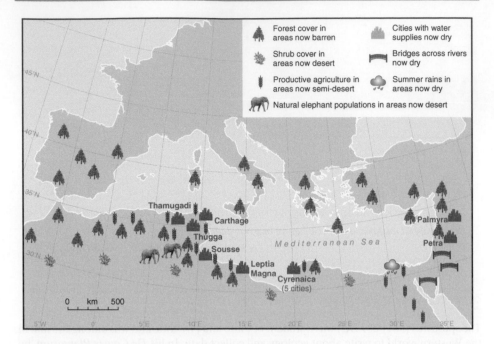

Figure 3.1 Evidence for land cover change during the last 2000 years, based on classical literature and archaeology. Source: Reale and Dirmeyer, 2000. Reproduced by permission of Elsevier

showing the change from classical times to the present (Figure 3.1). Their map records major loss of forest cover and savannah vegetation over the last 2000 years and shows areas where water supplies have become significantly reduced, with a consequent loss of agricultural production and impact on settlement patterns. Data compiled in this way are unlikely to match modern standards of data quality and often cannot be statistically analysed, but as long as the potential sources of error are identified and their consequences assessed, they allow interesting issues to be explored, such as what caused these observed changes of vegetation cover and water availability: was it over-exploitation of resources by people, natural climatic change or a combination of pressures? Do these changes have relevance for current and future ecosystem dynamics? These are themes that we explore later in the book, while in this chapter we review the types of data that are used in the study of ecosystem dynamics.

3.2 Which data are relevant?

We use two main types of data in this book: observations of ecosystems in time and space and information about the possible causes, forcing factors or drivers of ecosystem change. Driver is a simplified term, as these factors may also react to ecosystem change and take part in complex feedbacks. As the ecosystem is a broad and abstract concept, observations tend to be of ecosystem components or fluxes, such as species, soil properties, gas emissions, accumulation of dead material, DNA or structural units of vegetation, which can be resolved by remote sensing. The data can be directly observed or measured, but time series are usually short in relation to ecosystem dynamics, since scientists lose interest or retire and impatient funding agencies tend to favour experimental

Table 3.1 Types of data used for the study of millennial ecosystem dynamics and their drivers. Unfilled circles indicate irregular or fragmentary datasets

	Ecosystems	Drivers	Time series	Spatial array
Direct observation				
Short-term				
Remote sensing	●	●	●	●
National monitoring programmes	●	●	●	●
Field mapping projects	●		○	●
Population studies	●		●	●
Permanent plots	●		●	○
Long-term				
Chronosequences	●		●	
Historical accounts	●	●	○	○
Diaries	●	●	●	
Historical maps	●		○	●
Monastic records		●	●	○
Tax records	●		●	●
Photographs	●		●	○
Indirect proxy measurements				
Pollen	●	●	●	●
Plant macrofossils	●		●	○
Charcoal		●	●	●
Megafauna	●		○	●
Microfauna, insects	●		○	○
Packrat middens	●		●	●
Tree-lines	●	●	○	○
Ancient DNA	●		●	○
Stable isotopes		●	●	●
Ice cores		●	●	
Glacier mass balance		●	○	○
Rhizopods		●	●	○
Archaeological sites		●	○	●
Human population size		●		
Geochemistry		●	●	

research over monitoring. Datasets must therefore be combined if we are to develop an understanding of entire ecosystems. Direct, precise observations cover the recent past, but they can be complemented by longer-term, indirect or so-called proxy data, preserved in biological or geological archives, in order to provide a long-term perspective (Table 3.1, Figure 3.2).

3.3 Ecosystem dynamics: direct observation

3.3.1 Phenology

On 9 January 2012, scientists from the National Museum of Wales reported that 63 plant species were already in flower in Cardiff City, instead of the more usual 20–30 species. This type of phenological observation can give information about the effects of climate change and the Welsh scientists were confident that the exceptionally mild autumn and early winter conditions that year lay behind this flowering record. Observation and recording of phenological events is among the most common and complete records available, sometimes

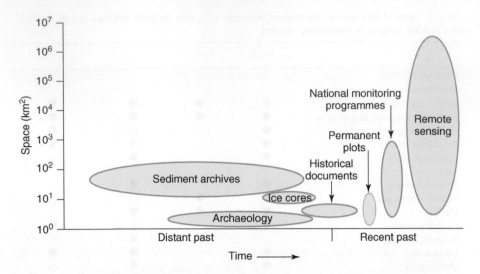

Figure 3.2 The spatial and temporal scales of ecosystem dynamics covered by different types of data collection and forms of archive

going back hundreds of years. However, such events need to be recorded over many years for any trend to be observed, as climate varies from year to year. In plants, phenological events are often easily observed occurrences, such as leaf emergence, flowering and leaf fall. In animals such as birds, they may involve changes in migratory patterns. Such careful recording has until relatively recently been the preserve of naturalists, and their long-term observations have provided direct evidence of climate-driven biological change (Sparks, 1999). Records include the full-flowering date of Japanese cherry trees, which have been observed for 732 years (Aono and Kazui, 2008), and the set of phenological observations dating back to 1736 made by five generations of the Marsham family in Norfolk, UK: 26 of the 27 monitored events taken to indicate the onset of spring in the Marsham series are significantly related to climatic variables, with oak (*Quercus*) coming into leaf and *Anemone* flowering earlier in response to increased temperatures through time. The first croaking of frogs and toads has proven to be a more reliable indicator of the arrival of spring weather than is nest building by rooks.

The timing of spring events, such as leafing out and flowering, is currently advancing at a rate of 2–3 days per decade, although there is large variation among taxa. Parmesan (2007) showed that most taxa have already responded phenologically to changing climate and that amphibians have responded the most, with a greater shift to earlier breeding than all other groups (Figure 3.3).

Specific examples of these sorts of change can be found throughout the recent literature; for example, in southern Sweden, at Gullsmyra, there exists a rare long-term record of the flowering times of 25 species, carried out by a local farmer from 1934 to 2006 (Bolmgren *et al.*, 2012). Recent analyses of this dataset show that all species responded to increased spring temperatures by flowering earlier, although trends were variable depending on the species and time period considered. The study also shows that the period 'mellan hägg och syren' (a romantic Swedish time when shoemakers were allowed to leave their guilds and go walkabout), between the flowering of bird cherry (*Prunus padus*) and that of lilac (*Syringa vulgaris*), has moved earlier over the last 40 years, although the relationship between the

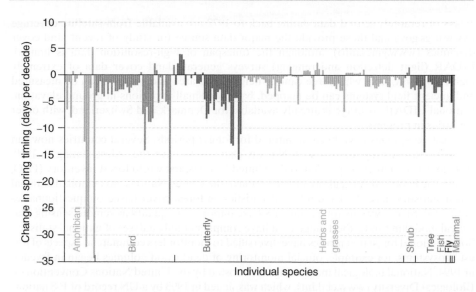

Figure 3.3 Changes in timing of spring events in days per decade. Each bar represents a species. Negative values indicate earlier event occurrence, positive values are delays (after Parmesan, 2007)

two species remains the same as they seem to respond in a similar way to the changing climate. Another study from 1980, using data from bird observatories at Falsterbo and Ottenby in south Sweden, shows that long-distance migratory birds, wintering south of the Sahara, have advanced their arrival in northern Europe to a greater extent than have short-distance migrants (Jonzén et al., 2006). Earlier egg laying is also often reported, sometimes in conjunction with earlier arrival in birds, sometimes not. In Finland, the common buzzard has begun hatching its eggs more than 4 days earlier per decade over the last 25 years. Autumn phenological events such as leaf fall are more complicated. Leaf fall involves warmer days, light levels across long days and the incidence of sharp frost. It is becoming later, but not at the same rate as spring events.

Phenological events occur across a range of species and trophic levels and mismatches can thus occur in the timing of these events (e.g. Visser & Rienks, 2003). For example, in the food chain relating the oak (*Quercus*), the winter moth (*Operophtera brumata*) and the great tit (*Parus major*), the bird feeds its young on caterpillars, which are available only for a short time in spring when they feed on new oak leaves. In the recent warmer springs, winter moth eggs have begun to hatch up to 3 weeks before bud burst in oak, and newly hatched caterpillars can survive only for a few days without food. Since species respond individually to their environment, especially on different trophic levels, climate changes are likely to promote disruptions of important relationships. These may be unpredictable, leading to a loss of biodiversity and possible ecosystem degradation. However, altered phenologies may not always be negative, and new relationships between species may well emerge as new combinations of species come to coexist.

3.3.2 Biological monitoring

Direct observations of phenology and other ecosystem dynamics and their drivers are now frequently made at a variety of spatial and temporal scales (Table 3.1). Enormous arrays of remotely sensed data dating back to 1972 are available from satellite coverage (www.usgs.gov) and these provide the major data source for study of recent land cover dynamics (www.gmes.info), although the timespan of high-resolution data is limited. LIDAR (light detection and ranging) surveys generate land cover data of extremely high resolution, from which tree canopy heights, leaf area and biomass can be measured and ancient field systems can be mapped, based on small variations in ground elevation. National LIDAR coverage is already available for Denmark and Switzerland, with other countries in progress.

Forest ecosystems have been monitored for longer periods. Several countries now run forest inventory programmes, with Norway claiming to have the oldest, dating to 1919. In 1946, 101 countries responded to the United Nations first world forest inventory (FAO, 1948). The Food and Agriculture Organisation of the United Nations has compiled national forest statistics since 1948 and released statistics on forest cover based on questionnaires until 1980. Since 1980, its forest statistics have improved in quality as remote sensing, statistical modelling and expert judgement have complemented analyses of national statistics. Some national forest inventories have diversified to monitor less commercial aspects of forest ecosystems; for example, official monitoring of dead wood volumes began in Sweden in 1994. National biological monitoring was boosted by the United Nations Convention on Biological Diversity (www.cbd.int), which was signed in 1993 by a UN record of 168 nations (from a contemporary membership of 184).

Finland has a proud history of biological monitoring and now runs programmes that collect data on changes taking place in ecosystems, habitats, species, communities, genes and genotypes (Niemi, 2009). Individual species monitoring focuses on nationally or internationally rare and endangered species. However, even in this well-organised country there are an estimated 60 different biodiversity-related schemes that monitor changes at the species, habitat and landscape scale in forests, peatlands, alpine fells, agricultural, marine and coastal habitats and inland waters, which has created a need for centralised, harmonised databases if these data are to be used for the systematic study of ecosystem change over large areas. The European Environment Agency seeks to take on part of this role for Europe (www.eea.europa.eu), while the International Long-Term Ecological Research (ILTER) Network (www.ilternet.edu), which developed from a successful US research initiative, coordinates field sites where systematic observations are made in over 30 countries, such as the classical agricultural experiments begun by John Lawes and Henry Gilbert at Rothamstead, UK in 1843, which are probably the longest-running field experiments in the world. These experiments were established to examine the effects of different manures and fertilisers on agricultural crop production, but their longevity now makes them valuable in a range of different types of study, including of carbon sequestration, pests and pollution: topics that would have astonished the scientists who began the original experiments. Another ILTER site is the permanent mixed hardwood forest plots at Harvard Forest, MA, USA, at which historical data and palaeoecological research are used to extend the period of ecosystem observation (Foster *et al.*, 1992). The Millennium Ecosystem Assessment (www.maweb.org), launched in 2001, was designed to summarise several global ecosystem dynamic processes and their drivers in order to assess the consequences of ecosystem change for human well-being. The data come chiefly from within a 100 year time period (1950–2050) and the focus on ecosystem services was an important development, to which we will return in Chapter 7.

A rapidly increasing volume of ecosystem observations are being made around the world, but there are rather few sets of observations that extend beyond a handful of decades, even though several important ecosystem processes operate on timescales of hundreds to thousands of years (Figure 3.4). Human activities have impacted on ecosystems for millennia, with hunters already systematically slaughtering wild horses in Germany about 400 000 years ago (Thieme, 1997) and humans using fire for cooking for perhaps almost 2 million

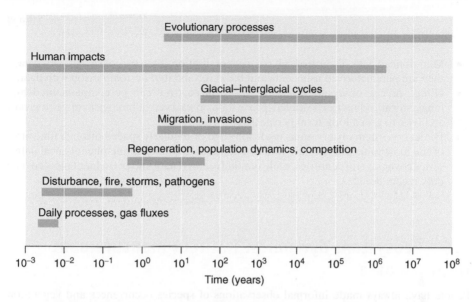

Figure 3.4 Temporal scale of some important ecosystem processes (after Miles, 1987)

years (although the first firm archaeological evidence for controlled fire currently dates from 1 million years ago; Berna *et al.*, 2012). Several other significant changes in plant communities, driven by sporadic disturbance, pathogen outbreaks, migration, gradual climate change and evolutionary processes, require observations over longer time periods than are covered by most current monitoring programmes. What type of data can extend the records now being collected?

Chronosequences have been a classic research tool in studies of succession and are included in many introductory textbooks. They are a series of sites that vary in age since formation or age since latest catastrophic disturbance and are used to infer dynamic ecosystem processes over millennial timescales. When compared with independent sources of evidence, their reliability as long-term sequences of vegetation dynamics is very doubtful, as the assumption of uniformity of nonsuccessional processes through time is frequently violated (Johnson and Miyanishi, 2008). A palaeoecological analysis of the classical chronosequence by the southern shore of Lake Michigan, USA shows that recent disturbance has influenced current vegetation more than once since habitat creation (Jackson *et al.*, 1988). It has been argued that chronosequences can contribute to understanding of long-term soil developmental processes, which are difficult to study using other methods (Wardle *et al.*, 2004). We thus include chronosequences in Table 3.1, but they come with a definite warning; direct observations of new land surfaces, such as the progressive colonisation of the volcanic island of Surtsey, which formed in 1967, provide more reliable insight into primary successional processes (Magnússon *et al.*, 2009).

Permanent plots and transects exist in several countries that are unsystematic or poorly known and which have not been incorporated into official databases. These have often been established by phytosociologists or foresters and can date back to the early 1900s, predating official government monitoring programmes. Such early records can complement historical documents, old photographs and maps and lead into the subjects of historical ecology, palaeoecology, archaeology and other indirect ecosystem observations, all of which have longer time perspectives but are harder to organise into databases.

Summary: Direct measurement of ecosystem dynamics

- Models need data. They are built from generalisations that are consistent with data, they are parameterised using data and they are validated by comparison with data.
- Direct, precise observations cover the recent past but can be complemented by longer-term, indirect or so-called 'proxy data', preserved in biological or geological archives, to gain a long-term perspective.
- Direct ecosystem observations made long ago or regularly spaced through time are of the most value but they are rare and of variable quality. Some phenological data series cover several centuries, while remotely sensed data and permanent plots cover only recent decades.

3.4   Ecosystem dynamics: indirect measurement or proxy data

People have always made informal observations of species occurrences and vegetation types; the remarkable 32 000-years-old paintings in the Chauvet Cave, France (Figure 3.5) are one of the earliest human records we can use to infer ecosystem dynamics and include a broad assemblage of animals that are mostly extinct or found far from France today, such as mammoth, rhinoceros, horse, aurochs, musk ox, cave bear and panther. The painting of lions hunting bison must have been based on accurate personal observation. The Chauvet cave paintings differ from later Palaeolithic cave art in focusing on species not generally hunted by people, so there is a sampling bias that cannot easily be assessed, which is a general feature of much of the proxy data outlined here. Prehistoric paintings are known from every continent except Antarctica and often depict hunting themes that include extinct animals. Ubirr in northern Australia has the longest sequence of rock art in the world and predates Chauvet by about 10 000 years, but little biological information can be obtained from those mysterious images.

Scientists extracting biological information from art bring two different research cultures close together in an exciting but provocative manner. This approach characterises the developing, related fields of environmental history and historical ecology, in which written records and maps are used to reconstruct past ecosystems. The editor of *Forest and Conservation History* (now *Environmental History*), which publishes papers in both ecology and history, analysed review comments and found that scientists consider the rather qualitative approach of historians to be imprecise, with exaggerated, poorly justified conclusions, while historians regard the quantitative methods of scientists to be over-mechanistic and to generate often trivial conclusions (Crumley, 2003). Many must share this experience and it argues for the constructive meeting of research approaches that we adopt in this book.

Historical ecology and palaeoecology, which carry the study of ecological systems back in time beyond written records, are vast but fragmentary disciplines that require the approach of the forensic scientist. Complex systems and processes are reconstructed from scraps of evidence left in deliberately created archives such as libraries, museums, galleries and graveyards, as well as 'accidental' archives such as most archaeological sites and biological archives in lake sediments and peatlands. The reconstructions are rarely as detailed as the modern ecologist might wish for, but the glimpses into the past they offer provide a valuable perspective on current dynamic ecosystem processes, such as the impacts of climate change and former land use, over timescales unachievable using modern ecological techniques.

Figure 3.5   A hunting pride of lions painted in the Chauvet Cave, France about 32 000 years ago. Reproduced by permission of Jean Clottes, Chauvet Cave Scientific Team

3.4.1   Historical ecology

Pehr Osbeck was a well-travelled student of Carl Linnaeus who ended his career as a parish priest in the county of Halland in southwest Sweden in the late eighteenth century. He made several striking observations of locally common species of plants and animals that have now not been seen in Halland for years. Birds like the European roller (*Coracias garrulus*), black stork (*Ciconia nigra*) and hoopoe (*Upupa epops*) would cause a sensation if they nested in Halland today, but they were apparently commonplace just 200 years ago (Larsson and Simonsson, 2003). Data such as these belong to the field of historical ecology, where useful information can derive from written records, old maps and photographs. Their interpretation requires specialist knowledge, particularly with regard to assessment of the accuracy of each source.

Witness trees – trees marked and recorded by North American land surveyors during the eighteenth and nineteenth centuries – have been widely used to map forest composition at the time of European settlement. However, the early surveyors had their own sampling biases, which must be taken into consideration, such as the underrecording of trees growing on steep slopes (Black and Abrams, 2001). These types of data have been combined with later, more detailed inventories to show the broad-scale vegetation dynamics driven by the farming and barrel production of European settlers in northwest Pennsylvania and the later abandonment of these practices (Figure 3.6; Whitney and DeCant, 2003).

European historical ecologists have longer and far more diverse written records to draw upon. Rackham (2003) exploits Anglo-Saxon charters from the ninth to the eleventh centuries, the Domesday Book from 1086 and the more continuous written records that exist from about 1250 onwards to estimate that about 15% of England was covered by woodland in 1086, making it one of the most deforested European countries at that time. Lincolnshire, in eastern England, had 432 individual woods in 1086, covering 4% of the county. By 1895 this woodland area had only reduced to 3% of the county, and many of the individual

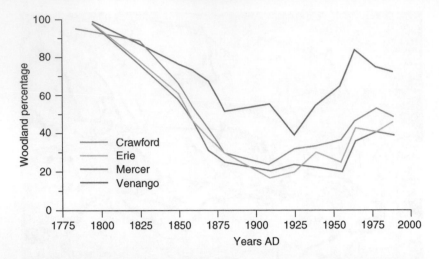

Figure 3.6 Changing woodland area in four counties of Pennsylvania, USA since European settlement.
Source: Witney and de Cant, 2003. Reproduced by permission of NRC Research Press, Canada

woods from 1086 can still be recognised today (Rackham, 2003). Sweden is a strictly administered country, with archived city council meeting minutes reaching back to 1381 and
continuous tax records from the 1530s onwards. Such data can be used to reconstruct long
records of change in types of land use. The Swedish land registry was established in 1628
and began work by compiling annotated maps showing agricultural land and its economic
value. A hand-drawn map from the province of Småland in 1637 shows the land of the
Huseby ironworks estate and the former village of Nya Barquara (Figure 3.7(a,b)). In 1637
there were clear distinctions between the in-fields and out-fields, with abundant meadows,
wood pasture and even a planted hazel (*Corylus*) grove. The traditional land use boundaries are still visible in an aerial photo from 2006 (Figure 3.7(c)), but cultivated fields have
considerably increased in area at the expense of meadows and open woodland, and Nya
Barquara is abandoned. The remaining wooded areas are now managed by foresters and
are no longer grazed by domestic animals as part of the agricultural system. In Malmöhus
County, divisions of the landscape were stable until the late 1700s, at which time a rapid
transformation took place to create the modern configuration (Figure 3.8). The current
balance between open and forested land in coastal regions of southern Sweden has its origins in the Late Bronze Age 3000 years ago, as is the case in large regions of northwestern
Europe (Odgaard and Rasmussen, 2000).

3.4.2 Palaeoecology

Historical ecologists use documents and maps for their ecosystem reconstructions, while
palaeoecologists study the past using materials that range from molecules to macrofossils
preserved in soft sediments, ice cores and biological archives. Every set of proxy data has
its own limitations, making their indiscriminate use risky without advice from experts in
each field. However, the various types of proxy ecosystem data preserved in geological
sediments mostly share the same assets and weaknesses, and together they make up the
discipline of palaeoecology. Data from this discipline have probably not yet reached their
full potential contribution to the biological analysis of ecosystem dynamics.

Biological proxy ecosystem data include: pollen, diatoms, plant macrofossils, tree
rings, stomata, phytoliths, charcoal, insects particularly coleoptera and chironomids,
molluscs, crustaceans (particularly ostracods), vertebrate remains, fungi, rhizopods and

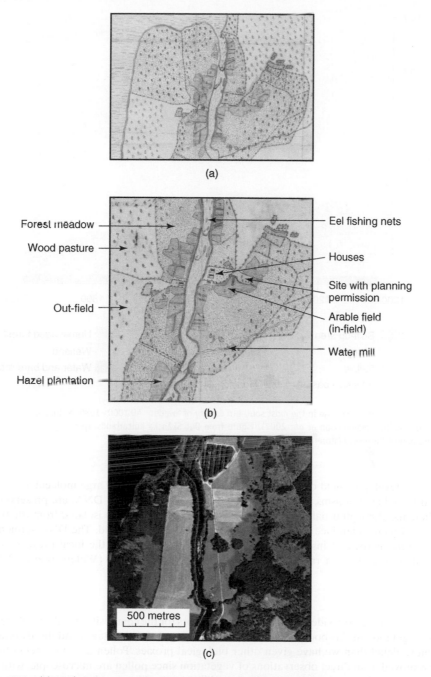

(a)

Forest meadow

Wood pasture

Out-field

Hazel plantation

Eel fishing nets

Houses

Site with planning permission

Arable field (in-field)

Water mill

(b)

500 metres

(c)

Figure 3.7   (a) Land registry map from 1637, showing the village of Nya Barquara, Sweden located in a forested landscape. (b) Part of Nya Barquara, showing the land use as recorded in the 1637 land registry records. Source: From GEORG Database, Courtesy of National Archives of Sweden. (c) Aerial photo of the area mapped in (a), taken in 2006. The traditional land use boundaries are still largely visible, but cultivated fields have considerably increased in area at the expense of meadows and open woodland, and the village of Nya Barquara is abandoned. Source: Image from Google Earth ©2006

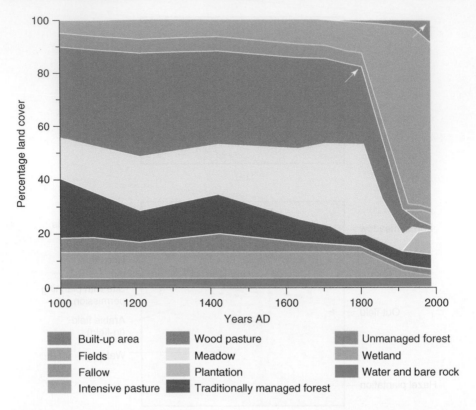

Figure 3.8   Land use change in the most southern county of Sweden, AD 1000–1980. Malmöhus län is over 5000 km² (after Emanuelsson *et al.*, 2002). Figure from Det Skånska kulturlandscapet, reproduced by kind permission of Naturskyddsföreningen i Skåne

the fast-developing field of ancient biomolecules, such as DNA. Large molecules synthesised by living organisms, including pigments, lipids, proteins and DNA, are preserved in sediments, although usually as fragments. Ancient DNA has been isolated from the basal layers of a Greenland ice core, which are at least 450 000 years old. The DNA sequences from trees, herbs and insects were specific enough to interpret the local presence of a boreal forest ecosystem, rather than the current arctic environment (Willerslev *et al.*, 2007).

3.4.3   Pollen analysis

Pollen and spores are widely used as proxy indicators for the composition and abundance of past vegetation in this book, so here we will examine their advantages and disadvantages in more detail than we have given other biological proxies. Pollen data may seem to be far removed from direct observations of vegetation since pollen are microscopic, with the largest grains being about one-tenth of a millimetre in diameter, but as with any proxy system, it is the source plants that are of interest and not the pollen grains themselves. The abundant production of pollen is one of the greatest assets of this proxy. The pollen are well mixed and widely dispersed, the walls of pollen grains are extremely resistant to decay processes and large sample sizes can be recorded. Pollen are preserved in soft sediments found in lakes and peatlands and at other sites with little decomposition. Small, wet forest hollows closely surrounded and overhung by trees can contain pollen concentrations of up to 1 million grains per cubic centimetre of sediment. Wind-pollinated tree species are the most abundantly recorded pollen types, with insect-pollinated herbs being very poorly represented. Around 100 different pollen types can typically be recorded from Holocene (the last c. 11 500 years) lake sediments in the boreal and temperate zones, usually to the level of the genus, so taxonomic resolution is limited and many families are underrecorded or absent from the record. However, these weaknesses of the pollen record are more than balanced by the unique records of vegetation change that can be traced back millions of years through the subfossil and into the fossil record.

Von Post (1916) and Godwin (1934) were among the first to speculate about the ways in which pollen analysis could contribute to the biological sciences. Following the initial enthusiasm and excitement from the pioneer days, a significant development was the routine application of reliable timescales to late Quaternary sedimentary sequences through the development of radiocarbon dating. This development took place during the 1960s and contributed to an increased level of precision in the reconstruction of the rate of ecosystem dynamics and the recognition of the synchronicity of major ecosystem 'events' during the last c. 40 000 years. It became possible to make meaningful comparisons within and between regions, and eventually to build databases that made pollen data accessible to a broad range of scientists. Quaternary pollen analysis developed from being merely a method of geological correlation into a useful tool for exploring vegetation dynamics.

Davis (1976) and Huntley and Birks (1983) published continental-scale maps of pollen abundances showing the Holocene spreading of tree taxa from the regions where they grew during the last ice age (in so-called 'glacial refugia') to their present distributional limits. Their work illustrated the dynamic nature of tree distributions that have undergone continuous change on a millennial timescale. The subsequent work of the Cooperative Holocene Mapping Project (Wright et al., 1993) demonstrated that vegetation was continuously tracking the shifting areas of favourable climate. At any single point in time, vegetation was adjusting to climate change, so periods of stability in which vegetation reached an equilibrium with local climate were exceptional during the last 11 500 years. This line of research, which originated from a strictly geological approach, has led to a paradigm shift in biology. The concepts of climax vegetation and traditional succession have been shown to be largely untenable under typical disturbance regimes with a constantly changing climate. This development in the subject is a good example of the results from one discipline leading to constructive development in another: the best possible outcome of interdisciplinary research.

Pollen diagrams have tended to increase in length and detail over the years, with more types identified and increased precision in all aspects of the work, which has transformed the nature of the subject. A fundamental issue, particularly when mapping pollen data, is how accurately pollen abundances reflect the source vegetation. When mapping large pollen datasets, the mapping unit is usually the percentage value for each taxon based on a count of a defined group of pollen types. Many threshold percentage values have been proposed to indicate the local presence of the taxon being mapped, but Lisitsyna et al. (2011) illustrated the level of accuracy with which pollen data can be expected to map plant distributions: they compared a European dataset of modern pollen distribution with recorded and modelled modern tree distributions and showed acceptable correspondence for trees such as *Quercus*, *Pinus*, *Picea*, *Betula*, *Alnus* and *Fagus*, while *Populus*, *Acer*, *Juglans* and *Tilia* pollen significantly underestimated the true distribution of trees on the landscape (Figure 3.9). In a landmark paper, Davis (1963) made an important attempt to relate pollen spectra to their source vegetation in a more quantitative manner, through the development of factors capable of adjusting pollen counts to make them reflect more closely the abundances of the source taxa. Her work was revised and given increased theoretical rigour by Prentice (1985), who laid the foundation for improved characterisation of the processes of pollen production, transport and preservation in sediments and of the modelling of pollen–vegetation relationships. This approach has now culminated in the development of practical tools for the quantitative reconstruction of vegetation at various scales using

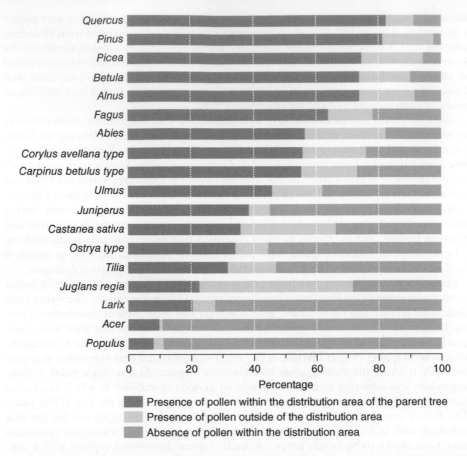

Figure 3.9   Relationship between presence of pollen and distribution of parent trees according to the Atlas Florae Europaeae (Jalas and Suominen, 1972–1994). Dark red bar: presence of pollen within the distribution area of the parent tree. Pale red bar: presence of pollen outside the distribution area. Blue bar: absence of pollen within the distribution area. Source: Lisitsyna, Giesecke and Hicks, 2011. Reproduced by permission of Elsevier

the 'Landscape Reconstruction Algorithm' (Sugita 2007a,2007b). This algorithm incorporates two mechanistic models: REVEALS (Regional Estimates of VEgetation Abundance from Large Sites), which estimates vegetation cover at a regional scale using fossil pollen from large sites, and LOVE (LOcal Vegetation Estimates), which reconstructs vegetation at a finer spatial scale using pollen data from smaller sites (Gaillard *et al.*, 2008). Pollen productivity data are critical model inputs; they can only be estimated under current conditions and are not globally available, but nevertheless this modelling approach presents pollen data in ways that are familiar to a wide variety of users and increases their scientific value through interdisciplinary cooperation.

Conventional pollen analysis from lakes and peatlands records vegetational change at spatial and temporal scales that are largely unfamiliar to the plant ecologist. Forest succession, or the predictable replacement of one group of plants by another, is usually directly observed at the scale of the woodland stand ($10^2 - 10^3$ m^2) during the lifetime of the ecologist, while pollen analysis of lakes and bogs reconstructs vegetation cover from far larger areas ($10^4 - 10^6$ m^2), typically over time periods of thousands of years. Forest hollows and

other 'closed-canopy' sites, by contrast, are dominated by pollen and spores that have travelled only a few metres from their source plants and thus resolve vegetation dynamics at
the scale of the woodland stand. Vegetation around these small sites can be reconstructed
with very high spatial resolution, which can be compared easily with plot-based surveys of
modern vegetation. Ecological processes and properties that can be more easily studied
using stand-scale rather than conventional palynology include: (1) the effects of disturbance on vegetation – either events such as fire, disease and storm damage or chronic
disturbances such as animal browsing; (2) species invasions, succession and its relationship
to climatic change and anthropogenic impact; and (3) vegetation structure and openness
(Bradshaw, 2013).

The Holocene has been most thoroughly studied by pollen analysts and the wide range of
sites available permits sampling flexibility at both spatial and temporal scales. Sampling frequency is flexible within limits imposed by the conditions of sedimentation. Annually laminated lake sediments occur where there are visible seasonal changes in sediment composition and no post-depositional mixing of sediment layers. These sediments can be sampled at
close, regular time intervals but tend to integrate events over broad regions. Peglar (1993)
used annually laminated sediments to show that a British elm (*Ulmus*) population collapsed over a 6 year period about 5700 years ago, which is consistent with a pathogen
outbreak such as Dutch elm disease (*Ophiostoma ulmi*). Small forest hollows permit high
spatial resolution, but temporal resolution depends on the depth of bioturbation or mixing
of the sediments. This versatility of scale in pollen data is an asset when matching data to
models at a variety of scales. Spruce (*Picea*) pollen data from both regional sites and small
forest hollows have been compared with climate-driven regional (STASH) and stand-scale
(FORSKA2) vegetation models to show that climatic change could largely account for the
observed changes in spruce distribution during recent millennia (Bradshaw *et al.*, 2000,
see Chapter 6).

Increased quantitative and spatial precision in vegetation reconstruction, combined with
genetic data, has led to exciting new discoveries in the disciplines of evolution, ecology
and conservation biology. An example is the identification of glacial refugia for beech
(*Fagus*) in Europe, white spruce (*Picea glauca*) in Alaska and *Asplenium hookerianum*
(a warmth-loving fern) in New Zealand far closer to ice sheets than was previously thought
possible (Hu *et al.*, 2009). Further, new details concerning migrational pathways have
shown that mountain ranges have rarely posed a barrier to the spread of species and that
the Holocene spread of beech in Europe was actually facilitated by mountain climates.

3.4.4 Charcoal and fire scars

Preserved charcoal fragments are another proxy system recorded from sediments and are
the major source of long-term information about past biomass burning. As with the reconstruction of vegetation, proxy data for past fires in the form of damaged tree rings and
charcoal can be linked with historical records and remote sensing to reconstruct global
fire history back to the origins of complex terrestrial ecosystems (Figure 3.10). The very
long relationship between fire and people, both hominins and humans, reaches far into the
past, so data covering long timescales are needed to understand the present situation. The
oldest known charcoal dates from the Late Silurian (c. 420 million years ago) but charcoal became more widespread during the Late Permian (c. 250 million years ago), when
atmospheric oxygen levels were sufficiently high to support widespread fire (Scott and
Glaspool, 2006).

Charcoal fragment records give reliable evidence for the presence and approximate timing of fire within terrestrial ecosystems but do not provide firm information about important characteristics of the fire regime, such as fire size, intensity and frequency. The long

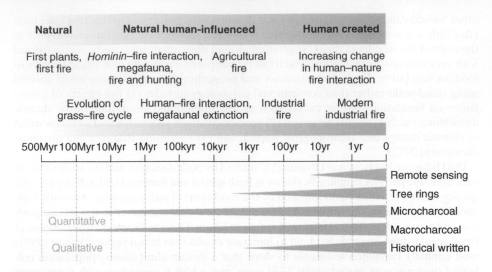

Figure 3.10 Summary of the major sources of information used to reconstruct the history of fire on Earth. Data sources decrease in quality as one goes further back in time (after Bowman *et al.*, 2011)

but imprecise time series of charcoal fragments preserved in sediments overlaps with the more detailed but shorter records obtained from the analysis of fire scars in trees (Falk *et al.*, 2011). Tree ring reconstructions of fire histories based on fire scars cover considerably shorter time periods than charcoal records – often just the last c. 300–800 years – but can generate detailed information on fire years, seasons, frequencies and locations (Niklasson *et al.*, 2010). Yves Bergeron has used fire scar data in a systematic manner to generate maps showing forest age following stand-replacing fires in boreal Canada (Bergeron *et al.*, 2004). In turn, fire scar data can complement and prolong official records of direct observations of forest fires, which in Sweden extend back to 1942, and can be statistically compared with climatic data in order to explore climate–fire relationships (Drobyshev *et al.*, 2012).

Charcoal fragments have been analysed using over 100 different recording methods, which has complicated intersite comparisons and the reconstruction of regional fire histories. As with pollen, charcoal records in lake sediments are the result of complex processes of production, dispersal and sedimentation. However, these processes have been modelled by Higuera *et al.* (2007) in order to reproduce several of the features observed in charcoal records, suggesting that we now have sufficient knowledge of charcoal representation to reconstruct with some confidence past fire 'regimes', or the frequency and extent of past burning. The fire regime in a particular region is recorded as total charcoal abundance per unit of sediment (which is proportional to the total biomass burnt in a given depositional environment) and as peaks in charcoal accumulation originating from individual fires (Power *et al.*, 2008). Data–model comparisons have suggested that key controls on macrocharcoal abundance (usually fragments retained in a 250 micron sieve) include the height that charcoal particles are thrown into the atmosphere during a fire, the wind direction and the area burnt within the potential charcoal source area. Typical source areas range between 300 m and 20 km, largely depending upon charcoal injection height (Higuera *et al.*, 2007). Secondary deposition of charcoal (that washed into the sampling basin after a fire) and sediment mixing processes can blur the signals of individual fires. Small forest hollows again appear to yield records with higher spatial resolution, because if a fire passes

Figure 3.11   Charcoal bands in sediment from a Norwegian small forest hollow. Each black band is the record of a local fire (photo Richard Bradshaw)

directly over the sampling site a discrete charcoal band is deposited, while charcoal from more distant fires is largely filtered out by the local vegetation (Figure 3.11). Intersite comparisons of lake sediment charcoal records chiefly reflect variations in large-scale climate and biome type, which exert the major controls on fire regimes at regional to continental scales. Mapped summaries of charcoal data show a coherence in observed changes in charcoal abundance on millennial timescales among locations with similar climate, vegetation and human impact (Power *et al.*, 2008).

Analyses of the biological components of sediments are naturally complemented by proxy physical data, which include inorganic and organic chemical and elemental analyses and isotopic determinations of carbon and oxygen, giving information on geochemical fluxes. All these analyses are thoroughly discussed in such texts as Elias (2013), Smol *et al.* (2001), Lowe and Walker (1997) and Berglund (1986).

Palaeoecological data from sediments have the advantage that spatial arrays of data can be collected at variable temporal resolutions. Single fires can be identified from charcoal remains in forest hollow sediments and post-fire successions can be traced using pollen and remains of seeds, buds or leaves, known as macrofossils (Figure 3.12). Similar analytical methods can be applied to lake sediments to generate regional, long time records covering millions of years. At Lake Baikal, 300 m of sediment covered the last 5 million years of Earth history and has yielded information about the cooling continental climate as the Tertiary was superseded by the Quaternary period, with its repeated glaciations (Williams *et al.*, 1997). This spatial and temporal flexibility is a major asset of the palaeoecological data held in sediments.

Palaeoecological archives are not uniformly distributed, however, with the most abundant evidence found in the lakes and landscape depressions left by retreating glaciers. Even within glaciated regions, sediments tend to occur in wetland or upland areas and may not be fully representative of entire landscapes. The data are usually only proxies for the species

Figure 3.12   A 6000-year-old poplar (*Populus tremula*) bract preserved in sediment from Germany, proving local occurrence of a tree that leaves a very weak pollen record. (Photo courtesy of Gina Hannon)

of interest (e.g. pollen and not plants) and many species and larger taxonomic units are not preserved in sediments. The transport and incorporation of living material into sediments introduces further bias as fragile materials and the soft parts of plants and animals may be broken, oxidised or poorly preserved. Several of these biases have been thoroughly investigated and are generally well understood. The vegetation history of the tropics is less well known than that of temperate and boreal zones because of a scarcity of study sites, the low pollen productivity of insect-pollinated plants and the probable poor preservation of ancient DNA in hot, moist climates. With modern data handling and presentation, bias in the palaeoecological record can be minimised, but it is important to keep it in mind when using data from the past.

Combining historical ecology with palaeoecology and archaeology in a multidisciplinary approach is powerful but can be labour intensive. The Swedish Ystad Project (Berglund, 1991) represented a decade of collaboration between the disciplines of palaeoecology, plant ecology, archaeology, history and human geography. 6000 years of cultural landscape development was described in unique detail, including unstable periods that culminated in socio-economic crises. The researchers concluded that social rather than environmental factors primarily precipitated these crises, although soil fertility exerted an ultimate control on population size. However, the broader impact of the project has been limited, as only part of the data is incorporated into digital databases and the study covered an area of less than 300 km². The upscaling of integrated palaeoecological data to continental scale is made possible through the establishment of databases and a broad acceptance of standardised methodologies, but it is a challenge when most previous research has been site-based, often focusing on single lake catchments. Integrated international projects and initiatives such as the Neotoma Palaeoecology Database and Community (www.neotomadb.org) are stimulating progress in both upscaling and the standardisation of data collection.

Summary: Indirect measurement of ecosystem dynamics

- Past observations of ecosystems come in several forms, including cave art, historical documents, maps and photographs.
- Historical ecology draws on human archives, while palaeoecology is the study of materials ranging from molecules to macrofossils preserved in sediments. Palaeoecology generates proxy data of past ecosystem dynamics.
- Pollen analysis is the major proxy for past vegetation dynamics. Quantitative reconstruction of land cover is possible using the landscape reconstruction algorithm.
- Charcoal fragments yield records of past fire regimes. Combinations of records in multiproxy studies give more complete ecosystem reconstructions.
- Sedimentary records of ecosystem dynamics are most abundant in formerly glaciated landscapes covering the Holocene.

3.5 Drivers of ecosystem dynamics

The major external drivers of ecosystem dynamics include climate change, greenhouse gases, human impacts, soil maturation, disease and disturbance. Several of these drivers have attracted considerable research in recent years, and climate change, greenhouse gases and human impacts have been directly observed and documented in detail for the recent past (www.ipcc.ch). As was the case for the ecosystem data, it is more challenging to characterise these drivers at millennial timescales as this involves linking proxy climatic archives to the instrumental record and merging archaeological, palaeoecological and historical data.

3.5.1 Palaeoclimates and greenhouse gases

Many of the same sources are used to reconstruct the pre-instrumental climate record as we outlined earlier for ecosystem dynamics. In fact, some of the same proxy observations of ecosystem dynamics are used to reconstruct palaeoclimate, so research in this field must avoid misleading circularities. Past records of oxygen isotopes are found throughout the world and can be analysed from marine and lake sediments, ice sheets and calcite deposits in caves (Heikkilä et al., 2010). Most of these records closely covary for recent glacial–interglacial cycles and appear to record truly global signals. What are they recording? The ratio of the rare, heavier ^{18}O isotope to the widespread, lighter ^{16}O isotope is influenced by the temperature of clouds at the time of snow formation and by the progressive rainout of the heavier isotope during the poleward transport of air (Heikkilä et al., 2010). The use of this isotopic ratio as a simple 'palaeothermometer' is complicated by the observation that changes in the source and seasonality of precipitation also influence oxygen isotope ratios. Also, when global ice volumes are large, the lighter ^{16}O is depleted in the oceans, although this effect is rather small on land. Nevertheless, oxygen isotope ice-core records are among the most widely used proxies for past temperature change during the last glacial–interglacial cycle (Johnsen et al., 2001). The boreholes from which ice cores are extracted retain a thermal memory of past ice surface temperatures, which can be used as an independent check on the isotopic temperature reconstructions (Dahl-Jensen et al., 1998), and both these methods suggest that Greenland temperatures 20000 years ago were about 20 °C colder than today (Johnsen et al., 2001). However, the Greenland palaeotemperature data give a weaker indication of the mid-Holocene warming than has been inferred from Scandinavian pollen data (Seppä and Birks, 2001). Perhaps we should not be surprised at this, as no one would use modern climate data from central Greenland to gauge the current length of the growing season in central Sweden.

Other terrestrial physical systems used for palaeoclimate reconstruction include: (1) speleothems (stalagmites and stalactites), which develop in unfrozen caves and also yield oxygen isotope ratios sensitive to cave temperature and moisture source; (2) glacier dynamics, which are particularly sensitive to summer temperature and winter precipitation (Nesje *et al.*, 2008); and (3) variations in thickness of annual layers in lake sediments (Wohlfarth *et al.*, 1998).

Several biological systems are used as proxies for past climatic data, even though these systems are inherently more complex and variable than the physical ones. For temperature, these include changes in assemblages of temperature-sensitive freshwater chironomids (non-biting midges), changes in tree lines, pollen and plant macrofossil data, beetles and variations in tree ring widths. All these approaches build on calibration using present-day temperatures and require specialist knowledge for a full assessment of their assumptions and reliability (e.g. Brewer *et al.*, 2013). Transfer functions have been calculated for pollen data based on modern pollen–climate relationships (Seppä and Birks, 2001) and these are discussed in some detail in Chapter 4. Bartlein *et al.* (2011) reviewed all the quantitative methods for climate reconstruction that use pollen or plant macrofossil data and generated useful global maps for the climate 6000 and 21 000 years ago.

Precipitation patterns are even harder to forecast or reconstruct than temperatures. The proxy systems most widely used include lake level changes reconstructed from sediments, glacier mass balance variations when summer temperatures are estimated, speleothem records (where these are not dominated by the temperature or isotopic properties of the water source) and the surface wetness of peatlands, where rhizopods or testate amoebae are proving to be useful indicators of average water table depth and hence drought variability (Booth, 2012). Lamb (1977) still provides a readable and authoritative approach to how different sources of data – biological, physical, historical and archaeological – may be combined to derive the most complete reconstructions of global palaeoclimates.

Considerable knowledge of the global climate system is incorporated into general circulation models, and through use of appropriate boundary conditions from the past, large sets of palaeoclimatic model 'data' can be generated (e.g. www.paleo.bris.ac.uk/ummodel/list _of_simulations.html). It is attractive to use these model data to drive dynamic vegetation models, although the output from models linked in this way can compound the estimates and inaccuracies found within each one. A data–model comparison using Holocene vegetation data from central Fennoscandia did however match data and models rather well and indicated that the length of the growing season exerted the dominant control on the changing distributions of temperate deciduous trees (Miller *et al.*, 2008).

Holocene records of greenhouse gas concentrations derive from two data sources: the continuous records of CO_2, CH_4 and N_2O obtained from gas bubbles trapped in glaciers and the more intermittent records of past CO_2 derived from stomatal densities measured in fossil leaves. The stomatal method assumes an inverse relationship between the stomatal density on a leaf surface and the atmospheric concentration of CO_2 in which that leaf developed (Beerling and Royer, 2002).

Wanner *et al.* (2008) combined proxy data describing palaeoclimate and natural climate forcings such as orbital variation, solar activity, volcanic eruptions, land cover and greenhouse gases with modelling to provide an explanatory framework for climate change from 6000 years ago up to the Industrial Revolution. Their analysis showed the benefits of data–model comparisons, indicating, for example, that the Little Ice Age cooling observed in the northern hemisphere between AD 1350 and 1850 was linked to lower summer insolation, a minimum in solar activity and several volcanic eruptions in the tropics. Their summary included land cover change as a significant forcing factor for climate change

(Wanner *et al.*, 2008), and data are needed on the role of people in Holocene vegetation change in order to evaluate the timing, type and extent of this land cover change.

3.5.2 Human impact on ecosystem dynamics

The data most relevant to assessing the impact of people on past ecosystems are how many people lived on the planet, where they lived and what they did to the ecosystems. This type of information is available from national records for the recent past, but for periods prior to national bookkeeping we have to look to the disciplines of history, archaeology, palaeoecology and, increasingly, genetics. Research into long-term human impact on ecosystems has been stimulated by controversies, where inadequate knowledge of the sizes and capabilities of ancient human populations have made it difficult to choose between competing hypotheses for the major causes of observed ecosystem change. One of the great debates about extinction is whether humans or climate change caused the global loss of about 100 of the 150 large animal species (or 'megafauna') weighing more than 44 kg, known from the fossil record to have lived 50 000 years ago (Barnosky *et al.*, 2004). The current consensus is that hunting by humans interacted with rapid climatic change on most continents to cause the extinctions, but more definite conclusions are hampered by a lack of detail about human population size and the paucity of investigated kill sites (Barnosky *et al.*, 2004; Lorenzen *et al.*, 2011; see Chapter 4).

The Ruddiman hypothesis has proposed that changes in land use and land cover significantly altered atmospheric concentrations of methane and carbon dioxide several millennia prior to industrial emissions (Ruddiman, 2003). This hypothesis has stimulated widespread debate (Kaplan *et al.*, 2010) and fundamental uncertainties about global population size and the timing and scale of anthropogenic land cover change during the Holocene are restricting attempts to test it (Doyle *et al.*, 2011). The accuracy of prehistoric population estimates are influenced by when anatomically 'modern' humans (you and me) first evolved, with estimates varying depending on whether archaeological or genetic dating systems are used (Endicott *et al.*, 2009). Our most recent common ancestor with the Neanderthals is believed to have lived between 300 000 and 450 000 years ago and ancestral 'Eve', the mother of all modern humans, lived in Africa between 100 000 and 200 000 years ago (Endicott *et al.*, 2009). We spread out of Africa between 54 000 and 84 000 years ago, and there is a weakly based but general consensus that by the start of the Holocene (11 500 years ago), world population lay at between 4 and 10 million (Boyle *et al.*, 2011). This figure had risen to between 170 and 300 million by AD 1, after which time records become more reliable.

These estimates of past population size have been combined with the area of land used by each inhabitant for food, either in the recent past or historically, to reconstruct past land cover change, which was typically conversion of forest to pasture and arable land. The History Database of the Global Environment (HYDE) has been widely used by modellers, but in its original form it only extended back to AD 1700, while its method of calculation would show insignificant land cover change prior to a major population increase around AD 1500. Kaplan *et al.* (2010) improved on HYDE by assuming that land was used more intensively as population density increased. Farms became more productive through time and a given area could feed a larger population. The current approach to reconstruction of the timing and extent of ecosystem conversion by people is still a population-based approach, containing these considerable uncertainties. Developments in the quantitative reconstructions of land cover based on pollen data (Sugita 2007a, 2007b), coupled with estimates of past fire frequencies, are likely to generate more direct reconstructions of anthropogenically induced land cover change, although upscaling of this approach to the global scale poses problems. The systematic use of archaeological records might generate better estimates of past human populations, or there may be new genetic tools that can help with population reconstructions by combining information from ancient DNA with genetic theory about population size, gene frequencies and migration (Wang and Whitlock, 2003). However, as with all studies of prehistory, we will never have the full picture.

3.6   Databases

Organising these disparate sources of data describing past and present ecosystem dynamics and their drivers into databases can make them accessible to a wider community. The ideal database will incorporate simple visualisation tools for the non-specialist and convenient access to the underlying data for the researcher. The developing Neotoma Palaeoecology Database (www.neotomadb.org) aims to provide 'an online hub for data research, education, and discussion about palaeoenvironments'. Neotoma covers the last 5 million years and is organising global data from pollen, plant macrofossils, pack-rat middens, diatoms, testate amoebae, vertebrates, insects, ostracods, geochemistry and geochronology. Neotoma is gradually replacing the USA's NOAA (National Oceanic and Atmospheric Administration) paleoclimatology programme site (www.ncdc.noaa.gov) as the most complete global data collection describing past ecosystems. This type of database is complemented by more static, stable archives such as Pangaea (www.pangaea.de), which stores georeferenced data from Earth system research with an emphasis on marine and polar data. Open-access databases such as these provide valuable support to the research outlined in this book.

3.7   Gaps in available data and approaches

The data record on past ecosystem dynamics is fragmentary by nature. Future research could fill some of these gaps, while others will prove to be less tractable. Past records have tended to give little information about the composition, abundance and activities of microbial communities, including bacteria, viruses, fungi and protozoa, and of organisms that have only soft tissues and do not preserve well in the palaeoecological archives. The past dynamics of insect communities is another poorly researched topic given their importance in modern ecosystems. The identification of fragmentary insect remains is a skilled and time-consuming task and there are far less specialists working with insects than with the abundant pollen and plant macrofossil remains of trees. Genetic barcoding, where diagnostic DNA sequences can be used to identify species and map active genes, indicating the occurrence of specific biochemical processes, are now technically possible, although these techniques generate intimidating amounts of data. Effective exploitation of these methods and data to produce focused analyses of key processes is still in its infancy.

There has been a recent strong research focus on the influence of climatic change and biological processes, such as competition, on ecosystem dynamics, and earlier theories concerning the influence of altered soil nutrients on vegetation composition have become neglected. Boyle (2007) applied a geochemical model to eight lake sediment records from Scandinavia, the UK and North America and showed that early Holocene lake acidification, as reconstructed from diatom records, could be fully accounted for by leaching of the calcium phosphate mineral apatite, at a rate controlled by the kinetics of dissolution. His model forecast rapid changes in lake acidification rates during the early Holocene that closely matched the reconstructions and he concluded that his inferred soil nutrient changes probably influenced vegetation dynamics (Boyle, 2007). There may well be other examples of soil maturation processes that are of sufficient significance to be explicitly incorporated into dynamic vegetation models, although climatic change was judged to be the dominant driver in a Hungarian study from the early Holocene, in which the palaeoecological record showed a soil transformation from a podzol into a brown earth, with an associated vegetation shift from coniferous to deciduous forest (Willis *et al.*, 1997).

Most of the data sources outlined in this chapter produce observations that are scattered in time and space, yet dynamic vegetation models can handle continuous data input and output, often on a global scale. Matching data to models requires further development of specialised statistical tools and measures of uncertainty in order to allow data points to be upscaled for data–model comparison exercises and model output to be downscaled to specific locations with appropriate uncertainty measures attached. Data are the lifeblood of the study of ecosystem dynamics. Without them there would be no subject, but datasets are vast and growing and are hard to summarise and interpret without tools. Models organise our understanding of ecosystems and help us interpret the signals contained within the data.

Summary: Data concerning drivers of ecosystem dynamics

- Oxygen isotope records from ice and calcite deposits are widely used to reconstruct past temperature. Glacial dynamics are another reliable physical system with sensitivity to a combination of precipitation and summer temperature.
- Biological systems are used as proxies for past climatic data but are inherently more complex and variable than the physical systems. Changes in assemblages of temperature-sensitive freshwater chironomids, beetles, pollen and plant macrofossil data and variations in tree ring widths and tree lines can all be used for palaeotemperature reconstruction.
- Changing greenhouse gas concentrations are reconstructed from air bubbles trapped in ice cores.
- Human activities have affected ecosystems for long periods of time but data on early Holocene population size are sparse.

4

Climate Change and Millennial Ecosystem Dynamics: A Complex Relationship

'The great tragedy of Science – the slaying of a beautiful hypothesis by an ugly fact.'

T.H. Huxley, 1825–1895

4.1   Introduction

Any map of global biomes, from arctic tundra to tropical forest, indicates that climate influences the geographical distribution of plants and animals. Yet these same distributions have been shown by palaeoecological research to alter through time. Consequently, the analysis of long-term dynamic relationships between organisms and climate has become a very active research area, with a particular focus upon records from the last 20 000 years (MacDonald *et al.*, 2008). Such analyses can tell us much about climates of the past and about the response of ecosystems to climate change, if we can assume that species distributions are in approximately equilibrium conditions with prevailing climate and if we can be confident that we understand the climatic controls on the distributions of organisms. There were two very large 'ifs' in that last sentence and they underlie a lively debate over the dominant controls of species distributions past, present and future. There is also potential circularity in this research area, because, as we saw in Chapter 3, the same biological systems can be used either to reconstruct past climates or to study the ecosystem impacts of climate change. The scientific literature probably contains more examples of climatic reconstruction from palaeodata than research into ecosystem dynamics. This chapter primarily deals with the latter activity, namely the influence of climate change on millennial ecosystem dynamics. It would be much easier to write if we had detailed knowledge of past and future climates that was fully independent of the biological systems under discussion. The general features of past climates are known for at least the last 25 000 years, but high-resolution details for

Ecosystem Dynamics: From the Past to the Future, First Edition.
Richard H.W. Bradshaw and Martin T. Sykes.
© 2014 John Wiley & Sons, Ltd. Published 2014 by John Wiley & Sons, Ltd.
Companion Website: www.wiley.com/go/bradshaw/sykes/ecosystem

individual sites from which biological data derive are not well documented. Researchers have tended to use either the oxygen isotope records from ice or marine cores, which can only weakly capture regional climatic variation, or simulated data from global or regional climate models run under past conditions, the accuracy of which is open to some debate.

The study of climate change and biological response has a long history. Shen Kua was a Chinese scientist with broad interests, including making the first known description of the magnetic compass, who lived during the eleventh century AD. Shen observed petrified bamboos, complete with roots and trunks, in a section of riverbank in northern China where no bamboos then grew. He deduced that the local climate must once have been wetter than it then was and became the first of many observers to reconstruct a past climate from part of a species distribution: 'Perhaps in very ancient times the climate was different so that the place was low, damp, gloomy, and suitable for bamboos' (translation in Needham, 1959). Shen made a valuable observation and quite logically assumed that the altered distribution of his bamboo over time was driven by climate change. Is this always the case? It would be useful to have other information about Shen's discovery before making a final assessment of his interpretation. Was the absence of bamboo in the area during Shen's time due to human activities and conversion of natural vegetation into an agricultural system? Perhaps bamboos were increasing their distribution when Shen was at the site and simply had not arrived there yet. Maybe the local climate was suitable for bamboo but dense forest grew there instead and bamboos were unable to compete. Or perhaps the fossil bamboo that Shen observed superficially looked like modern bamboo but was in fact an extinct form with a different physiology and climatic tolerance. Some of these alternative hypotheses may seem far-fetched in this case, but examples of all these situations have been described in recent literature, and they argue for caution in interpreting the relationships between climate and species distributions.

It is important to explore climate–species relationships because there is a large and growing body of literature that uses present-day species distributions and climatic data to forecast possible future changes in plant and animal distributions. Changes in these distributions are a key component of millennial ecosystem dynamics. It may feel safe to forecast the distant future as there are no data to challenge the conclusions, but it is prudent to test the methods in the past, where model forecasts can be compared to data, even if such data are incomplete.

4.2   Reconstructing climate from biological data

Using evidence of past distributions of species to reconstruct past climates is one of the most traditional and widely used applications of palaeobotanical data. Blytt and Sernander, two Scandinavian pioneers of Holocene palaeobotany, developed a widely used scheme to describe north European postglacial climate change based on the sequences of plant macrofossils in peat bogs. The reason for interest in climate reconstruction is probably because many more scientists and policy makers are interested in climate change than in fossil plants with difficult Latin names. Indeed, there has been an enormous increase of interest and research into palaeoecology since the 1970s, which is largely attributable to the developing research focus on climatic change. Tom Webb III was a pioneer in the use of pollen data for climatic reconstruction using transfer functions, which use present-day climate–plant relationships to reconstruct past climate. Others have questioned the physiological assumptions behind the 'black box' relationship between summer temperatures and, for example, the square root of pollen percentage values for *Carya* (hickory), *Fraxinus* (ash), Poaceae (grass) and *Tsuga* (hemlock) (Howe and Webb, 1983). As an eager graduate student on a conference field excursion in 1977, I (R.B.) listened to a lively discussion in a Scottish pub in which a highly sceptical Margaret Davis and John Birks pressed Tom

Webb to justify his techniques, while Herb Wright looked on with amused detachment. The transfer function approach has been widely used and is accepted as a standard method in many geological disciplines, but it is important to keep in mind its major underlying assumptions, which include that: (1) the climatic parameter to be reconstructed is either a controlling factor on the modern and fossil abundance of the taxon or is linearly related to another controlling factor; (2) the ecological responses of the taxon have not changed through time; (3) variations in present-day abundance of taxa are unaffected, or only negligibly affected, by factors other than the one to be reconstructed, for example competition or atmospheric CO_2 concentration; and (4) the taxa recorded in the modern calibration dataset must be in equilibrium with modern climate for the temporal and spatial scales being calibrated in the fossil data (Brewer *et al.*, 2013). Very few tests have been or even can be made of these assumptions, as the necessary experiments and observations for 1, 3 and 4 are unrealistically difficult, and 2 can only be indirectly explored. There are examples of some well-supported physiological relationships between select climatic variables and plant distributions (e.g. Iversen, 1944; Pigott and Huntley, 1981), but the same ones are cited in almost all reviews of the topic, illustrating a lack of breadth in available knowledge. Rather secure temperature–species relationships have been established for Chironomidae (non-biting midges), while Rhizopoda (testate amoebae) are becoming widely used for difficult species-derived wetness reconstructions.

The transfer function approach has increased in sophistication and mathematical complexity over time, using techniques such as partial least-squares analysis, canonical correspondence, weighted averaging and artificial neural networks (Brewer *et al.*, 2013). These are mathematically powerful approaches but are still subject to the key assumptions. The transfer function approach to climatic reconstruction employs similar techniques to the equilibrium statistical models discussed in Chapter 2. A slightly different approach to climatic reconstruction from palaeoecological data is based on matching the composition and abundance of entire fossil taxa assemblages to modern communities and transferring the climatic setting of the modern communities to the past. This approach is subject to the same assumptions as the transfer function approach and incorporates an even larger black box, containing all the species–climate relationships in the modern and fossil assemblages. Methods include mutual climatic range, modern analogues and the use of sophisticated response surfaces, which allow limited interpolation and extrapolation, thereby increasing the range of climates that can be reconstructed.

Despite these potentially limiting, largely untested assumptions, some striking and consistent details about past climate have been discovered. The Kråkenes Project studied in detail the population dynamics of several groups of organisms at a coastal lake in southwestern Norway between 14 000 and 9000 year ago – a period that covered the transition from glacial conditions to the Holocene (Birks and Ammann, 2000). Independent transfer functions applied to the dynamics of pollen, plant macrofossils, chironomids and cladocera all showed an impressive synchroneity and consistency of reconstructed mean July temperatures based on different sets of preserved body parts (Figure 4.1). The decrease in mean July temperature from the Allerød to the Younger Dryas was estimated at about 2 °C, with a cooling rate of 0.7 °C per 25 years, based on the chironomid reconstruction. All organism groups showed a steep temperature rise during the first 500 years of the Holocene, with a subsequently slower increase until present-day temperatures were reached and exceeded between 10 000 and 9500 years ago (Figure 4.1; Birks and Ammann, 2000). A development resulting from visual comparison between temperature records based on different proxy systems is the mathematical combination of these records to exploit the strengths of each (Helama *et al.*, 2012). Pollen data appear to reflect long-term trends in temperature, while tree ring data are more sensitive to interannual temperature variability. These two proxy systems were combined using pollen and tree ring data from several sites in subarctic Fennoscandia to reconstruct a long-term decline in mean summer temperature

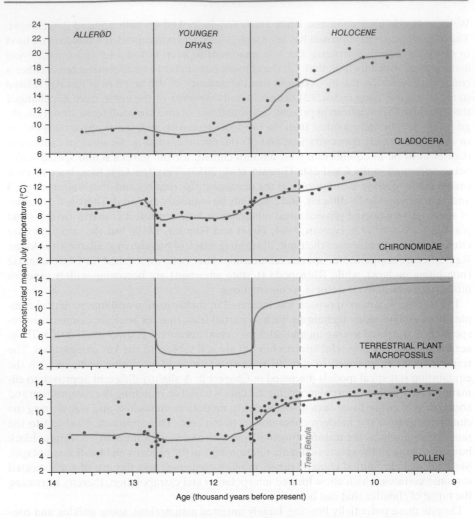

Figure 4.1   Mean July temperature reconstructions based on different proxy systems. The sequence begins in the Allerød interstadial, passes through the Younger Dryas stadial and ends in the Holocene. The dashed line indicates when tree birch macrofossils were first recorded at the site. Source: Birks and Ammann, 2000. © 2000, National Academy of Sciences, USA. Reproduced with permission

of 2.0 °C over the past 7500 years, with an interannual variability of as much as 10 °C (Helama *et al.*, 2012).

Transfer function temperature records can be usefully upscaled by combining multiple records to make regional temperature reconstructions (Davis *et al.*, 2003) or palaeoclimatic maps (Cheddadi *et al.*, 1997). Davis *et al.* (2003) combined pollen data from over 500 European sites and used a modified modern analogue matching technique to generate reconstructed temperature changes for the whole of Europe and for six regional subdivisions over the last 12 000 years (Figure 4.2). The reconstructed regional variation was striking, with a marked decline in late Holocene annual temperatures recorded from northern Europe that was not seen elsewhere. Davis *et al.* made separate reconstructions for summer and winter temperatures and showed that this north European mid-Holocene warming was principally confined to the summer.

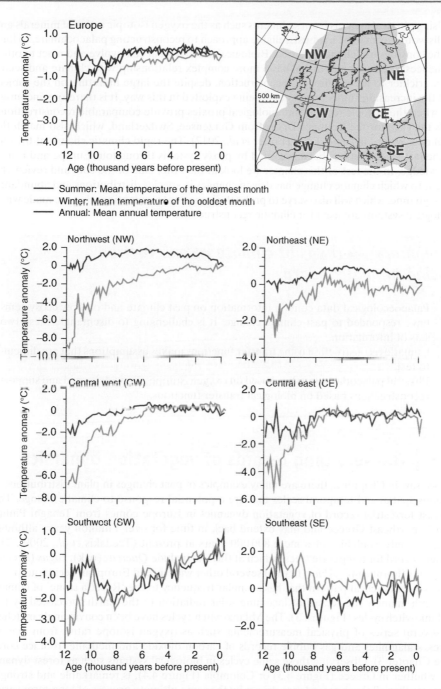

Figure 4.2 Reconstructed area-average temperature anomalies for the whole of Europe and for six regions during the Holocene. Source: Davis *et al*, 2003. Reproduced by permission of Elsevier

Using physical systems from sediments such as the oxygen isotopic ratios of minerals and shells would seem to be a robust, reliable approach to reconstructing palaeoclimate, so long as the controls are fully understood. Biological material that can evolve and react to other influences, such as competition, has a more complex relationship with climate and seems to be trickier to use in climatic reconstruction, despite the large literature on the subject and the increasing number of systems being exploited in this way. It is therefore encouraging when the isotopic records and biological proxies provide comparable reconstructions, as is the case for the Younger Dryas from Gerzensee, Switzerland, which also match the Kråkenes record in some detail (Lotter *et al.*, 2012). The rapid climatic change at the end of the last glaciation has been recorded by proxy systems throughout Europe and is firm knowledge. In the rest of this chapter we look at the other side of the coin and review the extent to which climatic change has driven changes in species distributions and abundance through time, which will also serve to probe the important assumptions that are made when biological systems are used for climatic reconstruction.

> ## Summary: Methods of climate reconstruction using biological data
>
> - Palaeoecological data contain information on past climate and on how ecosystems have responded to past climate change. It is challenging to disentangle these two sets of information.
> - Climatic reconstruction using transfer functions makes assumptions that are difficult to test.
> - Physical palaeothermometers based on oxygen isotopes are widely used and support reconstructions based on biological transfer functions.

4.3   The very long records of vegetation dynamics

As we saw in Chapter 2, there are many examples of past changes in plant distributions at various spatial and temporal scales that are a potential response to climatic change. The longest terrestrial record of vegetation dynamics in Europe comes from Tenaghi Philippon in northeast Greece and may extend back in time for over 1 million years, although data are only available for a mere 800 000 years at present (Tzedakis *et al.*, 2009). The pollen record for temperate trees (which at this site include *Quercus* (oak), *Fagus* (beech), *Carpinus* (hornbeam), *Abies* (fir) and several other widespread European deciduous trees) shows a semicyclical pattern that has a similar frequency to the 'Rosetta stone' of climate research, namely the record of incoming solar radiation to the Earth forecasted by the Milankovitch cycles (Figure 4.3). The Milankovitch cycles have been convincingly matched to several series of physical measurements, such as oxygen isotope ratios from marine cores, stalagmites and gas bubble records of carbon dioxide and methane from ice cores (see Chapter 3). Observing these same cycles in biological systems such as forest dynamics, whether in Greece (Figure 4.3) or Colombia (Figure 4.4), is remarkable and strongly suggests that all these records are driven by the same ultimate process of long-term variation in the amount and distribution of incoming solar radiation. Dating of these long cores of sediment and ice is difficult and it can be hard to identify 'leads' and 'lags' in the different records. There is thus room for speculation as to whether the long-term records of tree population dynamics are primarily driven by temperature change or by changes in the hydrological cycle that could also account for the variations in the methane record (Tzedakis *et al.*, 2009). A more detailed vegetation record for the last 133 000 years, from Ioannina, Greece (c. 310 km southwest of Tenaghi Philippon), shows rapid changes in tree

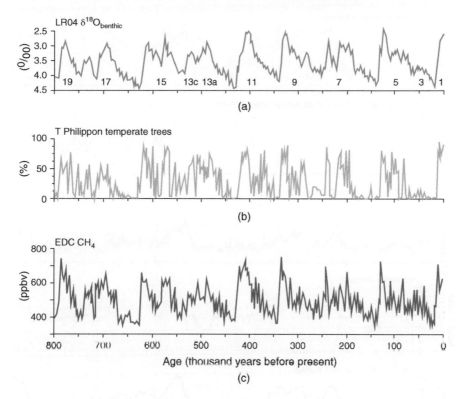

Figure 4.3 Comparison of (b) the temperate tree pollen sequence from Tenaghi Phillipon, Greece with (a) combined oxygen isotopic records from seabed foraminifera sampled from 57 globally distributed marine cores and (c) the methane record from the Antarctic EPICA Dome C ice core. Source: Tzedakis, Pälike, Roucoux and de Abreu, 2009. Reproduced by permission of Elsevier

populations on a millennial timescale, at a similar frequency to records of temperature variability obtained from marine cores in the North Atlantic (Tzedakis *et al.*, 2002). At Ioannina the lowest values for temperate trees occurred between 18 000 and 24 000 years ago, during the last glacial maximum (LGM). Tzedakis interpreted this forest minimum as a consequence of 'the combined effects of reduced annual precipitation and winter tempera-tures, a shorter growing season and also lower atmospheric CO_2 concentrations' (Tzedakis *et al.*, 2002). The rapid response of the trees strongly suggests local survival at the site and no extinction of trees during the period of the most extreme climate of the last 133 000 years. The Ioannina site in the Pindus Mountains is thus a true glacial refuge. Some tree species have been lost from Europe during the ice ages: *Liquidambar* (sweet gum), *Engel-hardia*, *Eucommia*, *Parrotia*, *Tsuga* (hemlock), *Carya* (hickory) and *Pterocarya* (wingnut) are all recorded from Europe during the last 1.5 million years but not the last 100 000 years (Magri and Palombo, 2012). The tree species most intolerant of harsh glaciations appear to have disappeared prior to about 600 000 years ago, at which time the European ice ages became more extensive and severe.

To move 10 000 km from Greece to Colombia and see the same long-term cyclicity of vegetation change is a powerful argument for global climatic control of long-term vege-tation dynamics. A 280 000 year pollen record from Lake Fúquene, high up in the eastern Cordillera of Colombia, shows just such a striking correspondence between the proportions of forest and nonforest biomes and the same 'stacked' global oxygen isotope record that was used by Tzedakis for comparison with the Tenaghi Philippon pollen data (Figure 4.4; Bogotá-A *et al.*, 2011). One of the climate records used for these comparisons was based

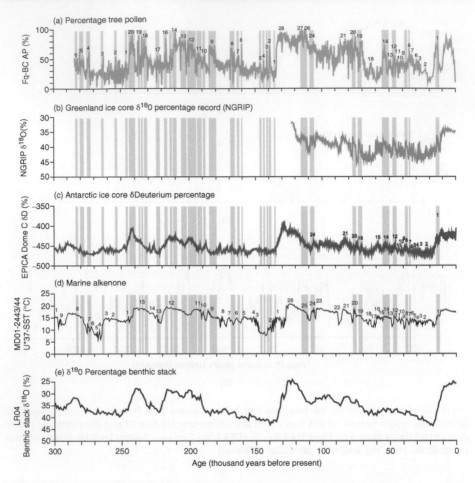

Figure 4.4 (a) Tree pollen record from Lake Fúquene, northern Andes. Compared with (b) the oxygen isotopic record from the NGRIP ice core, Greenland, (c) the Deuterium record from the Antarctic EPICA Dome C ice core, (d) a marine alkenone record from the Mediterranean, indicative of sea surface temperature and (e) combined oxygen isotopic records from seabed foraminifera sampled from 57 globally distributed marine cores. Source: Bogotá-A et al, 2011. Reproduced by permission of Elsevier.

on the isotopic content of shells from bottom-dwelling foraminifera preserved in multiple ocean cores from throughout the world (Lisiecki and Raymo, 2005). This isotopic record is believed chiefly to reflect global ice volume during the time periods covered by the pollen records. These pollen–isotope comparisons are based only on visual matching of curves, whose dating can be rather inaccurate. Also, the climate 'records' used for comparison are proxies for an often poorly defined combination of temperature at the site of foraminiferal shell growth and global ice volume. Similarly, the pollen data are also a proxy for the source vegetation from a poorly defined area. This sounds like rather inaccurate science and tells us little more than that vegetation is covarying with some aspect of climate. However, the climate record and its covarying ecosystem dynamics, with their characteristic 'Milankovitch' frequencies, are being found in so many different settings and systems around the world that they do seem to form part of a fundamental Earth system that has driven ice ages and movements of species throughout Earth history.

4.4   Holocene records

The records that cover very long time periods are informative but are too few in number to be used for spatial mapping of vegetation dynamics. The great abundance of sites now available from the Holocene for several regions of the world, when organised into databases, can be mapped to striking effect. The forest composition of Fennoscandia has changed continuously during the Holocene, with a maximum extent of broadleaved trees in the mid-Holocene and a rise to dominance of spruce-rich boreal forest during the late Holocene (Figure 4.5). While scientists have been mapping pollen data for many years and hypothesising about the major controlling influences (e.g. Rudolf, 1930), Margaret Davis' landmark paper of 1976 had a major impact on the biological and geological sciences within the English-speaking research community. She wrote that 'forest trees migrated to their present ranges from different source areas' and explored issues related to migration rates, climate–vegetation equilibria, the nature of succession and the influence of soil maturation and disturbance on spreading dynamics that are still eagerly debated (Davis, 1976). She reviewed data from both North America and Europe and prepared 'migration maps' that are as thought-provoking today as they were in 1976 (Figure 4.6). Her statement that 'clearly the sudden arrival of a species that is limited by migration speed cannot be attributed to climatic change, nor is its absence in older levels necessarily evidence

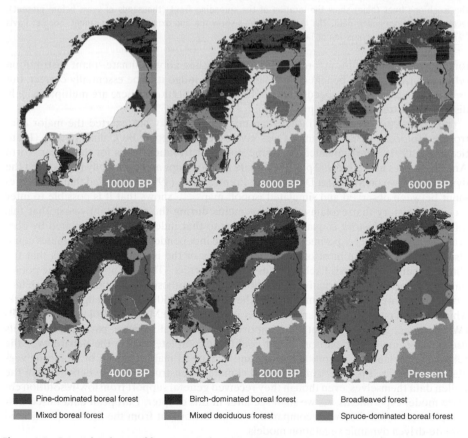

Figure 4.5   Interpolated maps of forest vegetation types at 2000 year intervals through the Holocene, based on 308 pollen diagrams from Fennoscandia

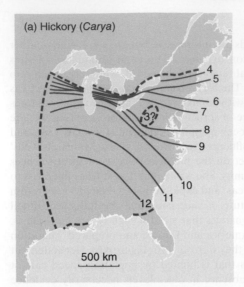
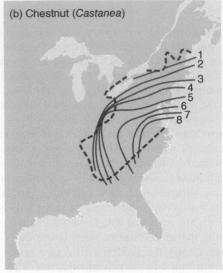

Figure 4.6 Holocene spreading patterns of hickory (*Carya*) and chestnut (*Castanea*) in North America. The isochrone lines (blue) show the location of the spreading front (thousands of years before present), reconstructed from pollen data. The red dashed lines show the approximate modern ranges. Source: Davis 1976. Reproduced by permission of Taylor & Francis

for unfavourable conditions' raises the central issues about climate–plant distribution relationships. With the benefit of hindsight, we can judge it to be essentially correct, but as we shall see migration speeds may not be such a limitation if there are multiple, widely dispersed centres for species spread.

The argument that climate rather than migration biology has exerted the major control on Holocene tree distributions was given elegant support by Colin Prentice (Prentice *et al.*, 1991), who simulated Holocene pollen distribution maps for the major North American trees using inferred pollen–climate relationships and essentially reproduced the dynamics shown on Margaret Davis' maps. Prentice *et al.* (1991) concluded that 'the agreement between observed and simulated isopoll maps … shows that it is possible to infer a climatic history that explains their movements during the past 18 000 years. That this inferred climatic history is qualitatively similar to that independently simulated by general circulation model experiments, and that these independently simulated climates could also correctly generate most of the major features of the isopoll maps, implies that the climatic change scenario is likely to be essentially correct.' They acknowledged the potential importance of the alternative 'migrational lag' hypothesis but proposed that such lags would be apparent in their comparison if they exceeded 1500 years in length. Their work, which formed part of the influential Cooperative Holocene Mapping Project (COHMAP) (Wright *et al.*, 1993), represented an important conceptual advance, particularly with its use of data–model comparison, but it could not completely close the debate over the relative importance of migration biology and climate as major controls of Holocene vegetation dynamics. Prentice *et al.*'s inferred climate reconstructions were derived from the pollen data themselves, even though they received general support from low-resolution climate model experiments. As we shall see later in this chapter, European data raised new complexities when they were compared with model output from the more sophisticated climate-driven dynamic vegetation models.

The discussion over the control of Holocene vegetation dynamics is highly dependent on the quality of the pollen data, and one weak link has been the dating control of individual sites used in the mapped summaries. Are major population expansions synchronous over large areas, suggesting climate-driven expansion of existing small populations, or asynchronous but spatially organised, suggesting that rates of migration might be a limiting factor? Giesecke *et al.* (2011) looked carefully at this issue using a selection of 58 high-quality, well-dated sites spread throughout Europe. Margaret Davis only had access to 26 North American sites for her work and her chronologies were less well constrained. Giesecke *et al.* found two types of synchronous change in vegetation within their European records. First, they detected synchronous changes involving several different species in different parts of Europe. The major changes occurred at the opening of the Holocene 11 500 years ago and again around 8300 and 3700 years ago. While a multitude of proposed climatic shifts or short-lived climatic events have been reported from the Holocene, these three dates do appear to have special climatic significance. The late glacial to Holocene transition (11 500 years ago) is well established as probably the most recent significant rapid climate change, with a typical estimate of 5 °C change in average temperature during a 50 year period obtained from Kråkenes in Norway (Birks and Birks, 2008). Previous research has also emphasised the significance of the climatic event 8200 years ago, when an enormous pulse of meltwater was injected into the North Atlantic following the final collapse of the major North American ice sheet and the catastrophic drainage of glacial Lake Agassiz. The effects were global, but European climate became considerably harsher for perhaps 200 years (Alley and Ágústsdóttir, 2005). One might have expected the next largest climatic event to have been the Little Ice Age, but Giesecke *et al.*'s analysis proposed a 3700-year-old event that they associated with a widespread onset of cooling and increased growth of European glaciers.

Giesecke *et al.* (2011) also identified two occasions where tree species populations (*Corylus* (hazel) and two species of *Alnus* (alder)) rapidly increased in size in synchrony over large regions of Europe. The most widespread species expansion of *Corylus* dates from 10 600 years ago, from 13 sites spread throughout central and western Europe, while 2 sites further east appear to show later population expansions (Figure 4.7). As Tsukada (1982a) and Bennett (1988) have postulated, regional pollen sites are rather insensitive to small tree populations and the rapid increases shown in Figure 4.7 are the later stages of exponential population increases that began just after the opening of the Holocene. The apparent later expansion at the eastern sites might simply reflect slower population increases under less favourable climatic conditions. This limitation of the sensitivity of regional pollen sites to the first arrival of species also restricts the possibility of calculating migration rates for trees using these data; small forest hollows prove to be more reliable in this respect (Bialozyt *et al.*, 2012). Nevertheless, the synchronously increasing pollen curves for *Corylus* and *Alnus* strongly suggest that climate change is the main driver of population expansion with no apparent migrational lag.

Linking vegetation and climate proxy records in this way does not provide conclusive proof of a dominating climatic control of vegetation dynamics, but pooling data from several sites across a continent summarises and complements the more detailed but local analyses and interpretations from single sites. How can we further explore hypotheses about the role of climate change on past vegetation dynamics, given this topic is of such importance today? A further step forward could surely be made by more detailed modelling of Holocene vegetation dynamics as available models became ever more sophisticated and reliable.

4.5 Modelling of Holocene vegetation dynamics to help understand pollen data

Fennoscandia has been a powerhouse of pollen analysis since the early days of the subject (Birks, 2005) and important progress in vegetation modelling was made at Uppsala and

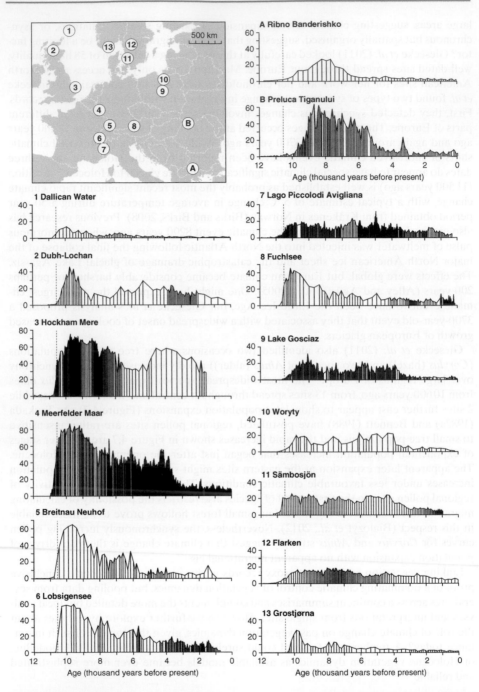

Figure 4.7 Comparison of *Corylus* (hazel) pollen records from Europe showing synchronous population expansions in 13 sites, with 2 sites in southeast Europe (A, B) expanding later. The data are percentages of a land pollen sum. Source: Giesecke *et al*, 2011. Reproduced by permission of Elsevier

Lund in Sweden under the leadership of Colin Prentice between 1983 and 1997. It was thus only natural that vegetation models should be used to explore Scandinavian pollen data in order to tease out some of the underlying controls on vegetation dynamics, in particular to estimate the importance of climatic control during the Holocene. As we have seen, pollen analysis has proved to be a remarkably versatile tool as it can both resolve vegetation dynamics at very fine spatial scales and track changing distributions of tree taxa at continental scales. A recurrent issue in the interpretation of Holocene vegetation dynamics at any spatial scale has been how to estimate the relative importance of climatic change, rate of spread and – particularly in Europe – anthropogenic activity as the dominant drivers of vegetation change. Pollen data alone yield descriptions of the rates and direction of vegetation change but give few clues as to the key drivers of this change. Behre (1988) reviewed approaches to distinguishing anthropogenic impact in pollen diagrams and suggested that this could more easily be achieved on a 'supraregional' scale, as 'climatic changes are normally geographically far-reaching', while 'prehistoric cultures are frequently regional in character' (Behre, 1988). His suggestion about the finer spatial scale of anthropogenic impact may be appropriate for prehistoric cultures but is likely to be less useful closer to the present time, as human impact increases in intensity and scale. Disentangling the effects of human activity from climate change in the recent past and at present is an urgent scientific task.

Davis and Botkin (1985) were among the first to use a dynamic, climate-driven vegetation model to help interpret pollen data. They modelled the potential effects of recent climatic change on temperate forests of the northeastern USA and concluded that the pollen record would yield a delayed response of vegetation to climatic events. This delayed response or inertia of existing vegetation to a gradual climatic change could be modified by disturbance, and in Chapter 6 we shall discuss another example of this process in the palaeoecological record. Following up on this earlier modelling of past climate–vegetation relationships, we began a new research programme into data–model comparison in order to increase our understanding of the influence of climatic change on long-term vegetation dynamics. As both climate models and modelled forecasts of future vegetation dynamics are rapidly developing areas of research, it seemed relevant to see how these models performed when hindcasting (see Chapters 2 and 3).

Summary: Long-term vegetation dynamics in the past

- There is a striking intercontinental correspondence between the global palaeoclimatic record and vegetation dynamics with Milankovitch frequencies, over at least 300 000 years.
- The density of Holocene pollen studies in Europe and North America permits detailed analysis of the underlying drivers. Climatic change drives dynamics, but migration biology can delay the response.
- Analysis of broad-scale synchronous vegetation changes in Europe has identified the opening of the Holocene (11 500 years ago), the catastrophic drainage of glacial Lake Agassiz (8200 years ago) and the onset of European neoglaciation (c. 3700 years ago) as events associated with widespread climatic change.
- Comparing output from climate-driven dynamic vegetation models with pollen data helps isolate the effects of climate change on vegetation dynamics.

4.5.1   Climate or people? The *Tilia–Fagus* transition in Draved Forest, Denmark

Our first case study looked at the change from mixed temperate forest to *Fagus* (beech)-dominated forest that took place over large regions of northwestern Europe in recent millennia. The decline in importance of *Tilia cordata* (the small-leaved lime tree) is a major feature of this change in forest composition and has been ascribed to anthropogenic impact using the arguments of regional nonsynchroneity and through association with pollen indicators of human activity (Turner, 1962). The decline in *Tilia* pollen values varies from site to site in rather an unsystematic manner and local pollen analyses from small forest hollows always record cultural activity coincident with the major decline in *Tilia* populations (Björse and Bradshaw, 1998). However, others believed that this change was largely a response to climate change, as *T. cordata* is sensitive to summer temperatures, which are believed to have been generally higher throughout northwestern Europe during the mid-Holocene. Estimating the area of Europe covered by *Tilia* using data from the European Pollen Database (www.europeanpollendatabase.net) shows maximum areas between 5000 and 8000 years ago, with a gradual decline in the last 5000 years – dynamics that can be interpreted as supporting a climatic-control hypothesis (Figure 4.8). These types of argument are common in the palaeoecological literature but are not fully convincing on their own. Independent evidence for the role of climate change would be valuable in helping choose between the anthropogenic and climatic hypotheses. We used the FORSKA forest-gap model (Chapter 2) to simulate the forest dynamics of a small plot within Draved Forest, Denmark over the last 1500 years, during which time beech (*Fagus sylvatica*) replaced *T. cordata* in places (Cowling *et al.*, 2001). This replacement took place during the Little Ice Age, which is widely regarded as one of the major climatic events of the Holocene. FORSKA simulates the establishment, growth and mortality of trees on 0.1 ha plots (Prentice *et al.*, 1993a); physiologically based processes such as establishment, growth and competition are subject to environmental constraints calculated from mean monthly climatic data; random disturbance by wind and fire is incorporated into the model but anthropogenic activities are not. A major difficulty in running forest simulation models for the past is obtaining palaeoclimatic data. These can either be derived from proxy systems such as tree rings or simulated using climate models, but both approaches incorporate uncertainties. There is, however, a growing consensus on palaeotemperature records for the last millennium, and in this study palaeotemperature anomalies from the present day were based on tree ring chronologies from northern Fennoscandia (Briffa *et al.*, 1992). Palaeoprecipitation data were derived

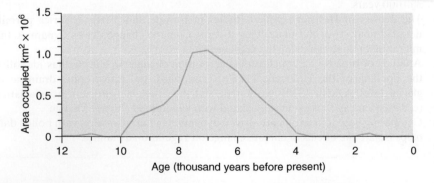

Figure 4.8 Estimated area of Europe covered by forest that includes *Tilia cordata* (small-leaved lime) during the Holocene. Reproduced by permission of Thomas Giesecke, European Pollen Database support groups

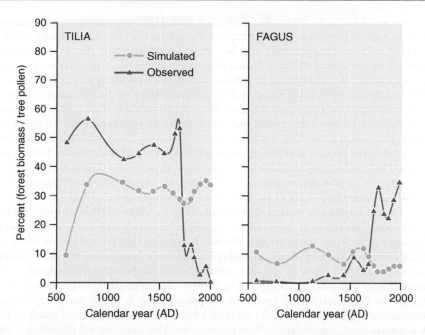

Figure 4.9 Data–model comparisons for *Tilia* (lime) and *Fagus* (beech) in Draved Forest, Denmark. Simulated tree abundance data are presented as percentage forest biomass. Observed tree abundance data are presented as percentage pollen, expressed as a percentage of a sum of all woody taxa (after Cowling *et al.*, 2001)

from Lamb (1967) but sensitivity tests showed that precipitation was not as important as temperature for the trees studied. The simulated biomass of *T. cordata* and *F. sylvatica* was compared with the trees' pollen percentage values in Draved during the last 1500 years (Figure 4.9; Aaby, 1983); the results showed that the observed stand scale replacement of *Tilia* by *Fagus* was not predicted by the model, even though the transition occurred during the coldest period of the Little Ice Age (Cowling *et al.*, 2001). Small reductions in biomass for both species were modelled, consistent with colder conditions, but the observed transfer of dominance was not predicted. This data–model comparison strongly suggests that a climate change was not the major driver for a stand-scale switch in dominant tree species. Aaby (1983) cites historical evidence of increased browsing pressure during the 1700s from cattle and horses. Cattle browsing on young shoots of *Tilia* during the early spring is known to restrict regeneration of the tree and favour the less palatable *Fagus*, and this replacement process may well have been accelerated by felling and planting activity. The model results in this case support the anthropogenic hypothesis: although it is likely that long-term trends in climate had made *Tilia* populations vulnerable to replacement, anthropogenic disturbance was the ultimate trigger for this change (see Chapter 6).

4.5.2 Climate or migration biology? The late-Holocene spread of *Picea* into southern Fennoscandia

Studies at the stand scale in southern Sweden have shown that late Holocene reductions in *Tilia* populations and the populations of other deciduous tree species were broadly coincident with increases in local populations of both *Fagus* and *Picea abies* (Norway spruce)

(Bradshaw and Lindbladh, 2005). The European-scale Holocene distribution dynamics of *Picea* show a gradual increase in area occupied throughout the Holocene, which could have been driven by climate change, migration biology or a combination of the two. Data–model comparison of the spread of *Picea* should help assess the relative importance of climate change as a driver in this regional to subcontinental late-Holocene process. We used the STASH bioclimatic model (see Chapter 2) to generate maps of the potential equilibrium distributional limits for *Picea* over the last 1500 years (Bradshaw *et al.*, 2000). The palaeo-climatic data used to drive this model were the same as those taken for the Draved Forest example. The pollen data show a southwesterly spread of *Picea*, which is the continuation of a spreading process that has occurred for most of the Holocene and that has been interpreted as being largely unaffected by anthropogenic activities (Giesecke and Bennett, 2004; Seppä *et al.*, 2009). A comparison of the observed range changes with the model output illustrated some important characteristics of the spreading process (Figure 4.10). The observed range lagged the modelled potential range by about 100 km at both 1000 and 500 years ago. This delay, or migrational lag, might be anticipated for a species extending its range in response to a changing climate; it suggests that the range was never fully in equilibrium with the continually changing local climatic conditions, but the lag did not appear to exceed 500 years in this case, which is close to the life expectancy of a *Picea* individual under favourable conditions. The data–model comparison also showed a recent change in the characteristics of spread in the present day, in which the observed range has overshot the potential range. At present, all of southern Sweden and much of Denmark supports large, regenerating *Picea* populations. This data–model mismatch is small but significant and reflects the widespread planting of *Picea* that has taken place during the last 300 years. Anthropogenic planting has bypassed natural regeneration processes and silvicultural management has removed competitive species in order to help *Picea* become dominant outside its potential natural range as modelled by STASH. It will be interesting to monitor the future of these populations. Many individuals have been destroyed during recent severe windstorms, although this type of disturbance was not explicitly modelled.

In this example there is a reasonable match between distributional data and the climate-driven model, but this is not of course conclusive proof of climatic control of the spread of *Picea*. The *Picea* population was expanding continuously during the 1500 year study period and this expansion may simply have been limited by the biological spreading rate: climate in this region may well have been suitable for *Picea* throughout the period. Dörte Lehsten ran LPJ-GUESS to answer just this question and showed that the model mapped most of Finland, Norway and Sweden as climatically suited for *Picea* for the last 9000 years, with Denmark also becoming suitable during recent millennia (Lehsten, personal communication). She argued that this was a strong indication that migration biology was delaying the spread of *Picea* into regions where climatic conditions appeared to be suitable. This migrational lag hypothesis has received further support from a data–model comparison using pollen data from small forest hollows and a diffusive spreading model that incorporates long-distance dispersal events (Bialozyt *et al.*, 2012), which shows that the data were fully consistent with a spreading rate for *Picea* of 250 m per year for at least the last 4000 years, which matches rather well with the observed timing from the most comprehensive summary of its Holocene spreading history (Giesecke and Bennett, 2004). So far this new story appears to be rather convincing. Most of Fennoscandia apparently became climatically suitable for *Picea* during the early Holocene, but the tree took several millennia to reach southern Sweden from glacial refugia in Russia, due to its maximum migration rate of 250 m per year. The two independent modelling exercises (by Dörte Lehsten and Bialozyt *et al.*, 2012) were consistent with much of the observed palaeoecological data and appear to support the migrational lag hypothesis, with a lag time for southern Sweden of almost 9000 years.

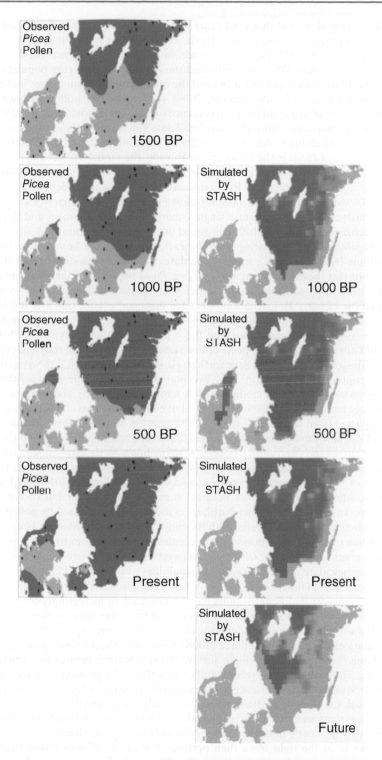

Figure 4.10 Observed and simulated *Picea abies* (Norway spruce) distributions in southern Scandinavia during the last 1500 years. The observed distributions are reconstructed from fossil pollen data, while the simulated distributions are generated by the bioclimatic model STASH. The forecasted future distribution assumes an atmospheric CO_2 composition of 730 ppm. Source: Bradshaw *et al*, 2000. Reproduced by permission of NRC Research Press, Canada

Migrational lags of several thousand years were widely discussed and accepted in earlier literature for many tree species (e.g. Davis, 1976) but they are falling out of favour as new evidence comes to light. Genetic evidence, in the form of both present-day population genetics and ancient DNA, has indicated the existence of outlying populations and even glacial refugia of tree species far beyond the main centres of abundance indicated by traditional pollen analysis (Anderson *et al.*, 2006). It turns out that pollen analyses of lake sediments do not reliably record the local presence of trees until those trees are really quite abundant on the landscape, although macrofossil data and fortuitously located small forest hollows do have the ability to detect small outlying populations (Overballe-Petersen *et al.*, 2012). In the case of *Picea* in the Holocene of Europe, the modern genetic evidence does not yet indicate any unique genetic variants in Norway and Sweden that do not also occur further east, close to the presumed Russian glacial refugial areas (Tollefsrud *et al.*, 2008). Ancient DNA evidence for full glacial survival of *Picea* in northwestern Norway has been reported (Parducci *et al.*, 2012), but a single occurrence based on novel and developing analytical techniques has not yet fully convinced the scientific community. A large number of wood megafossils of *Picea* dating from the early Holocene are known from the Scandes Mountains (Kullman, 2008); some of the dates have been questioned, but there can be little doubt that these finds represent outlying *Picea* populations far in advance of the major migrating front advancing at 250 m per year. These small populations could well be the outcome of early Holocene long-distance dispersal events, but why did they not become centres for early spreading themselves, given the vegetation models suggest that climate was suitable for *Picea* throughout the region for the majority of the Holocene? There would appear to be some aspects of this system that are not yet fully understood.

Thomas Giesecke has taken this issue further. He believes that LPJ-GUESS has not been adequately capturing *Picea* dynamics for two main reasons: first, the palaeoclimatic data that have been used to drive the model have low temporal resolution, and in the earlier simulations rather few climatic variables were used to characterise past climate; second, it has become apparent from several climate-driven vegetation models that we do not yet fully understand the bioclimatic controls on *Picea*. Giesecke *et al.* (2008) suggested that it was not sufficient to tell the vegetation models how many hours were warm enough for growth each year but that how those hours were distributed was also important. This temporal distribution of warmth for growth can give a 'continental' climate, with growth concentrated in the summer months, or an 'oceanic' climate, with a longer but cooler growing season. Giesecke showed that *Picea* distribution today, whether sensed by pollen or from direct tree observations, is associated with the most continental climates in Fennoscandia. He ranked five tree species from oceanic to continental distribution to give a gradient of increasingly continental species: *Ulmus glabra* (wych elm), *Corylus avellana* (hazel), *Quercus robur* (pedunculate oak), *Tilia cordata* (small-leaved lime) and *Picea abies* (Norway spruce) (Figure 4.11; Giesecke *et al.*, 2008). Giesecke also used transfer functions with pollen data to suggest that Fennoscandia had experienced an increasingly continental climate over the last 7000 years and that this aspect of climate may have exerted some control on the westward spread of spruce, hindering the early Holocene expansion of the outlying *Picea* populations described by Kullman (2008). Giesecke's theory could help explain how small outlying *Picea* populations might survive but not become centres for rapid population increase. These outliers may receive sufficient water and a growing season of sufficient length for survival but still experience a suboptimal distribution of favourable growing conditions through the seasons, so that young plants only rarely survive to maturity and the population size remains small. *Picea* is known to be particularly sensitive to late spring frost where growth begins early in the year, as it does in more oceanic climates.

If Giesecke is on the right track then perhaps climatic conditions rather than migration rate are the limiting factors for *Picea* after all and the European populations are in fact tracking changing climatic conditions rather closely. The debate, like many in science,

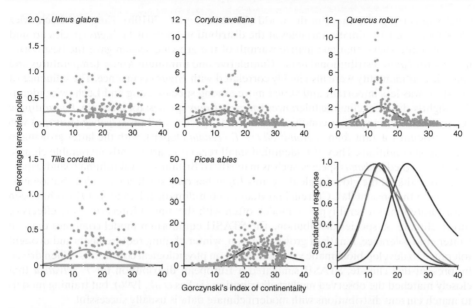

Figure 4.11 Scatter plots of pollen percentage versus Gorczynski's index of continentality. Response models are fitted to the data. Source: Giesecke *et al*, 2008. Reproduced by permission of Elsevier

swings back and forth in a slightly frustrating manner. This story shows that observations, even when supported by modelling, do not produce fully watertight explanations if the physiology of the system is not fully understood. Until we carry out more experiments on *Picea* trees in growth chambers and come to understand more about the physiology of climatic influence on the lifecycles of individual trees, we will really just be making educated guesses about the major controls on *Picea*'s spread during the Holocene. Modelling this system is an effective way of organising and integrating our knowledge in a logical manner, but the explanatory power of vegetation models is always limited by their initial parameterisation and how well they capture our understanding of plant physiology.

4.5.3 *Fagus* in Europe

Fagus sylvatica (European beech) is another tree that has increased its European abundance in the late Holocene and it has proved to be equally as awkward to model as *Picea*. Dynamic vegetation models do appear to have difficulties in controlling the abundance of European trees that only became really abundant during the late Holocene. At present, *Fagus* is a dominant tree in many European temperate forests. The factors that have led to its current abundance have often been debated, with Holocene climate change and human land use receiving most attention. It appears likely that several conditions must be fulfilled before a local *Fagus* population will expand to dominance. Ralska-Jasiewiczowa *et al.* (2003) have proposed that in northern Europe a change to a wetter and cooler climate, in association with human disturbance, led to species expansion. Similar predetermining factors have been recognised in southern Europe, with the addition of a decrease in the importance of fire, because *Fagus* is particularly sensitive to fire damage (Valsecchi *et al.*, 2008).

The arguments for the importance of an overriding climatic control on European *Fagus* distribution are strengthened by comparison of the European *Fagus sylvatica* with *Fagus* species occurring elsewhere in the world (Bradshaw *et al.*, 2010b). Fang and Lechowicz (2006) analysed 13 climatic variables at the distribution limits of 11 *Fagus* species around the world and found that cumulative warmth of the growing season gave the best correlation with *Fagus* distributional limits. Cumulative and minimum winter temperatures and climatic continentality were also highly correlated with *Fagus* occurrence. The influence of moisture was less important and varied more among species. Fang and Lechowicz identified slightly different climatic tolerances for each *Fagus* species; for example, *F. grandifolia* (USA) occurred in warmer conditions than other species, while *F. sylvatica* (Europe) was more drought-tolerant than *F. grandifolia* or *F. crenata* (Japan), with the latter growing in the wettest conditions. They also identified small regions of apparently favourable climate not yet occupied by *Fagus* species, such as in northern Britain and Yakushima Island, Japan, suggesting that at the global scale the genus *Fagus* has only small climate-distribution mismatches in the present day. Current knowledge of the climatic tolerances of *Fagus* has been incorporated into several types of model, often with the aim of forecasting the effects of future climate on species distribution. The STASH equilibrium model combines data on winter cold tolerance, length of growing season, winter chilling requirements and drought tolerance to develop bioclimatic variables based on physiological constraints. When driven by present-day climate, STASH simulated a European distribution for *F. sylvatica* that closely matched the observed modern distribution (Sykes *et al.*, 1996), but training models to match current distributions with modern climate data is usually successful.

The corresponding distribution of *Fagus* in Europe was simulated for 6000 years ago, again using STASH (Giesecke *et al.*, 2007; Figure 4.12(a)). This model run was driven by simulated climate data for 6000 years ago derived from atmospheric general circulation model (AGCM) output. The modelled distribution of *Fagus* based on climate data from three different AGCMs yielded distribution patterns broadly similar to the present-day distribution of the tree. Yet the palaeoecological record showed clearly that the regions in which *Fagus* trees were abundant 6000 years ago were confined to central and southern Europe, including the forelands of the Alps and parts of the Balkan Peninsula, and there is no evidence that the species had any significant presence in northwest Europe or Scandinavia (Figure 4.12(b)). This data–model mismatch was confirmed by Dörte Lehsten's LPJ-GUESS simulations for *Fagus* at 1000, 4000 and 9000 years ago (Figure 4.13), just as it had been for *Picea*.

There are several possible reasons for this data–model mismatch: the choice of climatic parameters used in the model and uncertainties in the palaeoclimatic estimates, particularly regarding possible changes in seasonality; the inability of regional pollen data to resolve small outlying populations of trees, meaning that *Fagus* may have occurred in the regions indicated by the model but in very small populations; or migrational lag (Giesecke *et al.*, 2007). As we saw for *Picea*, the migrational lag hypothesis is not fully consistent with the data for *Fagus*. *Fagus* survived the last glaciation in the Balkans and other locations around the Mediterranean but did not become a dominant species in Bulgaria or other regions close to established glacial refugia until late in the Holocene, so migration rate is unlikely to have controlled population dynamics (Tonkov, 2003).

Both Tsukada (1982a), working with *F. crenata* in Japan, and Bennett (1988), working with *F. grandifolia* in Canada, questioned the migrational lag concept for *Fagus* based on estimates of population doubling rates. An apparent northward spread of the species could be attributed to different rates of population increase among several dispersed population centres. Both the Japanese and the Canadian data indicated an early-Holocene widespread dispersal and establishment of *Fagus* populations, followed by differential expansion rates related to initial population size and latitude. It seems reasonable to propose that long-distance dispersal, establishment and initial population expansion were likely in open, nonforested landscapes but that subsequent population growth was dependent upon local climate and competitors. In Japan, the late-glacial rapid dispersal

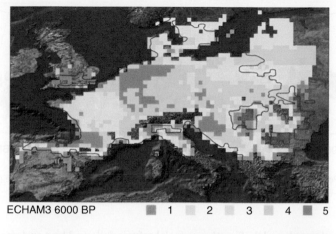

ECHAM3 6000 BP   1   2   3   4   5

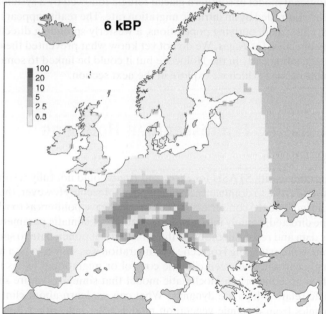

Figure 4.12   (a) STASH simulation of *Fagus sylvatica* distribution in Europe 6000 years ago, based on climate modelled by the ECHAM3 AGCM. Red lines show the modern distribution of *Fagus*. (b) Distribution of *Fagus* pollen percentages 6000 years ago, taken from the European Pollen Database and based on the dating strategy of Giesecke *et al.* (2014)

of *Fagus* is well documented on Honshu (Tsukada, 1982b). There is emerging evidence for this type of dispersal in Europe as well, with early Holocene records of very small *Fagus* populations being reported from locations in southern Britain, Denmark, northern Germany and Norway.

The European *Fagus* story follows similar lines to that of *Picea*. There are strong indications of an overall climatic control, but in large areas modelled as suitable for the trees during the early Holocene there are only small scattered populations that do not contribute to significant expansion of the range until late in the Holocene. Neither species follows

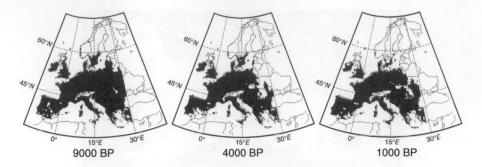

Figure 4.13   Potential distribution of *Fagus* in Europe at three different times (years before present), modelled using LPJ-GUESS (Lehsten, unpublished)

the simple, traditional model of southern glacial refugia followed by a systematic northern Holocene spread controlled by an intrinsic migration rate. The reality appears to have been more complex, with small, outlying populations, a westerly spreading direction for *Picea* and a northerly direction for *Fagus*. We do not yet know what prevented the outlying populations from expanding early in the Holocene but it could be linked to some more subtle aspects of climate change, which we explore in the next section.

4.6 Simulating Fennoscandian Holocene forest dynamics

Our pioneering efforts with STASH for *Picea* and *Fagus* could not fully resolve the debate as to why these trees rose to dominance so late in the Holocene. However, the data–model comparisons did indicate the complexity of the issue and gave pointers as to where research effort should be directed; namely into new combinations of climatic parameters, improved palaeoclimatic data and reconsideration of the sensitivity of pollen data to small, dispersed tree populations. Understanding Holocene tree migration was shown to be a more complex issue than simply choosing between climatic control or migrational lag as the dominant driver. STASH is an equilibrium bioclimatic model that simulates entire species ranges, yet tree–climate relationships are dynamic. What can we learn about climatic influence on forest dynamics from dynamic vegetation models? We saw earlier how the dynamic stand-scale simulator FORSKA was successful at distinguishing climate from land use in the shift from *Tilia* to *Fagus* dominance in a Danish forest. What could its more sophisticated successor LPJ-GUESS tell us about Holocene vegetation change at specific sites in Fennoscandia?

LPJ-GUESS was driven by climate model output and run for the last 10 000 years at four locations in Sweden and Finland on a north–south gradient. LPJ-GUESS estimates absolute plant biomass. We decided to compare its biomass estimates with pollen accumulation rates for the major trees recorded at the four sites, as pollen data in this form are closely related to biomass (Seppä and Hicks, 2006). Correction factors were applied to the pollen accumulation rates to account for species-specific differences in pollen production (Miller *et al.*, 2008). We found a rather good correspondence between modelled biomass and pollen accumulation rates for all four sites during the Holocene (Figure 4.14). Modelled and observed biomass are larger for the southern sites and build up rapidly as temperatures increase at the start of the Holocene. The pollen data, but not the simulated biomass, show decreasing values during the last 1000 years at the three more southern

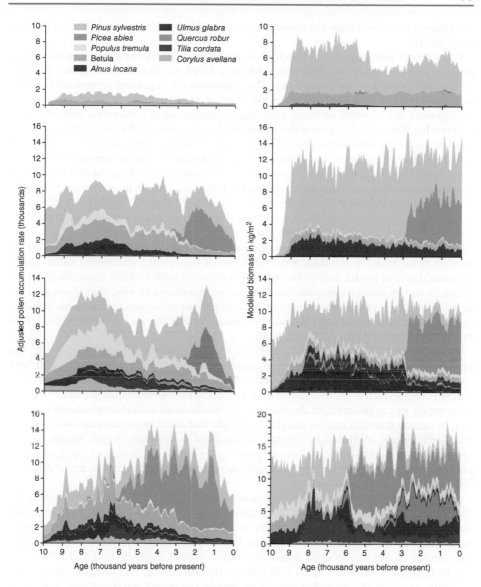

Figure 4.14 Adjusted pollen accumulation rate (left column) and modelled biomass (right column) for nine tree species at four sites on a north (top) to south (bottom) gradient in Fennoscandia (Miller *et al.*, 2008)

sites. This biomass reduction reflects deforestation for agriculture and forestry in southern and central Fennoscandia, which is not included in the model. Pollen data from the most northern site indicate declining biomass over the last 4000 years (Figure 4.14). The model does not capture this widely recognised retreat of Fennoscandian arctic treeline and we attribute this to an underestimation of regional late Holocene summer cooling in the modelled palaeoclimatic data (Seppä *et al.*, 2004). As expected from the earlier discussion, the model did not manage to exclude *Picea* from the simulation during the earlier Holocene, so it was artificially excluded until such time as the pollen data indicated presence. This piece of manual intervention, reflecting lack of knowledge rather than model failure, was only necessary for *Picea*, which is a considerable improvement over an earlier data–model comparison from Switzerland in which every tree had to be restricted in this way (Lischke *et al.*, 2002).

Pinus (Scots pine) and *Betula* (birch) dominated the modelled vegetation at the northernmost site throughout the Holocene, although *Alnus* was also more abundant prior to 4000 years ago than in the present day. These patterns are in general agreement with the pollen data. At present, the lake lies 30 km north of *Pinus*-dominated forest and *Alnus* is only present in very small amounts. Individual comparisons of pollen data and simulated biomass for *Pinus* and *Alnus* show that the model largely captures the observed trends, with an increase before 9000 years ago, maximum values between 8000 and 6000 years ago and a decline thereafter (Figure 4.14). These trends are driven by summer temperatures in the model (Miller *et al.*, 2008). The climate was more favourable for *Alnus* further south than at the northernmost site, and this is reflected in both data and the model output. A model experiment that excluded *Picea* from the system showed that the abundances of several deciduous trees, particularly *Alnus*, were sensitive to competition with *Picea* and that this competition effect could outweigh the direct effects of climate change.

The simulations for the two southern sites capture the overall species diversity and trends in abundance found in the data, although some of the relative abundance values differ between data and model, as is to be expected with proxy pollen values representing vegetation. Some model–data mismatches, such as the abundance of modelled *Quercus*, *Tilia* and *Ulmus* (elm) at the southernmost site during the last 3000 years, could be the result of these deciduous species being particularly sensitive to fire: the way in which LPJ-GUESS models fire is being developed as we write.

The real value of this modelling exercise is in exploring the potential controls on the observed forest dynamics, such as the separation of climatic (e.g. *Tilia–Quercus* ratio) from competitive (e.g. *Picea*–broadleaved trees) influences. We established that many of the significant changes seen in the four pollen diagrams could be approximately matched in model runs, which was an informative discovery as the model does not take into account migration or human activities and we can thus conclude that these processes may not have had a major influence on Holocene vegetation dynamics in the study region, except for the last 1000 years in the southern sites and possibly for the establishment of *Picea*, as previously discussed. Our analyses are consistent with climatic factors being the major driver of millennial vegetation dynamics.

As a further part of this study, we carried out a series of sensitivity experiments that included alteration of the disturbance frequency, removal of species-specific drought-limited establishment thresholds and variation of the interannual climatic variability. As we had no estimates as to how interannual climatic variability had changed through the Holocene, we applied the observed variability between 1901 and 1950 to the full Holocene record (Miller *et al.*, 2008). While the disturbance and drought experiments had little effect on the modelled vegetation trends, the model system proved to be very sensitive to changes in the amplitude of the interannual variability of meteorological parameters. Several of the trees in our study system were close to their bioclimatic distribution limits and doubling the level of interannual climatic variability completely removed the thermophilous *Tilia*, *Quercus* and *Ulmus* from the modelled system at one of the southern sites (Figure 4.15).

We followed up this discovery and used summer and winter varve thickness from a lake with annually laminated sediments throughout the Holocene as a proxy for interannual and seasonal climatic variability at a site in the Swedish southern boreal zone (Giesecke *et al.*, 2010). The results were conclusive. First, the varve data indicated that there had been systematic and significant trends and abrupt changes in past interannual climatic variability. Second, the model experiment showed that these changes in variability had the potential to influence vegetation composition and drive vegetation change without needing to invoke changes in average monthly or annual temperatures. This discovery supports

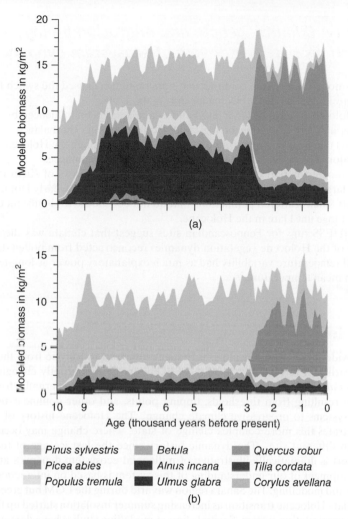

Figure 4.15 Modelled biomass for a southern Fennoscandian site when the interannual climatic variability is (a) halved or (b) doubled (Miller *et al.*, 2008)

earlier suggestions from modellers working with current and future systems (e.g. Notaro, 2008) and raises the key question of how much of the reconstructed vegetation change during the Holocene was caused by changes in interannual climate variability as opposed to changes in the mean values. This has implications for climate reconstructions from biological proxy data, which are generally aimed at reconstructing mean conditions using the modern relationship between mean climate and the abundance of a plant or animal. Of greater significance for society is the consequence that this conclusion also casts doubt upon projections of future plant distributions made by statistical vegetation models, because unlike the dynamic vegetation models used here, the statistical models do not take into account this important interannual climatic variability. An increasing number of studies suggest that future climate change will be associated with changes in variability of climate and weather. These changes need to be considered when making projections of future species distributions.

Summary: Climate, people and migration biology – interactions between drivers of ecosystem dynamics

• Data–model comparison confirms that human impact caused the switch from *Tilia* to *Fagus* in Draved Forest, Denmark during the 1700s.
• The Holocene spread of *Picea* into Fennoscandia has been influenced by small outlier populations, migration biology and recent planting. The balance of evidence suggests that an increasing continentality of climate through the Holocene has been the major influence on the changing pattern of *Picea* dominance.
• Data–model comparison indicates an overall climatic control of *Fagus* in Europe, but in large areas modelled as suitable for the trees during the early Holocene there are only small scattered populations that do not contribute to significant expansion of the range until late in the Holocene.
• LPJ-GUESS runs for Fennoscandian sites suggest that climate was the dominant driver of the Holocene vegetation dynamics reconstructed from pollen data. Interannual temperature variability had as much explanatory power as long-term variation in mean temperature.

4.6.1   Holocene dynamics of the Sahara

The three widespread abrupt ecological changes observed in Europe from the Holocene discussed earlier in this chapter were most likely driven by abruptly changing climates (Giesecke *et al.*, 2011). Williams *et al.* (2011) distinguish these 'externally' forced events from 'those resulting from thresholds, tipping points, and other nonlinear responses of ecological systems to progressive climate change'. The Holocene history of the Sahara desert illustrates this more complex change of state, where change may occur at different times in different sites. The dynamic history since the last ice age of the now arid Sahara desert is dramatically illustrative of millennial ecosystem dynamics and the continual development of our understanding of this system is based on interactions between palaeodata and modelling. The Sahara region was arid during the LGM but greened up during the glacial–Holocene transition as increasing summer insolation started up the seasonal monsoonal precipitation system. Several climate modelling studies have demonstrated the sensitivity of the monsoons to orbital changes in summer insolation. When climate models are linked to vegetation models, the Saharan climate can be made sufficiently humid to support the large lakes of the Holocene African Humid Period (between about 14 000 and 6000 years ago), for which there is good geological and biological evidence (deMenocal *et al.*, 2000).

A high-resolution marine sediment core taken from off the Mauritanian coast gave a 25 000 year record of sub-Saharan dust, of which about 400 million tons are annually transported out to sea at present. Present-day studies show a close linkage between the amount of dust transported into the atmosphere and drought severity in northwest Africa, so it is likely that this oceanic dust record can serve as a palaeodrought index. It shows three rapid changes in flux during the last 15 000 years, centred on 14 800, 11 400 and 5500 years ago (Figure 4.16). The 14 800-year-ago fall in dust flux coincided with evidence for the flooding of several subtropical African palaeolakes and the onset of the humid period. The flux of dust increased for about 1000 years around the time of the Younger Dryas period, which was a temporary return to near-glacial conditions at more northerly latitudes. This period ends rather abruptly – in less than 200 years – with a return to humid conditions, until an equally abrupt aridification 5500 years ago (deMenocal *et al.*, 2000).

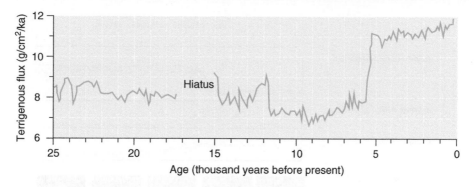

Figure 4.16   Terrestrial sediment flux, mostly dust, recorded in marine sediments off the west coast of North Africa. Source: deMenocal *et al*, 2000. Reproduced by permission of Elsevier

This system was modelled using coupled ocean–atmosphere–vegetation models to show how a gradual decrease in solar radiation might have given rise to a more rapid loss of vegetation and a consequent abrupt increase in the marine dust record (deMenocal *et al.*, 2000). Modelling indicated the potential importance of positive feedbacks from vegetation to albedo, lakes to precipitation and surface ocean temperature to moisture transport, which together could provide an explanation for the abrupt aridification (Claussen *et al.*, 1999; Krinner *et al.*, 2012). Detailed dynamic vegetation modelling showed that grass cover would rapidly disappear if rainfall were reduced to about 250 mm/year, so a long-term decline in precipitation could cause an abrupt vegetation change (Liu *et al.*, 2007). Thus the developing story was driven by observations from a marine core, supported by more fragmentary terrestrial data, with potential mechanisms indicated from detailed modelling of the system.

The next development in the story was the publication of a land-based sediment record from Lake Yoa, which is in the Sahara but is chiefly supplied from groundwater (Kröpelin *et al.*, 2008). This detailed multiproxy record contained in over 7 m of sediment the history of a regional terrestrial ecosystem development during the last 6000 years (Figure 4.17). The pollen data showed that originally the area was a grassy savannah with scattered tropical trees – a type of vegetation found 300 km further south today. Gradual drying out of this system began at least 5600 years ago as plants such as *Acacia* (indicating semi-desert conditions) began to replace the tropical trees. The grasslands, which are important in preventing soil erosion, continually reduced in area to reach minimum values of cover around 3800 years ago. As grassland became less abundant on the landscape, there was an increase in magnetic susceptibility of the sediments, indicating an increase in windblown dust. There was an abrupt increase in pollen deposition around 2700 years ago, which Kröpelin *et al.* (2008) interpret as indicating a change in regional wind regime to the year-round strong northeastern trade winds that still blow today. These winds bring fine sand and pollen from as far away as the Mediterranean region (Figure 4.17). The evidence from this site does not indicate the abrupt desertification that had been interpreted from the earlier marine core, but rather a gradual 2500 year reduction in the lusher tropical vegetation that preceded the subtropical desert conditions of the present day. There was an abrupt change in the hydrology of the lake system 4000 years ago when water conductivity increased as inflowing surface waters – perhaps a single river – ceased and the basin became hydrologically closed. Other terrestrial sites also indicate increasing aridification, but the timing and degree of abruptness varies from one to the next. The general pattern is a 'temporal mosaic' of local responses, both abrupt and gradual, to regional aridification. Each local

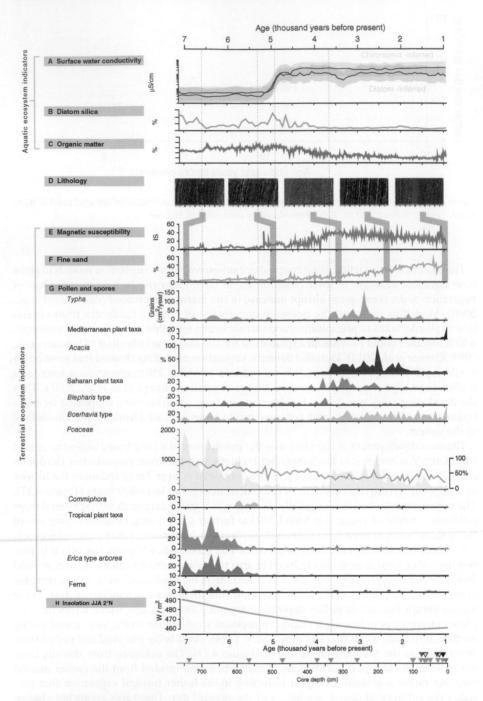

Figure 4.17   Evolution of aquatic and terrestrial components of Lake Yoa, Sahara during the last 6000 years. JJA, June July August. Triangles indicate dating points. Source: Kröpelin *et al*, 2008. Reproduced with permission of the American Association for the Advancement of Science

terrestrial record is influenced by site-specific thresholds of response to this aridification (Williams *et al.*, 2011).

This Saharan case study raises important issues concerning the strengths and limitations of data and modelling, and the rapidity of climate change and ecosystem response. Palaeoecological data are always fragmentary in space and time and it can be hard to judge how representative individual datasets are of larger regions. Data are also purely descriptive, and while authors are always keen to speculate on underlying driving forces and mechanisms, these hypotheses can rarely be adequately tested. In this example, the widespread applicability of the marine dataset, which indicates abrupt desertification 5500 years ago, is challenged by a well-dated, multiproxy lake core study, although the landscape area represented by the lake study is likely to be far smaller than the source area of the dust found in the marine core. The modelling work proposed plausible mechanisms for both abrupt and gradual ecosystem change and supported the claims that the desertification process was indeed a continental-scale process. The abrupt changes were dependent on feedback processes and the climatic thresholds of major vegetation types, while gradually declining precipitation could also have driven more gradual vegetation change. Both datasets could be simulated by models, although it proved more difficult to simulate the abrupt change. Distinguishing abrupt from gradual ecosystem change during periods of rapid climatic change is an important research challenge at the present time. Complex climate models have a doubtful ability to simulate threshold behaviour that might have occurred in the past (Valdes, 2011). In this example, the key lies with high resolution datasets that are representative for large areas, where sampling resolution is sufficient to measure rates of change and the proxy data recorded reflect ecosystem dynamics rather than changes in dispersal or sedimentation processes: so-called taphonomy, or the process of generation of material preserved in sediments. Data–model comparison of past climate-driven ecosystem dynamics is an important way of increasing our confidence in the value of complex models as forecasting tools.

4.7 Climate and megafaunal extinction

Zoological data have not been so extensively used as botanical data in the study of species–climate relationships, although there has been significant research on past climate–insect and –bird relationships (e.g. Huntley *et al.*, 2013). However recent developments in ancient and modern DNA analysis are affecting many fields of biology and have made an intriguing contribution to the classical debate about the causes of Late Quaternary megafaunal extinctions (Lorenzen *et al.*, 2011). Barnosky *et al.* (2004) had concluded that hunting by humans interacted with rapid climatic change on most continents to cause the extinctions, but ancient DNA studies are now able to contribute time series of estimations of species distribution area and population size, which help to disentangle human from climatic influence. Lorenzen *et al.* (2011) found positive correlations for horse, reindeer, bison and musk ox between genetically estimated population size and area of climatically suitable habitat in Eurasia and North America throughout the last 50000 years. As with our data–model research, they used climate model output to estimate past mean temperatures and annual precipitation, which incorporate considerable uncertainties. They concluded that climate change had been the major driving force for the population dynamics they observed (Figure 4.18). Lorenzen *et al.* (2011) also used their techniques to place extinctions into a temporal perspective. Populations of the now extinct woolly rhinoceros and woolly mammoth showed a five- to tenfold increase in size between 34000 and 19000 years ago, despite at least a 10000 year overlap with humans during this period, arguing against traditional hypotheses of a hunting 'overkill' or the spread of infectious disease following first human contact. Woolly rhinoceros disappeared from the

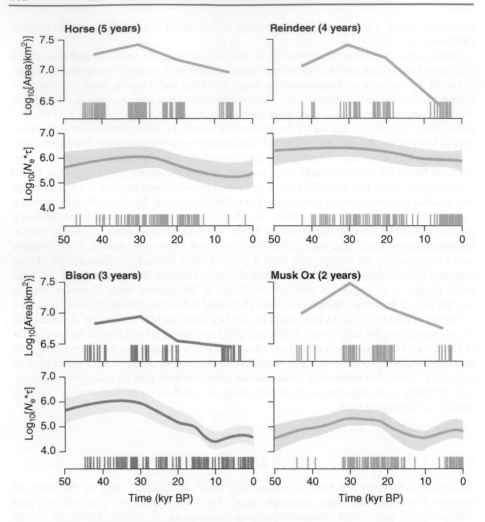

Figure 4.18 Temporal changes in global genetic diversity (a rough proxy for population size) and range size in horse, bison, reindeer and musk ox. Generation times are given in parentheses. Vertical lines represent dated individuals. Source: Lorenzen *et al*, 2011. Reproduced by permission of Nature Publications

fossil record about 14 000 years ago, following a period of genetic evidence for population fragmentation. There was little observed overlap between woolly rhinoceros and people in Siberia during the 6000 years before extinction, although there was such overlap in Europe just prior to extinction. The final extinction event for woolly rhinoceros was synchronous in both regions however, suggesting a primarily climatic control. Lorenzen *et al.* (2011) argue for significant climatic control of long-term population dynamics and regional extinctions for several large mammals, but acknowledge likely human influence in the decline of Eurasian horse and bison. They found no distinctive genetic changes or range dynamics preceding past extinction events, which has relevance for the forecasting of potential future extinctions (see Chapter 9).

The North American megafaunal extinctions were indirectly studied with high temporal resolution from sediments in a lake in Indiana, where spores of the dung fungus *Sporormiella*, which requires herbivore digestion to complete its life cycle, collapsed from abundance to rarity between 15 000 and 13 500 years ago (Gill *et al.*, 2009). These authors proposed a link between the loss of large herbivore populations and an observed restructuring of plant communities, with an increase in palatable broadleaved trees such as *Fraxinus* (ash), *Ostrya* (hop hornbeam), *Ulmus* (elm) and *Acer* (maple). Burning increased as a likely response to an increased fuel supply brought about by reduced herbivory. Gill *et al.* (2009) showed that the local megafaunal populations declined over several thousand years and that final extinctions could not be matched to any single events, such as abrupt climatic cooling, meteor impact or rapid 'blitzkrieg' overkill from hunting – all of which are hypotheses that had been previously proposed. So recent research into records of a dung fungus and ancient DNA are developing our understanding of processes that were previously viewed as driven by either climate change or human impact. These new data indicate the potential complexity of ecosystem interactions with a hierarchy of controls. Both papers argue for an overall climatic control that can be modified and accentuated both by human activity and by ecosystem processes such as fire and herbivory in order to generate a diversity of outcomes for population dynamics and extinctions. These examples show how technical developments in data collection allow more sophisticated ecological hypotheses to be evaluated from the past, and the outcomes have relevance for ecosystem management during the current period of rapid change.

4.7.1 Recent range shifts

Range shifts involve a species moving in some way to a more suitable environment. They can occur as changes in both latitude and altitude. As discussed earlier, there is strong evidence that species have responded to climate change in the past by range shifting. There is also evidence for range shifting under the current climate change. there are reports of shifts for amphibians, birds, butterflies and plants, for example. Pöyry *et al.* (2009) showed that 48 butterfly species in Finland had an average range shift northwards of almost 60 km between 1992–96 and 2000–04, with mobile species living on forest edges showing the greatest shift. Devictor *et al.* (2008), looking at the abundance of 105 different species of bird in France, found a shift northwards of 91 km in bird community composition between 1986 and 2006. Colonising new areas depends not only on suitable climate space but also on the ability of each species to disperse or migrate to new areas. Plants respond substantially more slowly than more mobile species such as butterflies and birds; most of the available evidence for plants relates to changes in altitude. In Sweden, treelines have moved upward at a rate of about 10 m per decade over the last 100 years (Kullman, 2001), in the European alps plants have moved by up to 4 m per decade over c. 90 years (Grabherr *et al.*, 1994) and in Spain beech has moved by 12 m per decade over the last 60 years (Figure 4.19; Peñuelas and Boada, 2003; Peñuelas *et al.*, 2007).

In the European Alps there is a trend towards increasing species richness on alpine summits, which suggests an upward shift in species ranges as the climate becomes more favourable. This is likely to lead to more specialised alpine species facing local extinction as they are outcompeted by migrating species. There are few reported examples of latitudinal shifts, although holly (*Ilex aquifolium*), a species well known for having a close link to winter temperatures, has a northern limit in Scandinavia closely associated with the location of the 0 °C isotherm. In the last 50 years this isotherm has moved north, and there has been a corresponding movement in the distribution of holly, which now occupies new climate space along the more southerly coasts of Sweden (Walther *et al.*, 2005).

4.8 So how important is climate change for future millennial ecosystem dynamics?

It would be misleading to regard climate–ecosystem relationships as just one-way traffic, with ecosystem dynamics passively responding to climate change. In the Saharan case study,

(a) *Fagus sylvatica* **(Beech)**

Upward shift to the top of the mountain

(b) *Quercus ilex* **(Holm oak)**

Heathland replacement by holm oak forest at medium altitudes

Figure 4.19   A global change-induced biome shift in the Montseny mountains (NE Spain). Upper: Fagus sylvatica (beech) (Peñuelas *et al.* 2007); Lower: Quercus ilex (holm oak) (Peñuelas & Boada, 2003). With thanks to Marti Boada and Josep Peñuelas

it was modelling that suggested the role of feedbacks from vegetation and water surfaces to climate, and the concept of feedbacks to climate is one of the conceptual advances to have arisen from data–model comparisons. Such feedbacks from vegetation dynamics can be significant, with a 13% increase in global warming estimated from this source if atmospheric CO_2 concentration should quadruple (O'ishi *et al.*, 2009). Modelling of likely future scenarios indicates that raised CO_2 levels fertilise vegetation, which increases terrestrial carbon storage, but that the associated warming also accelerates decomposition of soil organic matter, causing increased carbon loss to the atmosphere. The current consensus in the modelling community is that fertilisation is the dominant process and that terrestrial storage of carbon will increase in the future, at least if wildfires do not get out of control. This is an encouraging feedback from vegetation to the climate system, as the increased vegetation carbon storage can help control further build-up of atmospheric CO_2.

However, a model experiment that ran LPJ-GUESS coupled to an AGCM further into the future in order to allow biome distributions to adjust fully to higher CO_2 levels yielded an increased area of global vegetation cover, particularly in the semi-arid tropics. This increased vegetation mass enhanced absorption of short-wave radiation, which raised global temperatures further. When this extra heat was transferred polewards, to where enormous carbon stores are locked in frozen peatlands, the resulting decomposition of stored organic material was forecasted to raise atmospheric CO_2 concentrations still further (O'ishi *et al.*, 2009). Ecosystem feedbacks to the climate system of this type appear to have the potential to influence future climates in very significant ways; this is important as people are able to manage ecosystems, and to some extent influence these feedback processes (see Chapter 9).

Palaeoecological data provide limited but important validation of models. The Palaeoclimate Modelling Intercomparison Project consortium has used palaeodata to make comparisons with the output from coupled climate and vegetation models and has shown that models have tended to systematically underestimate reconstructed precipitation and temperature observations from the mid-Holocene and LGM (Figure 4.20; Braconnot *et al.*, 2012). The palaeoclimatic data used in these analyses are derived from pollen and plant macrofossil data based on a number of assumptions that are hard to test, as discussed earlier in this chapter. Direct experimental verification of model forecasts is potentially a more convincing way of validating models, but such experimental observations only cover rather short periods of time. Time series such as the 30 years of observation of tundra vegetation plots and associated climate data (Elmendorf *et al.*, 2012), the artificial warming experiments in open-top chambers of the International Tundra Experiment (ITEX) and the Free-Air CO_2 Enrichment (FACE) experiments (Leakey *et al.*, 2012) may lack the very long-term perspective contained in the palaeorecord but compensate for this with direct measures of vegetation and more accurate and less ambiguous measurements of temperature, precipitation and CO_2 concentrations than are found in palaeorecords.

Elmendorf *et al.* (2012) carried out a tundra study based on 158 plots from throughout the Arctic. They reported height increases of the plant canopies, an increased abundance of litter, a decreased area of bare ground and an increased abundance of evergreen, low-growing and tall shrubs, all of which are outcomes anticipated by vegetation modelling. They also found an association between the degree of summer warming and the monitored change in vascular plant abundance, but the heterogenous nature of the data limited the statistical confidence of their temperature–growth relationships (Figure 4.21). The ITEX experiments used 1.8 m² artificially warmed plots, with open-top chambers raising temperatures by 1–2 °C. Hudson *et al.* (2011) found significant and sustained increases in leaf size and plant height in a range of tundra species (*Dryas integrifolia*, *Salix arctica*, *Cassiope tetragona* and *Oxyria digyna*) after 16 years of continuous warming. The effects of elevated CO_2 have also been intensively studied, both in closed growth chamber experiments and in open-top chambers. Growth chamber experiments showed that an initial stimulus

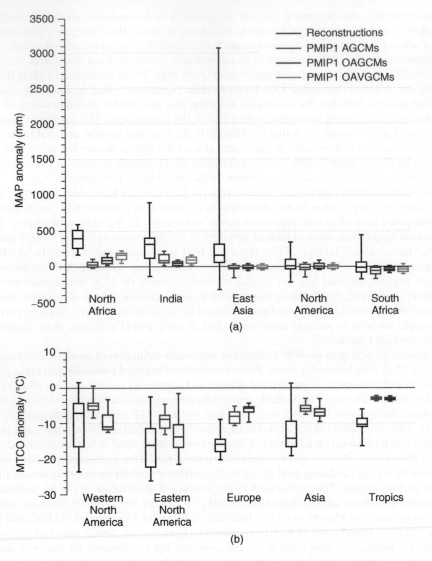

Figure 4.20 Comparison of reconstructed and simulated changes in regional climates during the mid-Holocene and LGM. (a) Change in mean annual precipitation (MAP) for the mid-Holocene. (b) Change in mean temperature of the coldest month (MTCO) for the LGM. The box-and-whisker plots show regional changes as given by AGCMs from the first phase of the Palaeoclimate Modelling Intercomparison Project (PMIP1 AGCM), coupled ocean–atmosphere and ocean–atmosphere–vegetation general circulation models from the second phase (PMIP2 OAGCM and OAVGCM) and reconstructions. The line in each box shows the median value from each set of measurements; the box shows the 25–75% range and the whiskers the total range. Source: Braconnot *et al*, 2012. Reproduced by permission of Nature Publications

to photosynthesis by elevated CO_2 concentrations was not sustained beyond a few weeks and that any increases in photosynthesis were not fully transferred into biomass increase (Leakey *et al*., 2012). Intriguingly, open-top chamber experiments, which are closer to field conditions, maintained increased rates of photosynthesis for at least 10 years, although the transfer of increased photosynthesis into increased biomass was also poor (Leakey *et al*., 2012). It is striking that there have been no tropical FACE experiments despite the

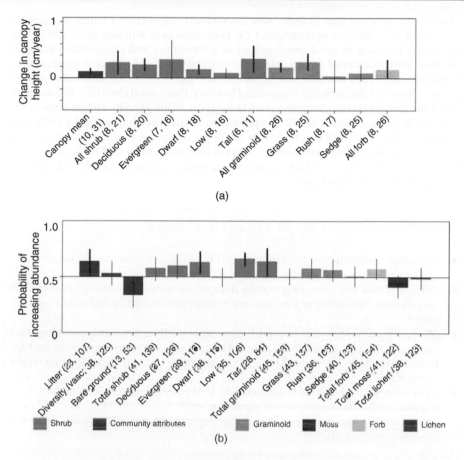

Figure 4.21 Biome-wide changes in (a) vegetation height and (b) abundance of several tundra plant groups. Error bars show ±2 s.e.m. and are in bold where mean change rates were significant at $P < 0:05$ using Wald tests. Sample sizes (number of studies, number of locations) and plant groups are shown on the x-axis. Source: Elmendorf *et al*, 2012. Reproduced by permission of Nature Publications

concentration of global net primary production in tropical regions and the importance for future food supplies of crop performance in the tropics. Vegetation models and palaeodata have a broader geographical coverage than plant physiological experiments.

Ugly experimental facts can add complexity to model building and the often simpli-fied interpretation of palaeodata, but there is a broad consensus across the fields of plant physiology, palaeoecology and vegetation modelling of a dominant and often immediate control of vegetation composition and dynamics by climatic factors. This consensus has important implications for the ecosystem impacts of the current rapid changes in global climate, which is well illustrated by Jack Williams' papers on novel climates and ecological surprises (Williams and Jackson, 2007). Jack is the Bryson Professor of Climate, People and the Environment and the Director of the Center for Climatic Research at the University of Wisconsin–Madison, which are high-profile positions to be held by a palaeoecologist. He argues convincingly that communities of plants and animals from the past that do not exist today were a consequence of unique combinations of climatic parameters that also do not occur today (Williams and Jackson, 2007). Unique combinations of trees and mam-mals are recorded from Alaska and eastern North America between 17 000 and 12 000 years ago, at the same time as there were anomalously large annual ranges of insolation and temperature and low atmospheric CO_2 concentrations. Williams *et al.* (2007) then map future forecasts of novel combinations of temperature and precipitation and show that these are concentrated in tropical and subtropical regions of high ecological complexity and diversity. They conclude that these climatic equivalents of uncharted waters will generate unpredictable ecological surprises. However, these novel climates will come at the cost of the disappearance of existing ones in hotspots of biodiversity and endemism such as the Andes, Mesoamerica, southern and eastern Africa, the Himalayas, the Philippines and parts of Indonesia, which will threaten existing ecosystem services, increase extinction risk and break up existing communities. There is thus much to be said for keeping global climates within the range of existing values, where models are more reliable and ecological surprises are kept to a minimum.

Summary: Responses to climatic change

- Holocene aridification of the Sahara desert is a probable example of a locally moderated response to a progressive climatic change.
- Ancient DNA and dung fungus data indicate an overall climatic control of past megafaunal extinction that is moderated by both human activity and ecosystem processes.
- Modern surveys show that several European butterfly species are extending their ranges northwards. Plants are responding more slowly but many are moving upslope in mountain areas. Latitudinal range shifts are so far rarely detectable using recent monitoring.
- Modelling shows that ecosystem feedbacks to the climate system have the potential to influence future climates in very significant ways. Some model forecasts for future ecosystems are supported by artificial warming experiments in long-term, open-top chamber experiments.

5

The Role of Episodic Events in Millennial Ecosystem Dynamics: Where the Wild Strawberries Grow

5.1 Introduction

Episodic events or disturbances are important features of millennial ecosystem dynamics. Long-term datasets and observation periods are required to detect and evaluate episodic events such as fires. The complex global climate models used to forecast the future have been 'built for stability' and have had limited success in modelling past abrupt climatic events, whose impact on ecosystems appears to have been considerable (Valdes, 2011). We chose not to use the word 'disturbance' in the chapter title here because it implies just a temporary interruption in a stable state or an orderly succession. We judge 'disturbance' or 'any relatively discrete event in time that disrupts ecosystem, community, or population structure and changes resource, substrate availability, or the physical environment' (as defined by Pickett and White, 1985) to be an integral feature of most global ecosystems and not simply a subordinate process that is an exception to the norm. Disturbances can range from very fine-scale events, such as those caused in a grassland by an animal's hoof, to widespread events covering many square kilometres, such as fires, insects, wind damage and timber harvesting. At this latter scale, the process can initiate a sequence of species dynamics or succession within the disturbed ecosystem. Hooves drumming in grassland or children running up and down grassy banks in southern Sweden can initiate a type of micro-succession, creating ideal conditions for regeneration or encouraging the clonal spread of wild strawberries.

Early in the twentieth century the importance of disturbance was more often recognised by foresters than plant ecologists. Disturbance of any origin was often regarded as incompatible with a virgin 'climax' successional state by ecologists. Nichols (1935), based in the USA, stated that climax forest was generally developed 'except where the natural conditions have been modified by fire or man'. There was a particularly heated debate in Sweden

Ecosystem Dynamics: From the Past to the Future, First Edition.
Richard H.W. Bradshaw and Martin T. Sykes.
© 2014 John Wiley & Sons, Ltd. Published 2014 by John Wiley & Sons, Ltd.
Companion Website: www.wiley.com/go/bradshaw/sykes/ecosystem

in the 1930s over Fiby Forest in the southern boreal zone, which lies close to Uppsala University. Professor Rutger Sernander of Uppsala wanted Fiby to be protected as a National Park. In order for it to qualify for this status, it was important that its modern state was not unduly affected by former human activities. Henrik Hesselman, a forester by training, was asked for an opinion and he pointed to various indicators of human disturbance and evidence for a formerly more open, grazed forest. He reckoned that the forest canopy had closed over quite recently and he presented historical evidence for local forest exploitation that had ceased about AD 1800. The end of a long period of human disturbance was followed by a dominating generation of light-demanding juniper (*Juniperus communis*) bushes, which is a frequent occurrence in Scandinavia following reduction in cattle browsing. Hesselman's observations were eventually confirmed for part of Fiby by pollen and charcoal analyses from a small hollow in the forest (Bradshaw and Hannon, 1992). These palaeoecological analyses showed how Fiby Forest had been in a state of continuous change for the last 4000 years, with episodic and chronic disturbances proving to be the rule rather than the exception. The episodic disturbances resulted in stepwise changes in forest composition, helping drive the system from a deciduous to a boreal state (Figure 5.1). It appears as if these episodic disturbance events ensured that forest composition tracked the changing climatic conditions, preventing the forest from maintaining a relict, stable composition through inertia, vegetative reproduction and maintenance of a buffered microclimate. In the absence of disturbance, the species in residence often have a competitive advantage, as they can potentially be the strongest seed sources or provide material for vegetative reproduction. In the case of trees, they also generate a forest microclimate that buffers extreme climatic conditions.

Palaeoecological reconstructions for Fennoscandia across most of the last 4000 years show how treelines moved downslope and the growing season gradually became shorter, which increasingly favoured hardy boreal species over temperate deciduous trees

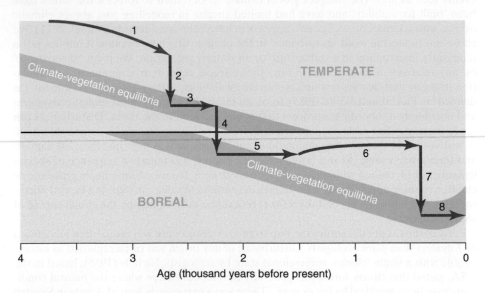

Figure 5.1 Conceptual model of dynamic climate–vegetation relationships in Fiby Forest, Sweden. (1) Chronic disturbance from burning deflects vegetation from equilibrium with climate. (2) Disturbing climatic event and burning stops. (3) Inertia as in situ vegetation retains dominance. (4) Disturbing climatic event drives vegetation switch to alternative stable state. (5) Inertia as in situ vegetation retains dominance. (6) Chronic disturbance from burning and grazing. (7) Storm and abrupt abandonment of burning and grazing. (8) Inertia as in situ vegetation retains dominance (after Bradshaw and Hannon, 1992)

(Miller *et al.*, 2008). Each discrete disturbance in Fiby Forest created gaps, which were filled by seedlings, the success of which depended on the prevailing climatic conditions. Of course, these gaps could occur through the 'natural' death of individual trees, as described in the widely held theory of 'gap phase' dynamics (Watt, 1947), but long-term observations from several boreal and temperate ecosystems show that external agencies such as fire, storm and disease have in practice often contributed more to ecosystem dynamics than the deaths of single, large trees. The rather regular fire cycle of the boreal region is a classical and widely accepted illustration (Heinselman, 1973). There is more debate about the role and dominance of episodic disturbance in temperate forest ecosystems.

Long-term observation series based on direct observation and palaeoecological proxies from protected temperate forest nature reserves such as Harvard Forest, USA (Foster and Boose, 1992), Lady Park Wood, UK (Peterken and Mountford, 1998) and Draved Forest, Denmark (Wolf *et al.*, 2004; see Figure 5.2) illustrate how disturbances from wind, disease and human activities have been dominant agencies controlling forest character in recent centuries. This can be an uncomfortable conclusion for conservation biologists, several of whom have used such sites to build a mental image of a temperate forest ecosystem reference state that may be dynamic and disturbance-driven but is largely unaffected by human activities and forms a potential target for restoration activities. Most if not all human activities are judged to have degraded the value of the ecosystem, just as Sernander felt in Sweden during the 1930s, and was outlined in the Swedish national survey of pristine forests carried out from 1978 to 1982. George Peterken has concluded that 'old-growth forest' is the most appropriate term for many of the most pristine forest reserves protected in several European countries and in the northeast USA, as it allows for limited human impact (Peterken, 1996).

While Fiby is an accessible forest close to a significant town, its story of longer-term human intervention is probably typical for many other old-growth European forests, which are now approaching close to 'natural' states after long periods of more intensive human

Figure 5.2   Richard Bradshaw inspecting storm damage in a Danish woodland

exploitation. The history of Fiby is fully consistent with Eustace Jones' (1945) classic review of virgin forest, based more upon field data than theory, in which he wisely concluded that 'it is conceivable that climax forest is a concept only, never existing in practice'. The Fiby story also illustrates one emerging issue of this book, namely that a long temporal perspective is needed in order to make judgements about the present state and direction of change of an ecosystem. Forests in particular can appear to be close to a natural condition and stable in the short term, but longer periods of observation will reveal the legacies of past episodic disturbance, both anthropogenic and natural, and provide a basis for future management (see Chapter 9).

Attiwill (1994) also reviewed disturbance of forest ecosystems and concluded that 'natural disturbance is fundamental to the development of structure and function of forest ecosystems'. He focused on wind, fire, insect pests and herbivory and observed that while wind will blow down an individual tree that is old and decayed, thereby contributing to a continuous internal ecosystem process, a hurricane will cause catastrophic damage and act as an episodic, external disturbance agency. His analysis shows how difficult it is to distinguish between internal and external ecosystem properties, posing problems for effective modelling of disturbance. Attiwill highlighted the importance of human activities in the humid tropics, again identifying a gradient of ecosystem influence, with shifting agriculture having fewer permanent effects on soils and ecosystem structure than improved pasture and uncontrolled logging. Our increasing knowledge of longer-term dynamics shows that the legacy of past human activities is widespread, including in the tropics (Willis *et al.*, 2004), and that the future challenge is to develop management strategies that balance the needs of people with long-term maintenance of key ecosystem values (see Chapter 9).

Episodic disturbance events can also trigger changes in ecosystem state, which may prove to be irreversible: a critical 'tipping point' has been crossed. Several regions of the world today have forest–grassland mosaics, but the origin and subsequent stability of this system has been most clearly demonstrated on the South Island of New Zealand (McWethy *et al.*, 2010). A rather small population of Polynesian Māori arrived on the island c. AD 1280 but managed to transform the vegetation of large areas within a short period of time using fire, with the eventual loss of 40% of the native forest area. Beech (*Nothofagus*) forests at high elevations and podocarps (*Podocarpus*) at lower altitudes were replaced by bracken (*Pteridium esculentum*), grasses and woody scrub, which European settlers subsequently turned into grazing lands. Analyses of sediments from 16 lakes along the Southern Alps showed a marked increase in post-settlement charcoal deposition at the 12 drier sites, with most fires during the first centuries after colonisation and subsidiary charcoal peaks following European settlement after AD 1840 (Figure 5.3). The records of magnetic susceptibility and concentrations of titanium, potassium and calcium in lake sediments indicated that burning increased the subsequent incidence of extensive soil erosion, whether this was triggered by local earthquakes or land use change. This well-documented example of people using fire to deliberately alter large areas of vegetation illustrates a type of ecosystem conversion that has probably taken place in many regions of the world. The increased susceptibility of burnt land to soil erosion in high-rainfall settings, particularly when grazing animals are introduced on to the cleared area, has contributed to the removal of trees, shrubs and soil from the uplands of northwestern Europe and north Atlantic islands, including the Faroes and Iceland, causing ecosystem change that has persisted for centuries in some cases and millennia in others.

Perhaps the most extreme example of human-driven ecosystem conversion comes from arid, east-central Australia, where Gifford Miller and colleagues have speculated that human intervention in the fire regime led to irreversible ecosystem reorganisation between 50 000 and 45 000 years ago (Miller *et al.*, 2005). Humans are believed to have first colonised Australia during this interval and almost all land mammals, reptiles and birds weighing more than 45 kg became extinct at approximately the same time, including

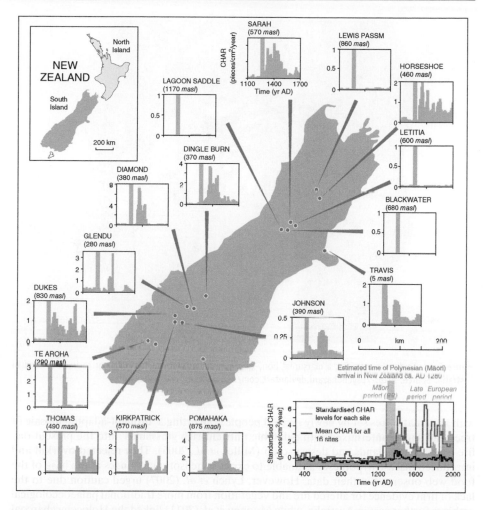

Figure 5.3 Charcoal accumulation rates (CHAR, pieces/cm²/year) at 16 sites on the South Island of New Zealand. Site names and elevations (metres above sea level, masl) are shown above plots. Small plots show variation in CHAR between AD 1100 and 1700. The inset in the bottom right shows a composite plot with standardised CHAR levels for each site (turquoise lines) and mean CHAR for all 16 sites (thick black line) for the period AD 300–2000. The grey band in both the small plots and the large one indicates the estimated time of Polynesian (Māori) arrival in New Zealand, ca. AD 1280. Source: McWethy *et al*, 2010. © 2010, National Academy of Sciences, USA. Reproduced with permission

Diprotodon optatum, the largest known marsupial, which was like a hippopotamus-sized wombat, and *Thylacoleo carnifex*, a marsupial lion, which was the largest carnivorous mammal ever to have existed in Australia (Figure 5.4; Roberts *et al.*, 2001). Miller and his colleagues analysed a 140 000 year record of eggshells from the emu (*Dromaius*) and an extinct flightless bird (*Genyornis*), together with wombat (*Vombatus*) teeth. Carbon isotopic (δ^{13}C) analysis of the *Dromaius* eggshells indicated altered isotopic signatures and a significant change in diet between 50 000 and 45 000 years ago (Figure 5.5). This change was consistent with a shift from a combination of C_4 grasses and C_3 shrubs and trees to a more restricted, C_3-dominated diet. A weakening monsoon after 45 000 years ago might have influenced this change, but the authors favoured a modified fire regime hypothesis,

Figure 5.4 *Thylacoleo carnifex*, a marsupial lion, the largest carnivorous mammal ever to have existed in Australia. Source: http://psychohazard.deviantart.com/. Courtesy of psychohazard

with systematic fires set by newly arrived people converting a 'drought-adapted mosaic of trees and shrubs intermixed with palatable nutrient-rich grasslands' into the present-day fire-adapted grasslands and desert scrub (Miller *et al.*, 2005). They rejected overkill and introduced disease hypotheses as unable to explain the isotopic changes at the base of the food web observed in their data. However, Lynch *et al.* (2007) urged caution due to the lack of firm evidence for altered fire and vegetation from more traditional palaeoecological archives further east in Australia, while Mooney *et al.* (2011) linked the Holocene charcoal record more closely to climatic data than to a composite record of dated archaeological finds in Australia. There are a number of plausible, competing hypotheses to account for developments in the long history of fire in Australia, which are actively debated among researchers. The data from the past are exciting and tantalising, but always require some imagination to fill the gaps. This type of speculation is what makes palaeoecology such a wonderful subject to teach but frustrating to use for the rigorous and conclusive testing of hypotheses. Modelling these systems can help with more thorough hypothesis evaluation, as we shall show later in this chapter.

In the following sections we look at specific types of disturbance that have been significant in the past and will affect us again in the future. We examine disturbance from fire, disease and other large, infrequent events such as hurricanes that cover large areas and leave significant biological legacies (Foster *et al.*, 1998a). Extreme climatic events such as drought were covered in Chapter 4. While people are implicated in the past dynamics of both fire and to a lesser extent disease, we make our major analysis of the direct effects of human impact in Chapter 6. Grazing and browsing animals have also disturbed vegetation in significant ways through time, although this type of disturbance tends to be more long-term than episodic. We return to this topic in Chapter 9, as animals are useful tools in conservation biology.

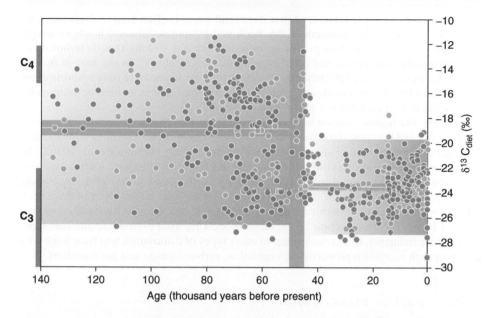

Figure 5.5 Time series of *Dromaius* dietary δ^{13}C, reconstructed from δ^{13}C of calcite crystals (red) and organic residues (blue) from individually dated eggshells from the Lake Eyre region of Australia. The vertical bar (50 000–45 000 years ago) defines the megafaunal extinction window, with its estimated uncertainty. *Dromaius'* vegetarian diet appears to have shifted in dominance from C₄ to C₃ plants (after Miller *et al.*, 2005)

Summary: Episodic events

- Episodic events or disturbances are integral features of most global ecosystems. Disturbances can range from very fine-scale events, such as those caused in a grassland by an animal's hoof, to widespread events covering many square kilometres, such as fires, insects, wind damage and timber harvesting.
- Frequent disturbance in most forest ecosystems renders the concept of successional climax of little practical value.
- Past disturbance has caused long-term changes of state for major ecosystems in Australia and New Zealand.

5.2 Fire

Fire is a global phenomenon, intrinsic to most terrestrial ecosystems, interacting with climate, vegetation distribution, biodiversity, land surface properties, atmosphere chemistry, the carbon cycle and human activities in a complex and still unresolved pattern (Bowman *et al.*, 2009, 2011). A world without fire would look very different, with an estimated doubling of the land surface covered by forest (Bond *et al.*, 2005). Fire is almost certainly a perfectly natural, episodic disturbance agency of many forest, savannah and grassland ecosystems that contributes to species dynamics, diversity, physical structure (e.g. biomass, leaf area index) and ecosystem function (e.g. nutrient cycling, water quality), but it has been exploited by people for so long that its natural status is often hard to determine (see Figure 3.10). One of the messages of this book is that human influence has been deeply embedded in many ecosystem processes, including fire, for so long that its removal can create something unfamiliar and often unwanted (see Chapter 9). In the case of fire, we are unlikely ever to have sufficient knowledge of the past to separate fully anthropogenic from natural burning, so the idea of 'restoring natural fire regimes without anthropogenic influence is neither possible nor useful' (Bowman *et al.*, 2011).

The term 'fire regime' is often used to describe important characteristics of interactions between fire and the ecosystem, but which characteristics are described, measured or modelled depends on the spatial and temporal scales under study (Whitlock *et al.*, 2010). The frequency, severity and extent of individual fires are important properties when assessing ecosystem impact. Oxygen, heat and fuel drive fire at the finest scales, while vegetation, climate and sources of ignition are key variables on longer time scales. Many of these characteristics have been incorporated into fire models and added to the dynamic ecosystem model LPJ-GUESS (see Section 5.2.5). These models are used to simulate different types of fire, their frequency, their relationship to other types of disturbance and their long-term effect on such ecosystem properties as vegetation, carbon storage and gas emissions.

5.2.1   Past to present fire

Fire has been influenced by different combinations of factors in space and time. For example, Europe and the Middle East are characterised by thousands of years of human settlement and fire regimes not only show a long-term sensitivity to climate but also reflect how fire has been an important land use tool for a long time. There is unequivocal archaeological evidence for people's use of fire in the home for at least the last 300 000 years (Roebroeks and Villa, 2011) but signs of sophisticated human activity keep being pushed further back in time. Berna *et al.* (2012) report 'unambiguous evidence' for burning in an occupied cave in South Africa 1 million years ago, which approaches the speculative age of 1.9 million years for the origin of cooking proposed by Wrangham (2009). There is also debate over when people began to use fire to modify ecosystems in significant ways. As discussed in Section 5.1, fire may well have been used for ecosystem management in Australia for 50 000 years, and current research may add detail to earlier indications that anthropogenic burning of African savannahs has a much longer history. We can write with more certainty about the Holocene in Europe and the USA, where archaeology and high-resolution pollen analysis with agricultural indicators have been linked to charcoal records.

Total fuel availability has limited burning in the past in dry periods or when growing seasons are short (Marlon *et al.*, 2006) and fuel type can also affect fire regime. Spruce (*Picea abies*) spread through Sweden and Norway during the late Holocene and as it established local dominance, burning became less frequent, but with perhaps an increased proportion of intensive crown fires (Ohlson *et al.*, 2011). Vegetation composition could prove to be an underestimated influence on fire regimes, at least in the boreal zone, where de Groot *et al.* (2013) have suggested that the greater abundance of open larch (*Larix sibirica*) stands in Russia has contributed to more smaller, low-intensity ground fires than are observed in Canadian boreal forests. Spruce trees, which are more abundant than larch in Canada, are much more effective 'fire ladders', drawing fire up into the canopies to create blazing, high-temperature crown fires. The influence of past climate change on fire regime has been investigated in rather general terms, often by comparing trends in proxy climatic records with charcoal data pooled from several sites within a region (Marlon *et al.*, 2013). Modelling the past opens new possibilities for hypothesis testing. In an elegant experiment, the dynamic vegetation model LPJ-GUESS was forced with modelled climate output for the Holocene in Europe and the outputs were compared with charcoal data from 156 European sites using regression analysis (Molinari *et al.*, 2013). The best predictor of biomass burning in Europe during the early Holocene was climate, with land use and vegetation dynamics subsequently becoming the dominant controls.

Satellite monitoring using thermal sensors can now detect fires as small as $50\,m^2$ under ideal, cloud-free conditions, but the normal sensitivity limit is about $900\,m^2$ (Giglio *et al.*, 2010). Composite satellite images give fantastic seasonal overviews of global fire and show that most burning currently occurs in the tropics, from fires set by people in savannah and forested ecosystems (Figure 5.6). Many heathland and grassland systems owe their origin and maintenance to burning, and fire is a key disturbance agency in Mediterranean and boreal biomes. Although our knowledge of past fires is weaker in the tropics, a strong case has been made for the fire-driven expansion of savannahs and grasslands in several parts of the world beginning about 10 million years ago during the late Miocene, clearly prior to any influence by people (Keeley and Rundel, 2005). Regular burning has continued to be a key factor in prairie and steppe systems throughout the Holocene, as has been elegantly demonstrated for Kettle Lake on the northern Great Plains of the USA (Brown *et al.*, 2005; Grimm *et al.*, 2011). Brown *et al.* (2005) and Grimm *et al.* (2011) illustrate the complexity of interactions between climate and fuel on control of the fire regime, even when human influence is absent or negligible. The first firm evidence for human impact in this study follows European settlement. Open spruce parkland covered the region 13 000–12 000 years ago and was associated with arid conditions and moderate charcoal abundance. This was the only period when trees formed part of the local vegetation. Both spruce and charcoal disappeared from the system at the opening of the Holocene about 11 500 years ago as temperature and moisture levels increased. The authors believe that the moisture suppressed fire, while spruce could not cope with the increasing temperatures. The early Holocene was characterised by relatively moist conditions, interspersed with periodic droughts at 100–500 year intervals. Charcoal flux gradually increased, favoured by the drought intervals and by abundant grassy fuels (Figure 5.7(a)). A 200 year 'megadrought' heralded a mid-Holocene arid period with insufficient fuel to support much fire, even though the geochemistry indicated high-amplitude moisture variability during this period. Amaranthaceae is a family of plants that includes several drought-tolerant species whose pollen was abundant throughout the mid-Holocene. Occasional charcoal peaks coincide with geochemical evidence for the brief moist periods that stimulated grass production,

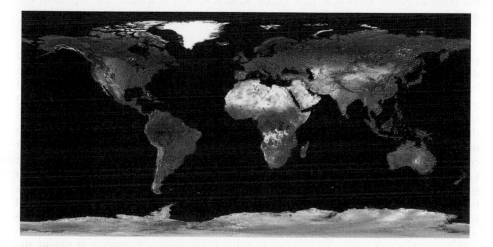

Figure 5.6 Remotely sensed locations of fires between 10 and 20 July 2011, detected by MODIS (NASA, no date). Source: http://lance-modis.eosdis.nasa.gov/cgi-bin/imagery/firemaps.cgi. Courtesy NASA

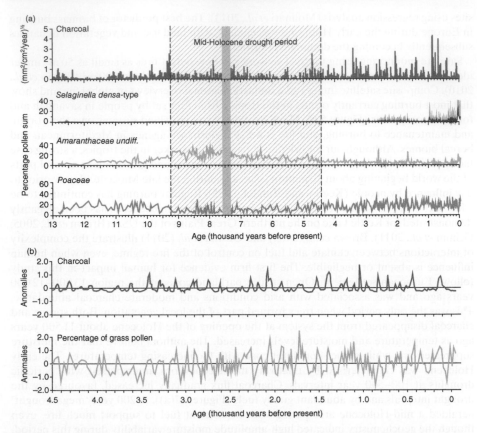

Figure 5.7 (a) Pollen and charcoal records from Kettle Lake, USA. Recurring drought episodes in the mid-Holocene resulted in less burning, due to lack of fuel. The grey bar marks a brief hiatus in the record and the other vertical lines are zone boundaries (Source: Grimm *et al*, 2011. Reproduced with permission from Elsevier). (b) Charcoal and Poaceae (grass) pollen trends, showing a tendency for coincident positive anomalies (Source: Brown *et al*, 2005. ©2005, National Academy of Sciences, USA. Reproduced with permission.)

generating temporarily high fuel loads. There was a shift to a generally wetter climate during the late Holocene but with rather regular periods of drought. Charcoal flux increased but was cyclical in nature, with a periodicity of about 160 years. The charcoal peaks coincided with moist intervals, when extensive grass (Poaceae) cover generated sufficient fuel for burning (Figure 5.7(b)). A nice twist in the tale is the entrance into this ecosystem of prairie clubmoss (*Selaginella densa*) during the last 4000 years, with maximum populations during recent centuries (Figure 5.7(a)); prairie clubmoss is a poor competitor for light and benefits from increased grazing pressure. Bison populations are known to have expanded at the same time and were replaced by cattle grazing after European settlement. This study illustrates how periods of relative stability and change, both gradual and abrupt, characterise Holocene climates, with consequent effects on fuel type and abundance. Add people and large fuel-eating animals to this mixture and the complexity becomes daunting.

Humans, fuel and climate have interacted closely to generate European Holocene fire regimes, while in the northwestern USA human influence has been judged to be minimal (Marlon *et al.*, 2006). Humans have most likely altered ignition frequencies in parts of the Mediterranean region for at least 6500 years, in conjunction with local agriculture

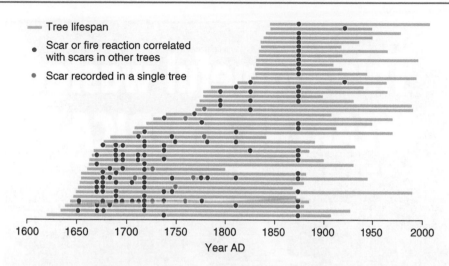

Figure 5.8 Fire history of a coniferous tree stand in Białowieża Forest, Poland, reconstructed by cross-dating *Pinus sylvestris* (pine) tree rings (Niklasson *et al.*, 2010)

(Kaltenrieder *et al.*, 2010), but regional syntheses suggest that humans only became the dominant factor in Mediterranean fire regimes during the last 2000–3000 years (Vannière *et al.*, 2011). Human impact on Fennoscandian fire regimes has been limited to perhaps the last 2000 years in the south, with the widespread use of slash-and-burn agriculture (Bradshaw *et al.*, 2010d), while human activities had little apparent influence on northern Scandinavian forests until recent centuries (Granström and Niklasson, 2008).

While the role of fire in both the boreal and Mediterranean biomes is widely known, its past status in temperate forest systems is more controversial. Białowieża Forest in Poland is a European flagship for temperate forest ecology because of its size, species diversity and presumed limited anthropogenic influence. Niklasson *et al.* (2010) sampled primarily Scots pine (*Pinus sylvestris*) trees, logs and stumps from Białowieza Forest in order to assess the past role of fire in Central European lowland forests. Previous analyses of charcoal in sediments showed that fire had been a part of this ecosystem (Mitchell and Cole, 1998), but Niklasson's analyses revealed just how frequent fires had been during recent centuries (Figure 5.8): single trees were experiencing fire events every 18 years on average from AD 1653 to 1780, but fires became much less common after 1781 and effectively ceased after 1874 (Niklasson *et al.*, 2010). People were almost certainly responsible for most of these fires, but our understanding of fire in temperate forest ecosystems is still far from complete. Better understanding of the timing of human exploitation of preexisting 'natural' fire regimes requires combining charcoal and fire-scar records and exploring their time course with ecosystem models.

Fire-scar studies in many systems have drawn attention to how we have removed or restricted burning in several coniferous ecosystems. Anthropogenic burning is now increasingly focused to the tropics and subtropics (see Figure 5.6). The collapse of burning in northern Scandinavia has been linked with both the rise of commercial forestry operations and changes in settlement patterns and land management. Similar trends can be seen in North America, where the Smokey Bear Wildfire Prevention campaign was created in 1944, becoming the longest-running public service advertising campaign in US history (Figure 5.9). Smokey caught the public imagination and was a key factor in the effective removal of human-caused fire from forested landscapes with long fire histories. This long-term development has several management and conservation implications, which are discussed in Chapter 9.

Figure 5.9   Smokey Bear at work. Source: USDA Forest Department

5.2.2   Present to future fire

Approximately 400 million hectares of the Earth burn each year, with 70% of fires located in Africa (Giglio *et al.*, 2010). It has been estimated that about 10–15 million hectares burn annually in the circumboreal forests of the northern hemisphere (Flannigan *et al.*, 2009). Boreal forest burnt area has increased over recent decades, at least in part due to a warming climate (Kasischke and Turetsky, 2006). This increase is likely to continue. At the same time, other ecosystems such as Arctic tundra, where fire has been mainly absent through the Holocene, are burning more frequently (Mack *et al.*, 2011). There is however not much evidence to date that changes in global peatland fires have contributed to changes in atmospheric greenhouse gases, although a review of field and experimental data suggests an increase in fire intensity and frequency caused by drier conditions and changes in drainage (Frolking *et al.*, 2011). Liu *et al.* (2010) used a drought index linked to outputs from various general circulation models (GCMs) to assess the potential for increasing wildfires globally. They considered the potential to be significant for many areas of the world, including the USA, South America, Central Asia, southern Europe, Africa and Australia.

Canada keeps some of the best statistics about forest fires in the world and the very effective Canadian firefighters use the well-tested Canadian Forest Fire Rating System, which predicts the occurrence, spread and intensity of fire in different fuels and takes as input a sophisticated fire weather index that is sensitive to temperature, rainfall, relative humidity and wind. Wotton *et al.* (2010) coupled future forecasts from two GCMs (from the Canadian Climate Centre and the UK Hadley Centre) to this fire rating system to produce a forecast for 2090 that has been described by a leading fire researcher as 'Armageddon', showing increases in fire activity of over 100% throughout much of southern Canada (Figure 5.10). Such increases in wildfire, especially in forests and peatlands, may eventually lead to a large carbon loss from the soil and vegetation as they feed back to the atmosphere (IPCC, 2007a). Human society is threatened with loss of control of wildfire, having mastered it for a few millennia at most. Fire is a good servant but a bad master. Recent research shows that fuel management is a better use of resources than fighting fires, which threaten to become rapidly uncontrollable in several regions of the world in the near future.

Summary: Fire past, present and future

- Fire is a perfectly natural, episodic disturbance agency of many ecosystems but has been exploited by people for so long that its natural status is often hard to determine.
- Humans, fuel and climate have closely interacted to generate fire regimes. While human activities are a major influence on ignition frequencies in many regions today, future forecasts for Canada indicate major increases in fire occurrence.

5.2.3   Modelling fire

While a rich variety of fire models are in use today, we will focus here on models that address broader-scale fire events, in line with the overall strategy of the book. A number of different models of fire spread and dynamics have been developed to simulate possible outcomes from wildfires, including carbon loss to the atmosphere, likely recovery time and future developments within an ecosystem. The modelling is not straightforward. For example, it is important to distinguish between those fire events that are local in scale and do not spread beyond the modelled grid cell (non-spatial) and those that spread into neighbouring areas or grid cells and beyond (spatial) (Li and Apps, 1996). Earlier forest community models

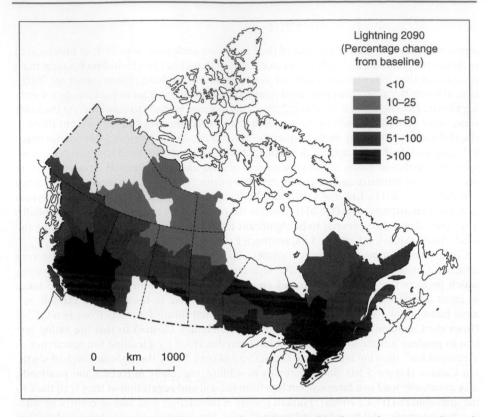

Figure 5.10   Percentage increase in fire occurrence from lightning ignition forecast for AD 2090, using modelled climate compared with fire occurrence during AD 1975–1990. Source: Wotton *et al*, 2010. Reproduced by permission of CSIRO

such as FORSKA2 (Prentice *et al.*, 1993a) did not take into account such contagion. Rather, they removed vegetation from just one of the replicated (0.1 ha) patches in a grid cell, as a representation of the disturbance, and then allowed secondary succession to begin within that patch. There was no contagion either to other patches or other grid cells. In the real world, contagion is very likely in areas where the average fire sizes are large. Fire size depends mainly on the landscape fragmentation and the causes of fire. The majority of fires in tropical areas are part of the landscape management and hence are confined to rather small areas. Large fires do occur in boreal areas; while low in number, they contribute significantly to the total burnt area. Stocks *et al.* (2002) estimated that Canadian fires larger than 200 ha were very small in number but accounted for up to 97% of the total burnt area. Most ecosystem models work on a grid-cell resolution and therefore need to include contagion. Other issues to be addressed include the mode of ignition, the likelihood of spread and the type of fire: ground, surface or crown.

5.2.4   Modelling ignition

Natural ignition is important in temperate and boreal zones, especially in regions of sparse human population. In Canada, 80% of fires are ignited by lighting (Stocks *et al.*, 2002). In regions where humans use fire to clear the land, such as for agriculture in parts of Africa, nearly all fires are ignited by humans (Saarnak, 2001). The density of gross domestic product (GDP), which is a convenient measure of economic activity, is the most important predictor at the regional scale of area burnt. The less economic activity in a region, the lower the GDP density and the greater the burnt area (Aldersley *et al.*, 2011). Ignitions of human origin tend to be either land clearance for pasture or agriculture or prescribed burning designed to reduce the fuel load in the landscape and prevent destructive fires; arson and negligence (through carelessly attended barbecues or camp fires) are also a major source. In southern Europe, 90% of fires are the result of human ignition, both deliberate and accidental. It is difficult to predict such human activity, but various statistical techniques have been used, at least at the regional scale, including logistic regression (e.g del Hoyo *et al.*, 2011) – which can be combined with a geographic information system (GIS) (Martínez *et al.*, 2009) – multiple regression (Syphard *et al.*, 2007) and Bayesian methods (Amaral-Turkman *et al.*, 2011). Methods such as these can be included in both land use and fire risk management at the regional scale.

Climate (high temperatures, average precipitation and especially long periods of low rainfall), rather than humans, influences the extent of wildfires at the global scale (Aldersley *et al.*, 2011). Generalised and more globally scaled fire models tend to ignore human ignition, although they do include lightning ignitions, which are modelled at least in part by linking to weather indices. Most lightning strikes occur when the weather is wet and stormy, so the likelihood of fire starting is therefore low. In those areas where the weather conditions are conducive to promoting fire, the result of a lightning strike can be a fire that remains in a smouldering phase for some days, which makes it difficult to relate the actual event to the subsequent fire. In an analysis of the drivers of fire, Archibald *et al.* (2009) ranked the different drivers of burnt area in southern-hemisphere Africa and found lightning to be one of the least important factors. At a global scale, biome burning can be ignition-limited, fuel-limited or limited by a moist climate. Savannah, a biome that has the highest fire return interval (typically 2–5 years), is not ignition-limited (Archibald *et al.*, 2009) as the fire system there is part of the landscape management: ignitions are provided at the right time by the humans inhabiting the area. Therefore, modelling of burnt areas in savannah does not require the effect of ignitions to be included. Even relatively simple modelling procedures such as logistic regression models can have relatively good predictive power, since land use practices are strongly related to the climate. In a study modelling burnt area in Africa, Lehsten *et al.* (2010) could predict the annual burnt area on a 1° longitude/latitude scale with a coefficient of determination of 0.7, using tree-cover variables, four different precipitation-related variables and population density as predictors.

While this approach works well in fuel-limited biomes, simulating fires in boreal regions is more complicated, since the fires are not solely determined by the climate but also by the random element of lightning. Boreal regions are the largest biome with pronounced fire activity and the majority of fires there are initiated by lightning strikes.

Most lightning strikes, however, occur in warmer climates such as the tropics due to hotter air, greater convection and higher numbers of thunderstorms (Figure 5.11). Although satellites that register single lighting flashes are currently available, linking lightning to recent fires is still a challenge, due to the fact that the satellites only record the lighting flash as such and not whether the bolt just goes from one cloud to another (which the majority do) or actually strikes the ground. A reasonable knowledge of the cloud-to-ground lightning ratio exists for the area of the USA, but this information is very speculative on a global scale, for which the currently available data are coarse. Satellites also miss a large number of strikes.

Simulating lightning in vegetation models is a somewhat hit-and-miss process. Various approaches are employed, usually involving weather variables. Prentice *et al.* (2011) used a spatial Poisson process to give an expected number of ignition events based on the number of observed lightning flashes in time and space and on monthly precipitation data, while Conedera *et al.* (2011) made use of Monte Carlo simulations and landscape aspects

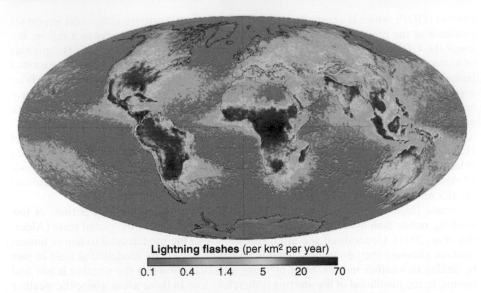

Lightning flashes (per km² per year)

0.1   0.4   1.4   5   20   70

Figure 5.11 Annual average number of lightning flashes per square kilometre in 1995–2002, using data from NASA satellites (IMAGE NASA). © 2005–2012 Geology.com. All Rights Reserved. http://geology.com /articles/lightning-map.shtml (last accessed 22 November 2013)

including vegetation type, elevation, aspect and slope. A more promising approach involves linking long-term trends in lightning activity to a climatic variable that can be modelled. For example, the upper-tropospheric (300 mb) humidity data from tropical Africa during the period 1950–2000 have been used as an acceptable proxy for lightning (Figure 5.12; Price and Asfur, 2006). Whether global warming might increase lightning activity is open to debate, but it may be possible to project future scenarios in climate models by using a proxy approach such as this.

5.2.5 Modelling fire spread

The science behind projecting the spread of fires is somewhat more developed that that behind ignition, perhaps due to the danger to life and property and the need for projections to aid fire fighting. It is based on empirical studies, statistical approaches and theory. Empirical studies at the landscape scale include data gathered on number of fires, weather and vegetation conditions. The total area burnt is related not only to the weather and vegetation but also to human activities, such as roads, population and cultivation (Archibald *et al.*, 2009), and is usually estimated in some statistical way. The rate of spread is much more difficult to estimate. Most models of spread are based on the Rothermel equation (Rothermel, 1972), using inputs such as fuel load, fuel depth and information on the fuel particles, including heat, moisture and mineral content. Environmental conditions such as the mean wind velocity and terrain slope are also required. BehavePlus (Andrews, 2007) is a widely used model suite that simulates fire spread and includes a number of fire behaviour systems, such as fire mapping at each point in the landscape, minimum travel times for fire spread, fire growth under different conditions varying in space and time and a fire spread probability system based on thousands of fire growth simulations according to weather. This model suite was developed for the USA but has been applied in Europe in many studies. However, it is designed for fire fighting and fire-prevention activities at the landscape scale and is not applicable to dynamic global vegetation models (DGVMs) because the input

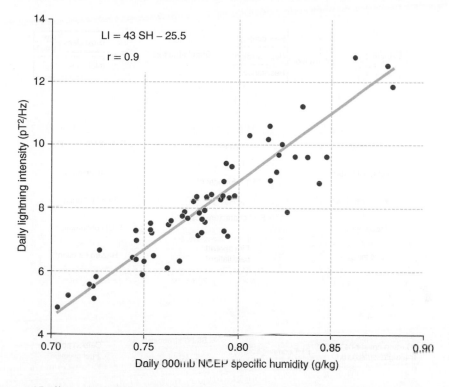

$$LI = 43\ SH - 25.5$$

$$r = 0.9$$

Figure 5.12   Upper-tropospheric (300 mb) humidity data from tropical Africa during the period 1960–2000 as a proxy for lightning. Source: Price and Asfur, 2006. Reproduced by permission of Terra Scientific Publishing Company (TERRAPUB)

parameters needed, such as slope, wind and precise fuel composition, are not available at the global scale.

In global-scale simulations, such as within DGVMs, the approach is more generalised and fire models are driven by fuel and various climate factors, such as wind, temperature and humidity. Thonicke *et al.* (2010) describe the features that should be considered if a process-based fire model is to be included in a DGVM (Figure 5.13), many of which they included in the SPITFIRE (SPread and InTensity of FIRE) model, which is part of the LPJ-DGVM (Chapter 2). SPITFIRE receives information on vegetation composition, fuel amount and its characteristics from LPJ-DGVM and then simulates the 'average fire size' that an ignition would cause and multiplies it by an estimate of the number of ignitions based on population density. A humpback relationship is assumed, with a low number of ignitions at both high and low population densities and a maximum number at medium density. Fire spread is simulated by checking whether there is enough dry fuel; if so, the fire spreads depending on a number of conditions, including wind speed. The rate of spread follows Rothermel equations. SPITFIRE also calculates the effects of fire on vegetation in terms of rates of mortality of individuals and loss of living biomass through combustion, as well as the effects on carbon fluxes through the ecosystem and into the atmosphere. Veiko Lehsten *et al.* (in preparation) used a variation of this model, LPJ-GUESS-SPITFIRE, to explore the dynamics of savannahs in Africa, particularly with regard to tree–grass ratios. LPJ-GUESS was applied as a gap model (see Chapter 2), simulating different age cohorts. Burnt area was prescribed from the MODIS satellite, while the fire effects on vegetation were calculated dynamically using the ecosystem model. They were dependent

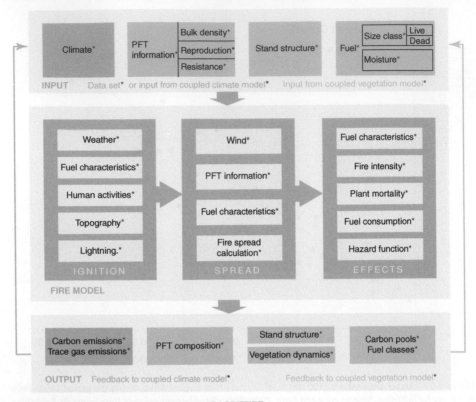

* Process, feedback or parameter captured in LPJ-SPITFIRE
* Process, feedback or parameter not included in LPJ-SPITFIRE

Figure 5.13 Model features required if a process-based fire model is to be included in a DGVM such as LPJ-DGVM. ©Thonicke *et al*, 2010. Reproduced by permission

on fire load, dryness, total plant height and plant-specific susceptibility to fire. Including wildfires improved the simulated distribution of biomes at the continental scale and the results showed that fire maintains savannah ecosystems that otherwise climatically would support forests. The impacts of climate change on African vegetation were also addressed in another modelling approach, this one using a different DGVM specifically developed for tropical vegetation (aDGVM) (Scheiter and Higgins, 2009). Fire was modelled as a disturbance using a semi-empirical fire model that defines fire by fuel biomass, fuel moisture and wind speed and simulates the effect of fire on single trees. Model projections and vegetation maps compared well, and also found that fire suppression strongly influenced tree dominance regionally but less so at the continental scale.

Prentice *et al*. (2011), using yet another version of the LPJ-SPITFIRE model, carried out a benchmarking exercise with the Land surface Processes and eXchanges (LPX) model and satellite data (the Global Fire Emission Database version 3, GFED3) (Giglio *et al*., 2010). This version excludes fires started by humans, as they proved too difficult to evaluate given the lack of information on the relationships between human ignition and population density, although it does take account of fires from lightning strikes. Data–model comparison showed that the model could simulate some of the key features of both variability in wildfire regimes and the global land–atmosphere fluxes of CO_2, although it did underestimate

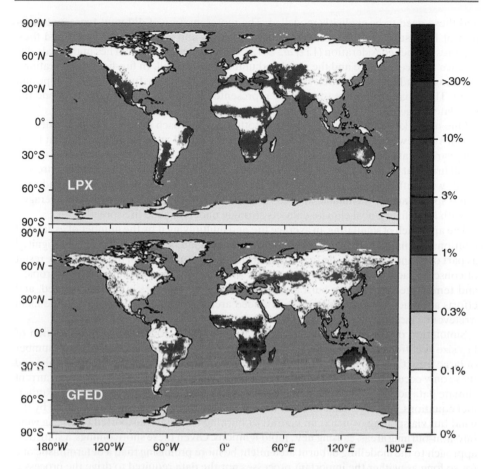

Figure 5.14 Comparison between annual fractional burnt area (average 1997–2005). Top: output from LPX-DGVM. Bottom: satellite data from GFED3 (Prentice *et al.*, 2011)

the burnt area in temperate and boreal forests (Figure 5.14) and overestimated total global CO_2 flux. The model was also too simplistic regarding the effects of land use on fire.

In another approach using a model of intermediate complexity (CLM-DGVM), Li *et al.* (2012) compared model performance at the global scale with data from the satellite GFED3. In this approach, fire occurrence, fire spread and fire impact were included and the burnt area in a grid cell was estimated using fire counts and average burnt area of each fire. The post-fire area was assumed to be elliptical in shape and the known properties of ellipses were applied to the fire-spread parameters. The results showed that simulated global totals, patterns of burnt area and fire carbon emissions were closer to GFED3 data than in earlier modelling efforts. It is suggested that these new parameterisations may improve other global models such as Earth system models (ESMs).

Modelling of fire, whether natural or human-induced, is difficult because of the many different factors at the range of scales involved. As with most models, a number of important aspects influence reliability, including the data used to drive and validate the model. Many aspects of fire modelling need to be improved. For example, the representation of human-driven ignition might only be related to population density in a region, whereas in reality other issues such as fire management policy and fire suppression are significant, and these need to be explicitly modelled. The representation of different types and duration of fires, such as those related to land use, crops and agriculture and peatland fires, requires further consideration (Li *et al.*, 2012). Another major problem lies in the scale difference between the fire and the simulated unit of the grid cell. In fire-fighting operations, fire danger and spread rate are evaluated using three parameters: fuel suitability, slope and wind. These parameters are very poorly represented in the available climate data and the simulated vegetation data.

One of the key parameters used by the Rothermel approach is the fuel bulk density, for which very few measurements exist, especially for tropical areas. Winds are known to be very variable at a local scale and are of major importance in driving fire and in transforming small fires that could be ignored into major ones. There is strong feedback between fire and wind, with sufficiently large fires generating uplift and hence increasing the spread rate. Wind speeds are only available at a 2.5° latitude/longitude resolution with daily averages, so with large fires global climate datasets strongly underestimate fire spread.

The approach of simulating an average fire and multiplying it by the number of estimated ignitions has limitations. It assumes an unfragmented landscape and that fires are ignited as part of a random process. In fact, the majority of the burnt area on the Earth is the result of conscious acts of landscape management. The aim with prescribed burning in tropical and temperate regions is to burn a defined area: if the fire goes out, it is reignited, and efforts are made to keep it within the intended area. Here climate factors are in the main irrelevant, since in every dry season a sufficiently dry period can usually be found.

Simulation of fires in boreal areas needs to take into account that the distribution of fire size is extremely skewed, with a large number of small fires and a very small number of large fires. Only the large to very large fires are of importance at the broader scale, and these only occur in exceptional weather conditions, which are hard to forecast given current climate data accuracy. The spread of megafires is also not adequately described by Rothermel equations, since the main mode of fire spread is not as a closed fire front driven by the wind but via spotting, which is an updraft of burning material transported by strong wind into non-burning areas, causing new, rapid ignitions. Given these shortcomings, a statistical approach to the modelling of burnt area might be more promising than a deterministic one for so long as neither the important processes nor the data required to drive the processes are sufficiently known.

5.2.6   Data−model comparison

One of the major challenges in understanding the role of fire in ecosystem dynamics is how to distinguish human influence from climatic and vegetation change as forcing factors. Modelling can help in this task, particularly as hindcasting to help interpret past datasets. In Europe, fire is most often associated with Mediterranean or boreal vegetation, yet some of the most detailed studies of Holocene fire regimes have been carried out in the Swiss Alps by Willy Tinner's research group based in Bern. Mountain areas tend to have plentiful lake basins that are ideal traps for charcoal and plant remains. Gouillé Roun is a 0.5 ha lake located 2343 m above sea level in the Rhone Valley, close to the modern limit for tree growth. Pollen data cannot easily resolve vegetation change close to treeline, so a palaeoecological study there was based on needles and leaves preserved in just 1 m of lake sediment, which recorded over 9000 years of lake catchment history from 11 800 to 2500 years ago (Kaltenrieder *et al.*, 2005). The major changes in forest composition through time were the rise to dominance of Arolla pine (*Pinus cembra*) between 9500 and 8500 years ago, apparently replacing larch (*Larix decidua*), the subsequent abrupt disappearance of Arolla pine 3700 years ago and the subsequent return to dominance of larch and expansion of juniper (*Juniperus communis*) and crowberry (*Empetrum hermaphroditum*)

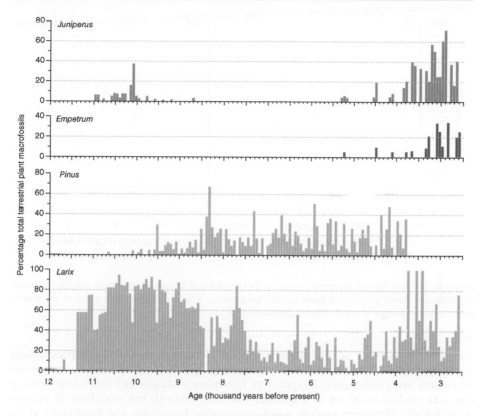

Figure 5.15 Needles from a Swiss mountain lake, recording vegetation change close to the treeline. Source: Kaltenrieder, P., W. Tinner, et al. (2005). "Zur Langzeitökologie des Lärchen-Arvengürtels in den südlichen Walliser Alpen." Botanica Helvetica 115: 137–154. Fig.4 p.151

(Figure 5.15) (Kaltenrieder et al., 2005). Colombaroli et al. (2010) used multiple-regression modelling to relate the major species in the system to charcoal abundance and summer temperatures, which were reconstructed from the fossil chironomid record (Figure 5.16). Their analysis suggested that Arolla pine was strongly favoured by warm summers but was disadvantaged by burning. Modelling of the system with the dynamic landscape model LANDCLIM (Schumacher et al., 2004) showed that the fire regime was more responsive to ignition frequencies than summer drought, and their dry, high-ignition scenario yielded burnt area records that best matched the late Holocene charcoal record (Colombaroli et al., 2010). Data–model comparison supported the hypothesis that with increased ignition frequency, human activities became the dominant influence on fire regime during the last 4000 years and contributed to the local loss of Arolla pine.

Colombaroli et al. (2009) also used redundancy analysis to estimate how much of the variance in past vegetation data could be accounted for by charcoal accumulation rates at four sites in the Mediterranean region, situated around the Adriatic Sea. At two of the sites, fire could account for between 25 and 35% of vegetation variance during parts of the Holocene. They also showed that holm oak (*Quercus ilex*) was surprisingly disadvantaged by fire, even though it favours relatively hot and arid conditions. These examples demonstrate the ability of modelling and quantitative techniques to discriminate between competing hypotheses in a more effective manner than simple visual comparison of records of vegetation dynamics, climate, proxies, human influence and charcoal records.

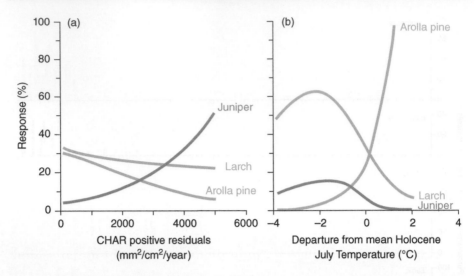

Figure 5.16 (a) Species response curves to CHAR (charcoal accumulation rates) positive residuals. (b) Response curves to Holocene mean summer temperature, which is approximately equal to the present mean July temperature of 8.1 °C. (after Colombaroli *et al.*, 2010)

Visual comparison of proxy records has been a standard approach in palaeoecology for many years but has limited value for testing hypotheses. Quantitative comparison of records using techniques of regression, redundancy analysis and time-series analysis can take understanding of past changes one step forward, but modelling of systems with dynamic ecosystem models, which incorporate knowledge and hypotheses about ecosystem dynamics, is an even more powerful analytical approach.

Data and modelling are showing how fire–vegetation relationships are altered by changing climate and that earlier fire regimes are now heavily modified by human activities, generating both biological and socio-economic problems. Intense or inappropriate fire can wreak enormous damage, and following recent extreme fire years in several regions of the world there is an urgent need for coordinated policies of fire management. Fuel management, through rigorous prescribed burning programmes, is only fully operational on a significant scale in southwestern Australia and in the southeastern USA. Models are the only tool we have for forecasting future conditions, and increased wildfire is a real concern. By using ecosystem models that incorporate fire to explore the past, we can increase our understanding of current ecosystem–fire relationships and prepare for likely future conditions.

Summary: Fire modelling

- The focus is on models that address broader-scale fire events.
- Natural ignition is important in temperate and boreal zones, especially in regions of sparse human population. In regions where humans use fire to clear the land, for example for agriculture, economic indicators are the most useful predictors of area burnt at the regional scale.
- Globally, climate (high temperatures, average precipitation and especially long periods of low rainfall), rather than humans, influences the extent of wildfires. Generalised and more globally scaled fire models tend to ignore human ignition, although they do include ignitions by lightning.

- Modelling the spread of fires is somewhat more developed. It is based on empirical studies, statistical approaches and theory. In global-scale simulations, such as within DGVMs, the approach is more generalised and fire models are driven by fuel and various climate factors, such as wind, temperature and humidity. However, there are many shortcomings, and a statistical approach to the modelling of burnt area might be more promising.
- Data–model comparison explores hypotheses about past changes in fire–vegetation relationships and builds confidence in the ability of models to construct realistic future scenarios.

5.3 Forest pathogens during the Holocene

Oliver Rackham (1980) has estimated that about 12% of trees in British forests were elm (*Ulmus*) prior to the well-documented 'elm decline', which has been widely found in the European pollen record dating from about 6000 years ago. The possible causes of the elm decline have been debated for many years and arguments have been made for the influence of human impact, climatic change and disease, or combinations of these factors. A typical sequence of events took place around the lake Diss Mere in eastern England (Peglar and Birks, 1993). The pollen record shows a 50 year period of local human influence, with evidence for cereal cultivation, local burning and opening of the forest canopy, at least 160 years prior to the elm decline. The collapse of the elm population took place over just 6 years, with a substantial reduction in the influx of elm pollen to the lake sediments (Figure 5.17). The time course of the collapse can be studied in detail as the Diss Mere sediments are visibly banded, with each band or lamination representing a single year (Peglar, 1993). The disease hypothesis has gained most supporting evidence because the insect vectors (*Scolytus* spp.) and elm wood containing both the characteristic tunnels cut by the mother beetles and the galleries where the beetle larvae develop have been found from the same time period as the elm decline (Figure 5.18). There have not however been any confirmed fossil finds of the pathogenic fungus *Ophiostoma ulmi*. The speed of the tree population collapse is typical for fungal pathogenic outbreaks. Secondary successions, including short periods with light-demanding shrubs and trees, often follow the loss of elm. Elm populations usually recover some time after the population declines, but rarely has elm ever been so abundant again, suggesting that the elm decline led to an irreversible change in European forest vegetation.

Parker *et al.* (2002) reviewed 139 sites from the British Isles containing records of the elm decline and they also favoured the disease hypothesis as it accounted for the abruptness of the decline. The uncertainty of radiocarbon dates did not allow for a reconstruction of the spread of the disease across the country, as the mean dates for the elm decline in England, Scotland, Wales and Ireland differ by just over 1000 radiocarbon years and lie well within the limits of dating error (Parker *et al.*, 2002). The recent outbreaks of Dutch elm disease in Britain also provide a modern analogue that shows some agreement with data from the mid-Holocene decline. In these recent outbreaks, adult elm has been virtually eliminated from southern England in just 20 years and ruderal species have increased in importance where elm once dominated, as is seen in several of the mid-Holocene records (Parker *et al.*, 2002).

However, the other debated factors of climate and human impact are certainly not out of the picture. Human impact has been implicated in the current European outbreak of Dutch elm disease through the inadvertent dispersal of the fungal pathogen and clearances in forests, which have facilitated the spread of the insect vector. These factors may

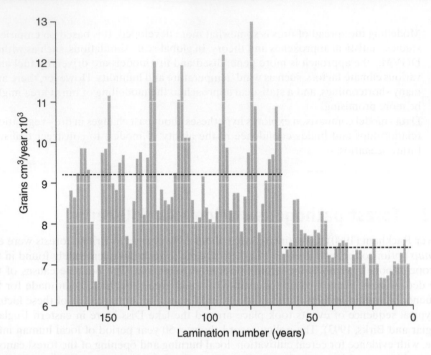

Figure 5.17   Sudden decline in *Ulmus* (elm) pollen flux into the sediments of Diss Mere, UK about 6000 years ago. The dashed lines show mean pollen flux before and after the elm decline. Source: Peglar, 1993. Reproduced by permission of SAGE Publications

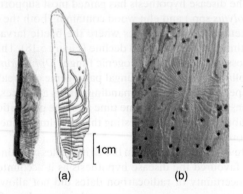

Figure 5.18   (a) Photograph and drawing of tunnels and galleries made by *Scolytus* sp. in ancient elm wood about 6000 years ago. (b) Photograph of modern infected elm wood. Source: Reproduced by permission of Peter Rasmussen

have operated in the past as well. The decline of elm pollen was more abrupt than subsequent late-Holocene declines in populations of *Corylus* (hazel), *Tilia* (lime) and *Quercus* (oak), which are all visible at a European scale and were caused by combinations of climate change and changes in land use. In northern Europe at least, the first major outbreak of elm disease may have struck just as climatic conditions were becoming less suitable for elm. LPJ-GUESS has been used to model Holocene vegetation development in Fennoscandia, and at two southern sites there are abrupt declines in the abundance of wych elm (*Ulmus glabra*) during the mid-Holocene (see Figure 4.14). These model runs were only forced by estimated past climate and are independent of human or pathogenic influence. The weight of evidence thus favours a dominant role for disease in the elm decline, but without the murder weapon – which in this case would be the elusive remains of the fungus itself – the case is not fully convincing. Ancient DNA techniques might be able to provide this critical evidence.

The North American hemlock (*Tsuga canadensis*) decline is another range-wide event recorded from the mid-Holocene (c. 5500 years ago). Davis (1981) reviewed a similar suite of hypotheses to those proposed for the European elm decline and also concluded that forest pathogens were the most likely cause. Hemlock needles chewed by the hemlock looper (*Lambdina fiscellaria*), together with head capsules from the looper and the spruce budworm (*Choristoneura fumiferana*), have been found in Canada by Bhiry and Filion (1996). Ring counts of hemlock trees at the same site showed growth abnormalities and premature death, providing strong evidence for the pathogen hypothesis. Palaeoecologists love to re-evaluate earlier interpretations reflecting the uncertainty that is always present in past reconstructions, and several papers subsequently presented evidence for widespread mid-Holocene drought events that were likely to have affected the vigour and survival of the moisture-sensitive hemlock (Shuman *et al.*, 2004; Oswald and Foster, 2012). Higher-resolution datasets and new analytical techniques can now detect the presence in the Holocene of abrupt climatic events whose impacts on vegetation can appear to be very similar to pathogen outbreaks (Shuman *et al.*, 2009). Simple, one-driver explanations are rare commodities in ecosystem dynamics, so perhaps drought events weakened the hemlock populations and made them more susceptible to disease. In the absence of direct evidence of large populations of pathogens, the disease hypotheses are under renewed scrutiny. Ecosystem modelling could help distinguish between the effects on vegetation of disease and climate, but further insights might also come from the study of contemporary systems.

The western USA and Canada support several magnificent coniferous forest communities, which often contain long-lived, giant trees. Parts of these systems have experienced rather little direct anthropogenic impact but have now been monitored for a few decades. Long-term monitoring data almost always reveal dynamics and there has been a tendency amongst foresters to be pessimistic about tree mortality in the wild, as they see exploitable natural resources going to waste. Van Mantgem *et al.* (2009) looked at records of nearly 60 000 trees located in 76 forest plots along the Pacific coast, using data extending back to 1960. They found that tree mortality rates had increased throughout the region from values below 0.5% per year to values over 1.5%, regardless of geographic region, altitude, stem size, tree species or fire history. Recruitment rates had not significantly increased over the same c. 50 year period to compensate for this increased mortality, so the authors were concerned for the future of these forests. Both these mortality values do lie within the typical range (0.5–3.0%) found from a literature survey of old-growth temperate, boreal and tropical forests (Wolf *et al.*, 2004), but the rate of increase appears to be unusually rapid. Van Mantgem *et al.* (2009) proposed a range of hypotheses to explain this increase but settled for a dominant contribution from regional warming, as mean annual temperatures had increased by up to 0.5 °C per decade throughout the period of study. They argued that warming would have increased drought stress and benefitted pathogens, exactly as has been proposed for the hemlock decline 5500 years earlier in the northeast USA. However, in this case widespread conifer mortality caused by bark beetles had also been observed in the same region by Raffa *et al.* (2008) and just about every other forest scientist who has recently visited any of the 13 million hectares of western Canadian lodgepole pine forest (*Pinus contorta*) affected by mountain pine beetle (*Dendroctonus ponderosae*) since 1999 (Figure 5.19).

Massive population explosions, or 'eruptions', of mountain pine beetle in western Canada were probably limited in the past by cooler summer and winter temperatures. In

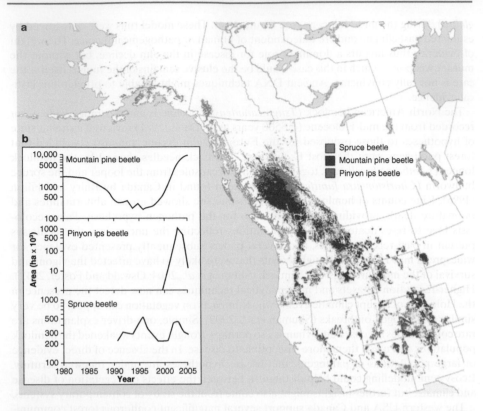

Figure 5.19   Recent mortality of major western conifer biomes to bark beetles. (a) Map of western North America showing regions of major eruptions by three species. (b) Sizes of conifer biome area affected by these three species over time. Data from the Canadian Forest Service, the British Columbia Ministry of Forests and Range and the US Forest Service. Source: Raffa *et al*, 2008. Reproduced by permission of the American Institute of Biological Sciences

response to warming, populations have expanded into more northerly latitudes and higher elevations than before (Raffa *et al.*, 2008). Mountain pine beetles have also attacked high-altitude whitebark pine (*Pinus albicaulis*) stands in the Rocky Mountains, which previously had suffered only occasional attacks during climatically favourable periods. Furthermore, the beetles now have potential access to the vast, mid-continental jack pine (*Pinus banksiana*) forests, which were previously protected by high-altitude regions of colder temperature where the beetles could not survive (Raffa *et al.*, 2008).

Human activities can also influence the course of pathogenic outbreaks, as has been suggested for the elm decline. In western North America, forest management has tended to increase the abundance of susceptible hosts. Lodgepole pine stands in interior western Canada have mostly originated from stand-replacing wildfires, but recent fire suppression has reduced this disturbance frequency and increased the proportion of older trees, which are more prone to beetle attack. This is a further factor that will contribute to the recent increased eruptions of bark beetle populations and subsequent tree mortality (Raffa *et al.*, 2008; van Mantgem *et al.*, 2009). We can see from these examples how climate change can interact closely with pathogenic insects and human impact to kill trees, where it would be misleading to identify one single cause of mortality. There are lessons here for the elm, hemlock and more recent chestnut 'declines' in the palaeoecological records of North America and Europe.

5.4 Hurricanes and wind damage

In November 1976 I (R.B.) was beginning the second year of research for my doctoral thesis and was feeling particularly receptive to well-written papers and innovative ideas about forest ecosystem dynamics. I read the latest copy of the *Journal of Ecology* in the library after lunch and was gripped by Douglas G. Sprugel's paper on balsam fir (*Abies balsamea*) forests in the northeastern USA, which summarised his own doctoral thesis (Sprugel, 1976). It was easy to read and entertaining, and it questioned some of the classical concepts of succession and introduced the importance of disturbance in long-term forest dynamics. Today it would most likely be criticised by referees for its informal style and bold speculation, but ecological scientific writing was more flamboyant and readable back then! Sprugel quoted from A.S. Watt's 1947 presidential address to the British Ecological Society: 'it is now half a century since the study of ecology was injected with the dynamic concept, yet in the vast output of literature stimulated by it, there is no record of an attempt to apply dynamic principles to the elucidation of the plant community itself.' Sprugel added that 'the intervening decades have not changed the situation much; except for Watt's own paper and a very few others, this field is still largely unexplored. The main difference is that advances in nearly all other areas of ecology have rendered our ignorance about long-term ecosystem dynamics even more striking by contrast' (Sprugel, 1976). His paper is certainly one of the stimuli for this book and this is the appropriate place to introduce his research on wind as a potential driver of long-term ecosystem dynamics.

Pure balsam fir forests are common in northeastern USA just below the upper limit to forest growth. On slopes facing strong prevailing winds, fir forests often have large 'crescent-shaped bands of dead trees whose exposed trunks show up silver' (Sprugel, 1976). 'These areas of dead and dying trees have frequently been thought to be small wind-fall areas, where particularly violent gusts of wind during a heavy storm flattened a few dozen trees and left a visible hole in the otherwise uniform canopy. Closer examination, however, shows a more systematic and remarkable pattern; each of the supposed wind-falls is actually an area of standing dead trees, with mature forest beyond it and an area of vigorous regeneration below it' (Figure 5.20; Sprugel, 1976). I have seen these 'Sprugel' waves in fir forests both on the windswept slopes of Mount Katahdin in Maine (Figure 5.21) and on the mountains of central Honshu, Japan. They look like the effects of wind blowing across a field of wheat and they move in the direction of the prevailing wind but at rates of only 1–3 m per year, so they take decades to move across the landscape. Sprugel proposed multiple hypotheses to explain how the waves could be initiated by the death of a line of trees, all related to wind stress. They include rime ice formation, winter desiccation and summer cooling of exposed needles. I heard several other suggestions made during the Mount Katahdin excursion. Once a stand has been linearly structured by age, the waves will be self-perpetuating in balsam fir, which is short-lived and can be senescent by 50–60 years of age in this environment. Rime ice formation has since been shown to be the most likely hypothesis for death (Foster, 1988).

Hurricanes and wind damage are probably the next most widespread type of disturbance after fire and pathogens to impact terrestrial ecosystems, affecting areas up to 100 000 km^2 (Table 5.1; Foster *et al.*, 1998a). Hurricanes originate over warm tropical oceans and lose energy as they move away from the tropics or travel inland. Many studies have examined the short-term impacts of hurricanes, but placing the effects in a longer-term perspective requires some luck. On 3 December 1999, winds were gusting up to 50 metres per second in western Denmark, and within a few hours over 3.6 million cubic metres of timber, comprising about 5% of the entire Danish forest volume, had been blown down, making this the most damaging storm and most expensive weather event on record in Denmark (Figure 5.22). Twelve days after the storm, I (R.B.) was in Draved Forest in southern Jutland with my colleagues measuring the effects of the hurricane on a mixed temperate deciduous

(a)

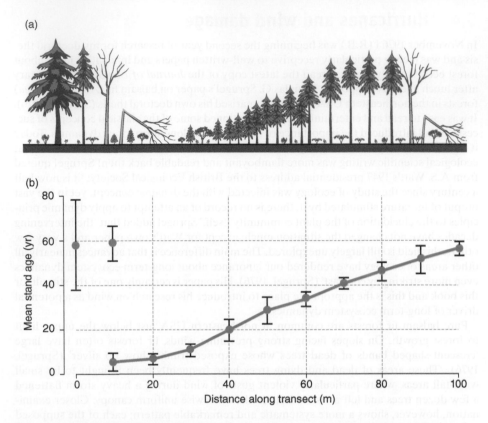

(b)

Figure 5.20   (a) Cross-section through a 'Sprugel' regeneration wave. (b) Mean stand age ±1 SD through a regeneration wave of balsam fir (*Abies balsamea*) (Sprugel, 1976)

Figure 5.21   Sprugel waves and avalanche tracks on Mount Katahdin, Maine, USA (photo Richard Bradshaw)

Table 5.1   Characteristics of large, infrequent disturbances (after Foster *et al.*, 1998a)

Characteristic	Volcanic eruptions	Forest pathogens	Tornadoes	Forest fires	Hurricanes	Riverine floods
Duration of event	Hours	Years	Minutes	Weeks	Hours	Weeks
Return interval (yrs)	$10^2 - 10^3$	$10^1 - 10^3$	100 – 300	75 – 500	60 – 200	50 – 100
Size of event (km^2)	5 – 100	1 – 100 000	5 – 100	50 – 20 000	50 – 100 00	50 – 50 000
Location	Volcanic mountains	Forest	Inland	Inland	Warm coasts	Riparian
Variables affecting severity	Volcanic eruptions	Forest pathogens	Tornadoes	Forest fires	Hurricanes	Riverine floods
Climatic factors	No	Yes	Yes	Yes	Yes	Yes
Season	Yes	Yes	Yes	Yes	Yes	Yes
Topography	Yes	No	No	Yes	Yes	Yes
Vegetation structure	Yes	Occasional	No	Yes	Yes	Yes

Source: Foster *et al,* 1998. Reproduced by permission of Springer

research forest, where 6000 trees had been monitored since 1948. Draved is a near-natural forest with varied structure and composition, lying only 21 km from the Atlantic coast and therefore often exposed to high winds. Draved felt the full force of the hurricane, yet only 4% of the monitored trees were broken or uprooted; many of these had resprouted a year after the event, so eventual mortality was considerably lower. Overall tree mortality in Draved (at least those trees over 10 cm in diameter, measured 1.2 m above the ground) had varied between 5 and 10% of trees per decade since 1948 and wind had been the presumed cause of death in up to 70% of these cases. The 1999 storm was an extreme event, but the 10% overall mortality for the decade 1991–2000 was only slightly higher than the 5–7% recorded from the earlier decades (Wolf *et al.*, 2004). In the country as a whole, seven times more exotic conifers blew down than native deciduous trees. The damage and mortality in uniform, neighbouring exotic conifer forest plantations was far more severe than in the near-natural Draved, which makes an important case for copying appropriate natural features in silviculture (Chapter 9). As with pathogens, forest management can inadvertently increase the susceptibility of forest ecosystems to wind damage, and even-aged conifer plantations have proved to be a risky strategy in several wind-stressed north European environments.

Wind storms have certainly influenced the age and size structure of Draved Forest, but they also appear to have affected species dynamics. Wind damage tends to result in less tree mortality than hot fires or pathogen outbreaks, so the subsequent forest regeneration can originate more from the recovery of damaged stems or release of suppressed individuals than from seedling recruitment, and the eventual impact on vegetation composition will be less (Martin and Ogden, 2006). In Draved, birch (*Betula pubescens*), beech (*Fagus sylvatica*) and lime (*Tilia cordata*) have suffered greater mortality from wind damage than oak (*Quercus robur*), alder (*Alnus glutinosa*) and ash (*Fraxinus excelsior*). Beech has been recruiting from seeds and lime from vegetative sprouts in Draved, so their populations have been stable. Birch, however, is disappearing from the forest due to a combination of mortality and lack of recruitment since 1839–1863, when German forestry operations created

Figure 5.22   The most severe hurricane to affect Denmark to date, 3 December 1999. Source: Cappelen, 2011. National Oceanic and Atmospheric Administration

larger gaps in the forest (Wolf *et al.*, 2004). Wind has been the major disturbance agency in recent decades – fire has not been an important factor in Draved for several centuries.

Harvard Forest, across the Atlantic from Draved, is another research forest in a coastal region that has periodically experienced hurricane-force winds. The New England hurricane of 21 September 1938 was one of the most catastrophic in the history of the USA, with sustained wind speeds over 50 metres per second and gusts of over 80 metres per second recorded in Massachusetts. The hurricane caused variable damage in a strip of New England 100 km wide and 300 km long (Foster *et al.*, 1998b). While the weaker Danish hurricane of 1999 raced over a flat landscape, slope and aspect influenced where the New England one wreaked most havoc (Foster and Boose, 1992). Uprooting was the main type of damage and was correlated with both tree height and species composition. As in Denmark, evergreen conifers such as white pine (*Pinus strobus*) suffered greater damage than deciduous trees, which included maples (*Acer*), red oak (*Quercus rubra*) and paper birch (*Betula papyrifera*) (Foster and Boose, 1992). Hurricanes in temperate regions tend to occur after deciduous trees have lost their leaves, which is a factor in their greater storm resistance. The biological legacies left by hurricanes in the northeastern USA include a patchwork of age and height structures, standing broken trees, coarse and fine woody debris (which may become a fire hazard) and soil disturbance such as wind-throw mounds and mass movement of soil following storm-related intensive precipitation events (Foster *et al.*, 1998b).

Hurricane damage may appear to be disorganised, yet Boose *et al.* (2001) combined a meteorological model with a topographic exposure model to compare modelled and observed forest damage for 67 historical hurricanes recorded in the New England region between 1620 and 1997. Their data–model comparisons for past hurricanes showed how frequently different degrees of forest damage had been inflicted in a regional gradient across New England. The most frequent severe damage was hindcasted for the southeastern coastal region where most hurricanes first come ashore (Figure 5.23). In Massachusetts, Rhode Island and eastern Connecticut, extensive blowdown of trees had most likely recurred every 85 years on average since 1620, yet local topography had provided

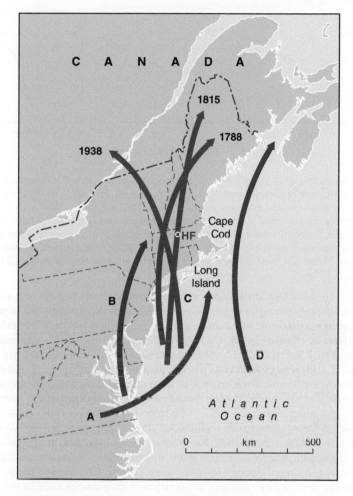

Figure 5.23 Northeastern USA, showing the four generalised pathways (A–D) that hurricanes follow into the region and the historical tracks of the hurricanes of 1788, 1815 and 1938 (pathway C). The location of Harvard Forest (HF) is indicated (Foster and Boose, 1992)

some long-term protection for small areas as hurricanes tended to travel in the same direction. In northern New England, extensive blowdown was far less frequent, recurring only every 400 years or more. Neither observations nor model results showed that all trees had been blown down over large areas in New England; damage had instead been focused on to multiple small patches, which typically experienced treefall, broken branches and defoliation (Boose *et al.*, 2001).

5.5 Conclusion

Our survey of the major forest disturbance agencies over longer time periods provides a background and reference to reports of drastic increases in disturbance frequency and severity throughout the northern hemisphere during recent decades (Seidl *et al.*, 2011). Seidl and co-authors collated forest inventory, disturbance and climatic data for 23 European countries covering the period 1958–2001. They used structural equation modelling

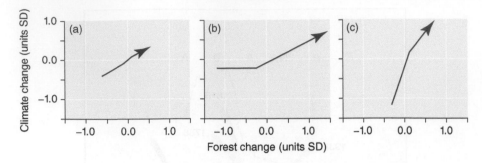

Figure 5.24   Trends from 1958 to 2001 in forest and climate drivers of natural disturbances in European forests: (a) wind, (b) bark beetles, (c) wildfire. Temporal trajectories are shown with the lines representing a lowest fit in the phase space of forest and climate change (standardised to units of standard deviation). A vertical trajectory upwards would indicate that climate change drivers facilitating disturbances increased over the study period while the influence of forest extent, structure and composition remained constant. A 45° trajectory from the lower left to the upper right, on the other hand, signifies that the intensifying influence of both climate and forest change drivers increased equally over time (Seidl *et al.*, 2011)

(Grace, 2006) to show that disturbance related to wind, bark beetles and wildfire had increased considerably throughout the study period, largely driven by interactions between changes in forest management and climate (Figure 5.24). Increased wildfire incidence was primarily driven by changes in precipitation and temperature, while increasing damage from wind and bark beetles was a combined effect of climatic change and altered forest management, acting through modified age class distributions and the increased proportion of conifers on the landscape. Their analysis confirms earlier indications about interactions between human and climatic influence on different types of disturbance, particularly in Europe, with its long, documented history of human impact on ecosystems. We do appear to be in a period where disturbance regimes are rapidly changing, driven by interactions between climate and land use change. Knowledge of the longer-term ecosystem dynamics shows that fire and pathogenic effects have altered rapidly in the past as well and that ecosystems have responded through compositional change. Climate and people have also interacted in the past to drive changes in disturbance frequencies such as fire dynamics. However, the current pace of climatic change and the intensity of management interventions, at least in Europe, are such that irreversible ecosystem change and modification of key ecosystem services are to be anticipated, as we discuss in the next chapter.

Summary:Pathogens and wind damage

- The European elm decline from about 6000 years ago provides an informative analogue for recent outbreaks of Dutch elm disease. Human activities and climatic factors are likely to have influenced the spread of forest pathogens earlier during the Holocene and in the present day.
- Hurricanes and wind storms influence the age, size structure and species dynamics of both tropical and temperate forest ecosystems. 'Sprugel' waves – wind-generated waves of regeneration – are found in the most wind-stressed montane forest systems.
- A European analysis covering AD 1958–2001 indicates that climate change and forest management interacted to affect disturbance regimes, with climatic change the dominant influence on wildfire frequency, forest management the dominant influence on bark beetle outbreaks and both factors of comparable importance for wind damage.

6

The Impact of Past and Future Human Exploitation on Terrestrial Ecosystem Dynamics

'So long, and thanks for all the fish'

So Long, and Thanks for All the Fish, *Douglas Adams, 1984*

6.1 Introduction

Agriculture now occupies 38% of the global terrestrial surface and human influence is detectable in almost all of the remaining 62% (Foley *et al.*, 2011). Human influence on ecosystems has increased through time to match and even overreach climate as an agent of ecosystem change, reflecting population growth and technological development. We reviewed the central but complex relationships between climate change and millennial ecosystem dynamics in Chapter 4, and in this chapter we explore the course of human impact from its origins to the start of the present critical state, with this planet rapidly approaching its carrying capacity for humans. The developing human influence on ecosystem dynamics is closely linked to the topics of ecosystem services and conservation, which are covered in the next three chapters.

In previous chapters we referred to two of the earliest documented records of human impact on ecosystems: the approximately 1 million-year-old date for the first controlled use of fire by humans in South Africa (Berna *et al.*, 2012) and the systematic slaughter of wild horses in Germany about 400 000 years ago (Thieme, 1997). Both of these examples are likely to be replaced soon by older finds in this fast-developing field of research. As the ink dried on that last sentence, the date of the first stone-tipped spears was pushed back 200 000 years to 500 000 years ago by new finds from South Africa (Wilkins *et al.*, 2012). Thieme (1997) described three sophisticated wooden throwing spears found with numerous butchered horse bone fragments and stone tools (Figure 6.1). The spears are

Ecosystem Dynamics: From the Past to the Future, First Edition.
Richard H.W. Bradshaw and Martin T. Sykes.
© 2014 John Wiley & Sons, Ltd. Published 2014 by John Wiley & Sons, Ltd.
Companion Website: www.wiley.com/go/bradshaw/sykes/ecosystem

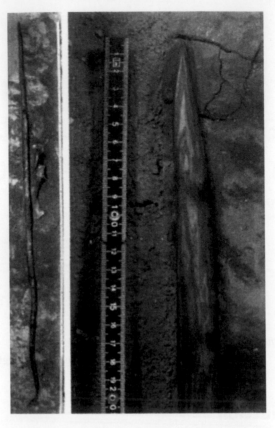

Figure 6.1 Tip of a spear used for killing horses c. 400 000 years ago in Germany. Source: Thieme, 1997. Reproduced by permission of Nature Publications

made from individual spruce trees, with the tips carved from the tough tree bases. They are over 2 m in length, are weighted like modern javelins and are the oldest complete hunting weapons known to science. These well-preserved archaeological finds (from an open-cast, brown coal mine) show that organised hunting, involving foresight, planning and the use of appropriate technology, was already taking place in Europe hundreds of thousands of years ago. The hunters were premodern hominins, but the scale of operation suggests that their activities caused noticeable ecosystem impact.

This German site is not the oldest in northern Europe to provide evidence of tools that could be used to modify ecosystems. In Norfolk, UK, 32 worked flints have been dated to between 680 000 and 750 000 years ago, demonstrating tool-making activities over 200 000 years earlier than the German horse slaughtering site (Parfitt *et al.*, 2005). At this time, Norfolk had a comfortable, seasonally dry Mediterranean climate and held a variety of tasty browsing and grazing mammals, including steppe mammoth (*Mammuthus trogontherii*), rhinoceros (*Stephanorhinus hundsheimensis*), large deer (*Megaloceros savini, M. dawkinsi*) and bison (*Bison* cf. *schoetensacki*). Stone tools of a similar age have also been found in Chinese loess deposits dating from 700 000 to 800 000 years ago (Lu *et al.*, 2011) and there is growing genetic evidence for the early coexistence of several types of hominin outside Africa, even at northern latitudes (e.g. Krause *et al.*, 2010). Early hominins probably lived in small family groups and had a very limited impact on their surroundings but there is growing evidence for a very wide geographical dispersal of hominins at increasingly old dates, and although the supporting evidence for ecosystem impact from the oldest archaeological sites is rare, this will doubtlessly increase with the use of new research technologies and the discovery of further populations.

While archaeological sites typically give snapshots of local evidence for human activities, sediment cores allow a more continuous, long-term assessment of the scale of landscape impact of human activities. As discussed in Chapter 5, a few people can potentially modify local fire regimes to cause considerable ecosystem impact and the regular burning of African savannahs for game may well prove to have a very long history, linked to the early evolution of hominins in Africa. A marked increase in the charcoal content of marine sediment cores around 50 000 years ago, collected near Papua New Guinea and Australia, matches rather well to the terrestrial Australian fire record and its possible links with human colonisation (Figure 6.2; Thevenon et al., 2004). However, charcoal records do not indicate how the fires began, so independent evidence is needed to connect the fires to people.

There is an intriguing opening of the forest canopy by fire and a subsequent grassland episode about 300 years in length recorded from two sites in eastern England during an earlier interglacial period, dating from about 400 000 years ago (Turner, 1970). Archaeological remains show that early humans were active in the area, but there is insufficient forensic evidence to link people directly to the forest clearance, even if they almost certainly ambushed the animals that enjoyed the rich grazing close to good drinking holes. Human ecosystem impact can be convincingly demonstrated when forest clearance, by fire or other means, occurs at the same time as evidence for the introduction of cultivated plants. However, other types of impact, such as low-intensity cattle grazing and selective felling of trees, do not leave a strong signature in the palaeoecological record. Past human impact in nonforested regions can be even more difficult to detect. The most secure evidence for widespread human impact on ecosystems comes from the Holocene, from palaeocological records of cereal cultivation on land cleared of forest, but debate rages about the timing and scale of human impact prior to the development of these intensive agricultural systems.

Cultural periods such as the Mesolithic and Neolithic are slightly slippery concepts for the layman as their definitions are based on the types of archaeological material they contain, rather than on set time periods. So at any one point in time, different parts of the world might be classified into different cultural periods. As with any classification, the focus is on periods of relative stability and the interesting transition zones between categories can be difficult to handle. The Mesolithic or Middle Stone Age period is characterised by stone

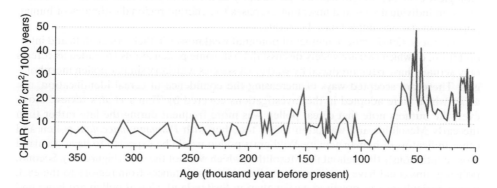

Figure 6.2   Record of charcoal accumulation rate in a marine core near Papua New Guinea. Source: Thevenon et al, 2004. Reproduced by permission of Elsevier

tools, weapons with small blades and the first settled communities. The Neolithic or New Stone Age directly follows the Mesolithic and is identified by the onset of agriculture, the first pottery, woven materials and polished stone tools. This classification gives the impression that the onset of agriculture was an abrupt and easily recognised transition, but this may not have been the case. In China, for example, finds of ancient pottery now date back to 20 000 years ago (Wu *et al.*, 2012), which has broken, at least for the time being, the formerly fixed link between the origins of agriculture, which in China are currently dated to about 10 000 years ago, and the use of fired ceramic containers for the storage and processing of food and for cooking, which appears to be twice as old.

As the onset of agriculture represents a significant change in the type and intensity of human impact on ecosystems, it is worth examining this transition from Mesolithic hunter-gatherer communities to Neolithic farmers in more detail. In Switzerland, it is believed to date from about 5500 BC, or 7500 years ago, yet Tinner *et al.* (2007) have found pollen from cereals and the classic weedy indicator of human activities, *Plantago lanceolata*, at several sites prior to the appearance of pottery, clearly placed within the Swiss Mesolithic period (Figure 6.3). Their conclusions are supported by similar finds by palaeoecologists from northern Italy, southern France, Germany, the British Isles, southern Scandinavia and the Baltic States. Tinner *et al.* (2007) argue for 'a slow and gradual change towards sessile agriculture' and 'agricultural adoption by indigenous hunter-gatherers as opposed to wholesale immigration of agriculturalists', which, if correct, indicates a gradual increase of human impact throughout the early–mid-Holocene rather than a dramatic 'Neolithic revolution' brought on by the sudden advent of agriculture. In the Po Valley, Italy there is a close relationship between the Holocene charcoal record and plants growing on cultivated fields and pastures, as burning was used to prepare the ground for agriculture (Kaltenrieder *et al.*, 2010). The palaeoecological record shows initial human impact between 8500 and 6500 years ago, which includes early forms of agriculture, with some cereal cultivation, followed by a significant and sustained increase in farming activities beginning 6500 years ago (Kaltenrieder *et al.*, 2010). Archaeologists place the Mesolithic–Neolithic transition in northern Italy at 7450 years ago, but the continuous palaeoecological records show a more gradual, phased transition into agriculture that contributes to the growing consensus from several regions of the world about the nature of the transitions from hunting and gathering to agriculture. This debate would not be the first example of palaeoecological evidence indicating a slightly different development in human–ecosystem relationships than that proposed by archaeologists. Archaeologists and palaeoecologists come from different research traditions and are familiar with, and place their trust in, different types of data. As our topic here is ecosystem impact, the biological evidence weighs heavier, particularly as archaeological research has tended to focus on individual sites and upscaling exercises to generate regional estimates of human population size and impact have been rare.

Tinner *et al.* (2007) raise a number of potential weaknesses in their pollen data and interpretation, of which two are widely discussed. First, some pollen grains identified as cereals may actually have been produced by wild grass species with very similar pollen morphologies. There are accepted ways of increasing the confidence of cereal identification, not all of which were adopted in the investigations covered by Tinner *et al.*'s review. However, the authors note that cereal pollen was more frequent during the late rather than the early Mesolithic and was often found together with *Plantago lanceolata* pollen and other indicators of human activity. Wild grass pollen would be expected to have occurred more continuously throughout the Mesolithic, which was not the case (Figure 6.3). Second, pollen grains could have been transported across long distances from regions to the east, where agriculture was practised earlier than in Switzerland. Cereal pollen are large and heavy and would only be transported long distances in the company of other 'exotic' pollen types, which also was not the case. Recent methodological advances in the interpretation

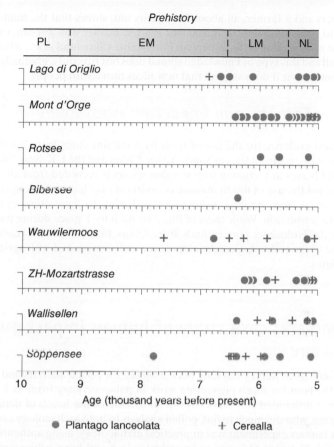

Figure 6.3 Early records of pollen from *Plantago lanceolata* and Cerealia from Switzerland. PL, Palaeolithic; EM, Early Mesolithic; LM, Late Mesolithic; NL, Neolithic. Source: Tinner *et al* 2007. Reproduced by permission of Elsevier

of pollen data further de-emphasise the significance of long-distance pollen transport of weak pollen producers, particularly when the majority of the landscape is covered by forest. The case for pre-Neolithic cereal cultivation would be clinched by the discovery of cereal macrofossil remains from an archaeological site. However, plant macrofossils of any kind are very rare in late Mesolithic archaeological sites in Switzerland, so it might be better to look elsewhere for this type of evidence. The provisional conclusion from this Swiss example is that organised agriculture gradually increased the human impact on ecosystems, which might already have had their fire regimes artificially modified. This example is probably representative of how agriculture spread throughout much of northwestern Europe. Although there is less detailed evidence from Asia, with less secure chronologies, a comparable course of events has been proposed for the spread of agriculture through China and India (Fuller *et al.*, 2011). There is evidence from these regions for the introduction of fire – or a modification of the natural fire regime – as early as 14000 years ago in the Ganges region of India. This is followed by palaeoecological evidence for the cultivation of millet and rice dating from the early Holocene. Subsequently, intensive livestock management was the final component of agriculture to become widely established (Fuller *et al.*, 2011). This model of a gradual development of agricultural systems may be modified by developing ancient DNA evidence from humans: DNA extracted from three Swedish hunter-gatherers and a farmer, all about 5000 years old, shows that the hunter-gatherers were related to present-day north Europeans but the farmer was a southern European, indicating that new farming skills came with immigrants (Skoglund *et al.*, 2012). The sample is very small and this type of knowledge spread does not rule out a gradually developing agricultural system, but it does suggest that new ideas move with people.

Summary: Early human impact on ecosystems

- The earliest evidence for the use of tools by hominins dates from 700 000 to over 1 million years ago, from sites in South Africa, China and the UK. Systematic slaughter of wild horses in Germany with wooden spears is recorded from about 400 000 years ago but the use of fire to manage ecosystems may have a far longer history.
- The transition from hunting, fishing and gathering to agriculture caused major ecosystem conversion. While most of this transition took place during the first 6000 years of the Holocene, pottery finds from China that are 20 000 years old show that some of the technological development usually associated with agriculture had early origins.

6.2   Denmark: case study of human impact during the Holocene

The Danish Geological Survey has long been a Mecca for pollen analysts and its 'bog laboratory' was the base for much pioneering work in palaeoecology from its foundation in 1896 until it was disbanded 106 years later in 2002. One of the heads of department was Johannes Iversen, who did much to link pollen analysis to both archaeology and plant ecology. His most famous experiment was in practical archaeology, using authentic stone axes from the Danish National Museum to investigate slash-and-burn cultivation techniques in Draved Forest (Iversen, 1956). Some of Iversen's interests were developed by his successors, including Svend Thorkild Andersen, Bent Odgaard and Peter Rasmussen, so that we now probably know more about the effects of human impact on vegetation in Denmark than anywhere else in the world. Odgaard and colleagues have combined site-based surveys to upscale results to the whole country, even if this is still only a paltry 42 500 km^2.

Dallund Sø lies near the geographical centre of Denmark and its Holocene sediments have been studied in some detail. Denmark's archaeological record is well documented and there are numerous finds from the Mesolithic period up till the present day within 5 km of the lake (Rasmussen and Bradshaw, 2005). This local archaeological record has been linked very elegantly with pollen, diatom and soil-erosion records analysed from the lake sediments (Figure 6.4). The pollen-based evidence for potential human impact on forest cover at Dallund Sø begins with changes linked to the elm decline, which dates from 5900 years ago at this site (see Chapter 5). There are however no clear signs of cultivation at this time and the small adjustments in tree proportions probably just reflect the local loss of elm, which had been a major tree in the forest (Rasmussen, 2005). Only one archaeological find is recorded from the area during the Danish Late Mesolithic period (7000–5900 years ago), suggesting little human activity.

The first cereal pollen grain is recorded from 5460 years ago, at the start of the period of megalithic grave construction in Denmark. It has been estimated that within a 300 year period (5500–5200 years ago) as many as 25 000 dolmens and 5000 passage graves (Figures 6.5 and 6.6) were built in Denmark alone as part of an enormous construction boom throughout Western Europe, and 12 of these were close to Dallund Sø. This activity

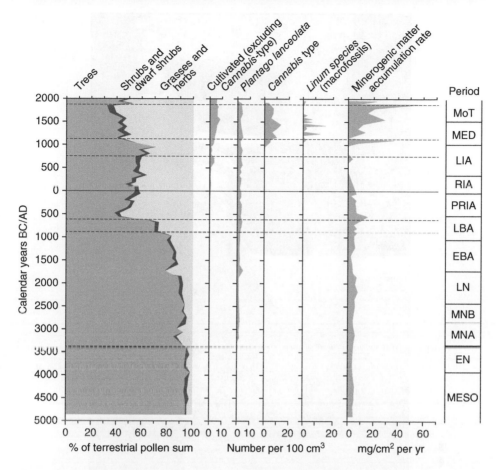

Figure 6.4 Summary of palaeoecological data describing the landscape development around Dallund Sø, Denmark. MESO, Mesolithic; EN, Early Neolithic; MNA, Middle Neolithic A; MNB, Middle Neolithic B; LN, Late Neolithic; EBA, Early Bronze Age; LBA, Late Bronze Age; PRIA, Pre-Roman Iron Age; RIA, Roman Iron Age; LIA, Late Iron Age; MED, Mediaeval; MoT, modern time. Source: Bradshaw *et al*, 2005. Reproduced by permission of SAGE Publications

is recorded in the lake sediments as a decrease in the proportion of tree pollen and corresponding increases in grasses and *Plantago lanceolata*, reflecting an opening of the forest canopy and small-scale grazing of farm animals, together with cereal cultivation using the ard as a cultivating tool (Rasmussen, 2005). The megalithic tombs were most likely designed to be seen from a distance (Figure 6.7), but pollen data from Ireland (Keeling *et al*., 1989) and Denmark (Nielsen *et al*., 2012) show that the open, cultivated areas were close to the tombs and the landscape as a whole was still 60–70% forest covered. This must have been a golden period in northwestern Europe, with pleasant summer weather on Atlantic coasts and novel food supplies from the new agricultural sources. However, the activities associated with the megalith construction period collapsed as quickly as they began and the tomb-building bubble had burst in Denmark by 5200 years ago (Andersen, 2000).

A 2000 year period with few archaeological remains around Dallund Sø follows, and lime (*Tilia*) forests reclaimed some of the abandoned agricultural land during the Late Neolithic (4400–3700 years ago in Denmark). The pollen record initially shows a reduced level of

Figure 6.5 Professor Frank Mitchell stands at the entrance to New Grange passage grave, Ireland (photo Richard Bradshaw)

agricultural activity, but at the end of the Neolithic period there was a short c. 130 year period with renewed forest clearance and pollen evidence for both crop cultivation and grazing (Rasmussen, 2005). This pattern is typical for early agricultural human impact in Europe: periods of plenty are followed by periods of abandonment covering rather large areas. Charcoal fragments recovered from Dallund Sø sediments show that fire was in use, but the system of cultivation differed from the small-scale, shifting slash-and-burn agricultural system recorded later from the boreal zone and tropical regions, in which local abandonment was driven by a rapid deterioration in soil fertility following the fertilising effects of the original burn. Iversen had slash-and-burn in mind when he carried out his archaeological experiment with stone axes in Draved Forest (Iversen, 1956), but new, quantitative approaches to reconstruction of vegetation suggest that these early Danish agricultural systems covered larger land areas for longer time periods than Iversen had envisaged. The subsequent Neolithic agricultural abandonments recorded from sediments in Denmark, which are typical for a broad region of the European temperate forest zone,

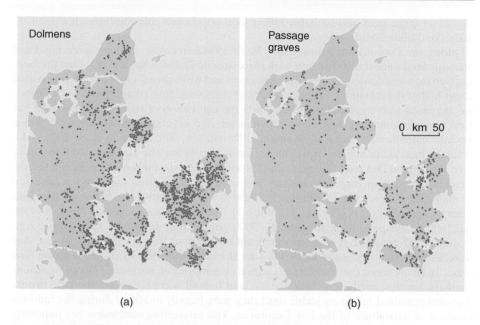

(a)   (b)

Figure 6.6   Distribution of (a) dolmens and (b) passage graves in Denmark. Source: Jensen, 2001. Reproduced with permission of Gyldendal, Copenhagen

Figure 6.7   Poulnabrone portal tomb, Ireland (photo Richard Bradshaw)

also cover large areas and probably reflect local population collapse through altered climatic conditions, famine or war.

During the Danish Late Bronze Age (3000–2500 years ago), an even more rapid and dramatic landscape transformation took place around Dallund Sø, and this time the landscape change was destined to be more permanent and widespread. Over a c. 170 year period about half of the remaining forests in the catchment were cleared and there was evidence for increased arable agriculture, particularly the cultivation of barley, together with hay meadows and animal grazing; these activities affected vegetation on all soil types within the lake catchment in one way or another (Rasmussen, 2005; Nielsen *et al.*, 2012). This dramatic increase in human impact on the landscape during the Late Bronze Age can be seen throughout Denmark. Odgaard and Rasmussen (2000) summarised pollen data from 10 sites spread throughout the country and used a sophisticated multivariate statistical technique to compare the distribution of forest in AD 1800 with 11 archaeological periods from the last 6500 years (Figure 6.8, Table 6.1). They could trace essential features of the AD 1800 landscape to their origin nearly 3000 years earlier during the Late Bronze Age. Regions of Denmark that retained significant forest cover through the Late Bronze Age were also forested regions in AD 1800, with the same long history of open, agricultural areas. In many ways, the traditional 'cultural' landscapes of northern Europe, including southern Sweden, northern France, the UK and Ireland, were created in the Late Bronze Age and remained relatively stable until they were heavily modified during the industrialisation of agriculture of the last 2 centuries. This interesting conclusion has important consequences for current conservation policies within these countries (Chapter 9).

Western Denmark today supports some of the most extensive heathland systems in Western Europe. Heathlands are open landscapes typically dominated by heather

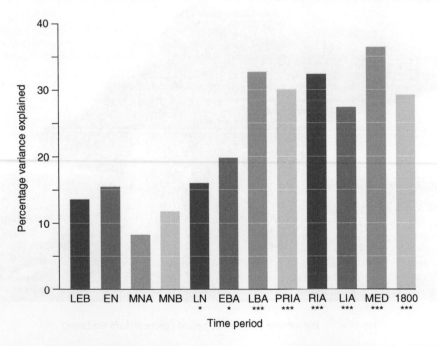

Figure 6.8 Variance in fossil pollen data from 10 Danish sites accounted for by woodland cover at AD 1800. LEB, Late Ertebølle; EN, Early Neolithic; MNA, Middle Neolithic A; MNB, Middle Neolithic B; LN, Late Neolithic; EBA, Early Bronze Age; LBA, Late Bronze Age; PRIA, Pre-Roman Iron Age; RIA, Roman Iron Age; LIA, Late Iron Age; MED, Mediaeval; *P < 0.05; ***P < 0.001 (Odgaard and Rasmussen, 2000)

Table 6.1   Danish archaeological periods (Odgaard and Rasmussen, 2000)

Period	Abbrevation	Start	End
Late Ertebølle	LEB	4500 BC	3900 BC
Early Neolithic	EN	3900 BC	3300 BC
Middle Neolithic period A	MNA	3300 BC	2800 BC
Middle Neolithic period B	MNB	2800 BC	2400 BC
Late Neolithic	LN	2400 BC	1700 BC
Early Bronze Age	EBA	1700 BC	1000 BC
Late Bronze Age	LBA	1000 BC	500 BC
Pre-Roman Iron Age	PRIA	500 BC	0
Roman Iron Age	RIA	0	AD 400
Late Iron Age	LIA	AD 400	AD 1050
Mediaeval Time	MED	AD 1050	AD 1536

(*Calluna vulgaris*), other members of the heather family (Ericaceae) together with gorse (*Ulex*) and grasses. Heathlands are found throughout Western Europe on sandy, acidic soils of low fertility and are especially common along Atlantic coastlines. They are closely linked with human activities and fire and there has been some debate about the extent of human influence on their origin and their Holocene development. Bent Odgaard's doctoral thesis was a study of the Holocene vegetational history of northwest Jutland, Denmark and his careful work still provides the best background to understanding the origin and current status of heathland (Odgaard, 1994). He convincingly demonstrated that western Jutland was largely forest covered from 10 000 to 6000 years ago but that within the forests there were open patches of heathland maintained by fire and windthrow in a 'moving mosaic' (Odgaard, 1994). He found close correlations between charcoal fragments and *Calluna* pollen in his sites and speculated that while the early Holocene fire regime could have been influenced by people, natural wildfires were probably dominant. Heathland expanded at the expense of forest, in step with agricultural development. The driving force was anthropogenic burning to increase heathland area for winter grazing of domestic stock. Danish heathlands reached their maximum extent during the AD 1600–1700s and their reforestation or conversion into agricultural land began after AD 1800. Similar sequences of events are likely to have occurred in western Norway, Sweden, northwest Germany and the British Isles. A key insight from Odgaard's thesis is the recognition of natural heathland nuclei in western Denmark, which expanded under human influence to cover large areas and then contracted again as land use practices changed.

There might well have been a limited expansion of heathlands in Western Europe even without human interference. There are popular theories of how changes in soil chemistry driven by leaching could sufficiently reduce soil fertility on sandy, acidic soils to limit tree growth. There is good evidence for the development of a type of heathland at Derrynadivva, western Ireland towards the end of a previous interglacial period, although separation of soil effects from the climatic limitation of tree growth has often been based on guesswork, at least until dynamic vegetation models could offer a more reliable approach to hypothesis testing (Coxon *et al.*, 1994). Odgaard's heathland study thus makes a strong case for the 'natural' status of heathland as a rare vegetation type whose distribution and composition have been heavily influenced by human activity during the Holocene. Palaeoecological research shows that almost the entire present land surface of Denmark is an outcome of millennia of human impact of variable intensity. By reconstructing the time course of human activities and their impact, it becomes possible to identify the surviving natural features and vegetation types that form the focus of attention for nature conservation activities (Chapter 9).

6.3   Islands: sensitive indicators of human impact

The Danish case study demonstrates that the development of agriculture exerted a major impact on terrestrial ecosystems and has left a larger legacy than the rather minor disturbances associated with the earlier initial human colonisation of the continents. However, human settlement on remote islands occurred much later, often long after the regional development of agriculture, and island systems tend to be particularly sensitive to human impact. David Steadman has dedicated his career to the study of dead birds; as of 2005, he had identified over 26 000 individuals from bones preserved on over 60 islands spread across Oceania in the tropical Pacific Ocean (Steadman and Martin, 2003). He has identified over 1000 specimens from just five islands and believes that in these places his record covers at least 80% of the total diversity of bird faunas, giving a reliable picture of what happened during the period of first human colonisation. The Tongan island of Eua, which is located 4200 km east of Cairns, Australia, was colonised about 3000 years ago. Today it supports 13 indigenous bird species, but Steadman found bones in cave deposits from 27 species of non-marine birds that had died natural deaths and were preserved before the arrival of humans (Steadman, 1993). Among these were several extinct species new to science, including a heron, a megapode, two types of rail, a dove, a parrot and a white-eye. Younger post-settlement deposits contained signs of 'civilisation', including charcoal, tools and bones of chickens, pigs, dogs, rats and people, but 14 of the native bird species had been wiped out. Once people occupy an island, human predation, habitat loss and introduced predators, competitors and pathogens exert enormous pressure on the unprepared native species (Steadman, 1995). One might think that the birds would fly to freedom, but many island species are reluctant fliers or even flightless and the islands that make up the Kingdom of Tonga lie far from any large land mass. Mass extinction was the inevitable outcome of human settlement, even 3000 years ago. Steadman concludes that 'the arrival of humans has influenced the Tongan avifauna more than any climatic, tectonic, or biological event of the past 400 000 years' (Steadman, 1993), showing that on island systems human impact can be dramatic and is easily detectable in the palaeoecological record.

Steadman knows the Pacific well and has also worked 9300 km east of Eua on the iconic Galápagos Islands, which first experienced documented human impact less than 500 years ago in AD 1535 – much later than most other islands in the equatorial Pacific (Steadman et al., 1991). He outdid Darwin by collecting a remarkable 500 000 bones of reptiles, birds and mammals during 37 weeks of productive fossil hunting. Most of the bones came from lava deposits on these tectonically active islands, having either been dropped by owls that roost or nest in the cave-like lava tubes or just fallen down holes into the tubes. Steadman collected his bones from five different Galápagos Islands, and when he dated them he found that 90% of his collection predated human arrival. His group found that over 20 populations of vertebrate species had disappeared from one island or another, and they described three closely related small rodents new to science that appeared to have gone extinct (Steadman et al., 1991). Remarkably, almost all of these losses occurred after human arrival. There may even have been no loss of diversity at all during a 4000–8000 year period prior to settlement, as some of the rodents presumed to be extinct may have subsequently been rediscovered (Dowler et al., 2000). In any case, the extinction rate prior to human settlement appears to have been negligible and 'when undisturbed by humans, the natural processes of dispersal, colonisation, and evolution may result in a very low rate of extinction for vertebrates on tropical oceanic islands' (Steadman, 1995). Steadman's remarkable studies question some of the evidence that has been used to test the theory of island biogeography developed by MacArthur and Wilson (1967): extinction rates and species area curves may well be affected by island area and distance from the mainland, but history – in this case the effects of human settlement – can be the dominant influence on species richness of birds, reptiles and mammals.

The 'rediscovery' of species formerly presumed extinct provides some welcome relief from the widespread gloom associated with the current sixth mass extinction event on this planet, in which high extinction rates, which can be directly linked to human activities, exceed geological 'baselines' (Barnosky *et al.*, 2011).

Species introduced by people during colonisation, either deliberately or by accident, can also pose problems for the indigenous biota and for conservation biologists (Chapter 9). The Galápagos Islands have had their share of introductions, but pollen analysis has shown the risk of jumping to hasty conclusions. Van Leeuwen *et al.* (2008) found continuous pollen records up to 8000 years old from six presumed non-native or doubtfully native species from Santa Cruz Island in the Galápagos. *Hibiscus diversifolius* had been labelled as a potentially threatening invasive species capable of transforming habitats, but the pollen data suggested that it had been more abundant in the past than it is today (Figure 6.9). The pollen record for *Ageratum conyzoides* is supported by seed fragments, providing clear evidence of local presence. So, while human settlement on these tropical islands can be devastating for diversity, not all of the species that increase in abundance are aliens.

Dates of island settlement are usually based on archaeological evidence, but given the sensitivity of many island ecosystems to human impact, one can argue that the continuous records contained in the sediments might prove more effective in demonstrating the effects of colonisation. The eastern Pacific island of Yap lies 1225 km north of New Guinea and the earliest known archaeological evidence for human settlement dates from about 2000 years ago (Dodson and Intoh, 1999). Pollen and charcoal evidence indicate burning and replacement of forest by a savannah with a grassland and fern understorey beneath a scattered tree canopy. This change in vegetation, which surely reflects settlement, occurred 3300 years ago, pushing back the estimated age of first human occupation by over 1000 years (Dodson and Intoh, 1999). Furthermore, the palaeoecological data argue for an anthropogenic origin of savannah vegetation in this setting. Palaeoecological evidence is more readily accepted as evidence for colonisation in the island than the continental context in the Pacific region (Head, 1999), but the reverse has been the case for the Faroe Islands in the North Atlantic.

There has been some controversy over the settlement dates for Iceland and the Faroe Islands as different estimates have been derived from the three available sources: historical documents and archaeological and palaeoecological data. As in the Pacific example, the palaeoecological data provide the earliest evidence for human impact, dating it to between AD 500 and 800 (Hannon *et al.*, 2001). The introduction of fire to this cold, wet and windy ecosystem is a consequence of human activity and leaves a clear charcoal record in the sediments (Figure 6.10). Pieces of wood preserved in peat profiles throughout the Faroe Islands show that trees and shrubs (*Betula* (birch), *Salix* (willow) and *Juniperus* (juniper)) grew there in sparse groves prior to settlement (Hannon *et al.*, 2005), and pollen evidence indicates that *Corylus* (hazel) can be added to this list (Bradshaw *et al.*, 2010c). However, initial human impact removed all woody vegetation from the Faroe Islands until replanting began in recent times.

Gina Hannon has studied the seeds, fruits and pollen preserved from an early farm on the Faroe Islands and shown how the flora changed at settlement (Table 6.2, Figures 6.10 and 6.11). An original damp fen increased its diversity with the addition of several wet grassland and ruderal species, appearing together with cereal pollen and charcoal (Hannon and Bradshaw, 2000). Apart from the cereals, these species almost certainly grew elsewhere on the islands before settlement, but they show substantial increases in population size with the disturbance from cultivation and grazing animals. Domestic animals were introduced to the Faroe Islands by people and contributed both to the loss of woody plants and to the spread of species-rich open communities. A similar situation has been proposed for Iceland, where sheep continue to have a major regional influence on the vegetation. Human impact on these islands in the North Atlantic is easy to detect in sites close to houses and farms (Lawson *et al.*, 2008), but it did not drive the major extinction events recorded from many

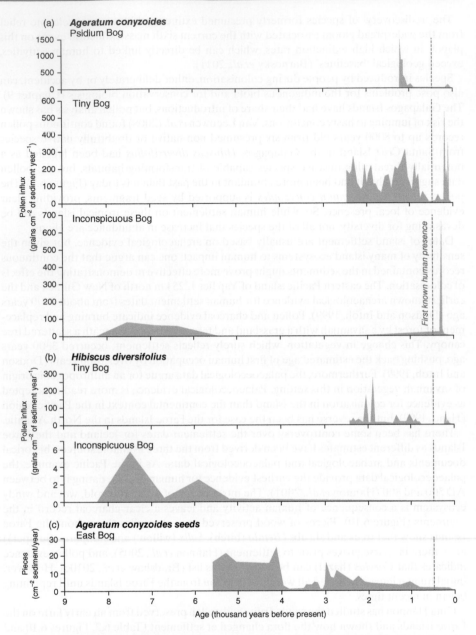

Figure 6.9 Pollen from (a) *Ageratum conyzoides* and (b) *Hibiscus diversifolius* and seed fragments from (c) *A. conyzoides* recorded from sites on the Galápagos Islands, providing evidence for the native status of plant species (after van Leeuwen *et al.*, 2008)

Pacific islands. Human impact has reduced forest cover but it appears never to have been more than 25% in any case and it probably supported a rare 'natural' heathland habitat during the Holocene. Excessive sheep grazing has led to large-scale soil erosion on Iceland and the Faroe Islands, but replanting programmes are helping to begin a restoration process.

These stories illustrate the greater sensitivity of small islands to human impact as compared to the mainland. Just as Darwin observed on the Galápagos Islands, the isolation

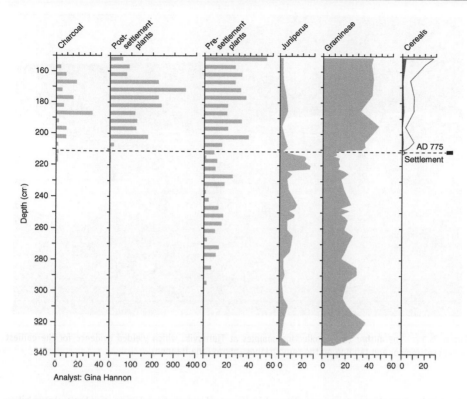

Analyst: Gina Hannon

Figure 6.10 Pollen percentages (continuous curves), plant macrofossil and charcoal concentrations per 30 ml sediment (bars) from Tjørnuvik, Faroe Islands, illustrating the impact of first human settlement. The horizontal line indicates the first cereal pollen. The hollowed graph beside the cereal pollen is a factor 10 scale exaggeration. Source: Hannon *et al.*, 2009. Reproduced by permission of University of California Press

Table 6.2 Pre- and post-settlement plants recorded as macrofossils on a Faroese farm (Hannon and Bradshaw, 2000)

Pre-settlement plants	Post-settlement plants
Equisetum palustre (marsh horsetail)	*Caltha palustris* (marsh marigold)
Hypericum pulchrum (slender St. John's wort)	*Cardamine flexuosa* (wavy bittercress)
Pinguicula vulgaris (common butterwort)	*Chenopodium album* (white goosefoot)
Ranunculus flammula (spearwort)	*Filipendula ulmaria* (meadowsweet)
Silene dioica (red campion)	*Galeopsis speciosa* (large flowered hemp nettle)
Viola riviniana (dog violet)	*Linum catharticum* (purging flax)
	Lychnis flos-cuculi (ragged robin)
	Montia fontana (blinks)
	Plantago sp. (plantain)
	Ranunculus repens (creeping buttercup)
	Sagina procumbens (pearlwort)
	Stellaria media (chickweed)
	Stellaria alsine (bog stitchwort)
	Rumex cf. *longifolius* (northern dock)

Figure 6.11 The author (R.B.) collecting samples at Tjørnuvik, which yielded evidence for the earliest human settlement and farming on the Faroe Islands (photo Gina Hannon)

of islands contributes to unusual combinations of species and a particular vulnerability to new introductions. Goats and sheep, introduced by people, have been particularly destructive when their populations are unmanaged. They can alter vegetation composition quite rapidly, as sheep managed to do on the Faroe Islands, which today support about twice as many sheep as people. Introduced predators and rats have been disastrous for many island-based ground-nesting birds. Islands are useful indicators of the spread of human impact around the world and provide instructive examples of the consequences of over-exploitation of finite resources, indicating where many heavily exploited mainland systems may be heading.

Summary: Human impact case studies

- Danish researchers have pioneered studies of human impact on landscapes during the Holocene. A shifting slash-and-burn agricultural system developed during the Late Bronze Age into more settled systems incorporating the cultivation of barley, the use of hay meadows and the grazing of animals. The first Danish cereal pollen grain is recorded from 5460 years ago, at the start of the period of megalithic grave construction in Denmark. It has been estimated that within a 300 year period as many as 25 000 dolmens and 5000 passage graves were built in Denmark alone as part of an enormous construction boom throughout Western Europe.
- Islands prove to be particularly sensitive to the effects of initial human settlement. Mass extinction was the inevitable outcome of human settlement, even 3000 years ago on remote but species-rich Pacific islands, while introduced alien species out-competed local natives on species-poor northern islands such as Iceland and the Faroe Islands.

6.4   Human influence on Mediterranean, temperate and boreal forests

Prior to the origin of agriculture, most of mainland Europe was covered by forest. The long-term history of human impact has to a large extent been mirrored by the extent of forest cover. How much of Europe would have been forest covered by the mid-Holocene if people had neither used fire nor developed agriculture? Pigott and Walters (1954) posed a similar question almost 60 years ago, and their answer has stood the test of time well. They listed seven habitats in the British Isles, which today are characterised by a rich herbaceous flora and are considered never likely to have carried closed woodland over their full extent. These were: (1) mountains above the tree limit (which was higher than it is today); (2) inland cliffs and screes; (3) sea cliffs; (4) river gorges, eroded river banks, river shingle and alluvium; (5) sand dunes and dune 'slacks'; (6) shallow soils over chalk and limestone, especially on steep slopes; and (7) certain marsh and fen communities and lake shores. Palaeoecological data have generally lacked the spatial resolution to test the openness of these habitats through the Holocene, except for the chalk and limestone habitats. The raised limestone plateau of the Burren and the currently open landscapes of the Aran Isles, western Ireland (Figure 6.12) appear to have been largely forest covered during the mid-Holocene, although the forest seems to have been rather open, with evidence of grassy glades (Molloy and O'Connell, 2004). Similarly significant quantities of grass pollen have also been found prior to extensive human impact from calcareous soils on sloping ground in Yorkshire (Bush and Flenley, 1987) and Dorset (Waton, 1983). However, the present, largely treeless state of these habitats in the British Isles is clearly a result of intensive human activity during the last 6000 years (Molloy and O'Connell, 2004). Estimating the quantity and composition of forest cover prior to significant human impact for large regions is still pretty much informed guesswork, but it nevertheless provides

Figure 6.12   Intricate wall system on a treeless landscape, Inis Óirr, Aran Islands, Ireland (photo Richard Bradshaw)

a useful base from which to assess the current status of terrestrial ecosystems. Peterken (2000) estimated that England was 90% forest covered during the early Holocene, but by AD 1086, when the Domesday Book was completed, forest cover had fallen to just 15% of land area (Rackham, 2003).

The history of the last 5000 years of European forest cover is primarily one of deforestation, which has been largely driven by the course of agricultural development. Reliable quantitative data from national forest inventories are only available for the last few decades, but generalised deforestation trajectories can be estimated for several countries by combining palaeoecological and historical sources. Some general features emerge (Figure 6.13): the deforestation trajectories fall to minimum values of national forest cover and then reforestation takes place, usually driven by changes in agricultural practice, forest legislation and the establishment of organised national forest agencies. The timing and minimum value of forest cover reached varies somewhat between regions. For Sweden as a whole, the minimum value was probably reached between AD 1850 and 1900, and whereas the minimum percentage cover was considerably greater than in other European countries, most of this lay in the relatively inaccessible and sparsely populated western and northern regions. The trajectory for Denmark is rather typical for northwest Europe, with a minimum forest cover of c. 3% of land area reached around AD 1800. A detailed survey was made of the Danish landscape between AD 1780 and 1820, which generated a map of forest area (Figure 6.14). Forests that existed in AD 1820 and are still present today are likely to have a history of long continuity through the Holocene, as they survived the bottleneck of minimum forest cover and consequently have a form of genetic continuity with the first postglacial forests. As discussed previously, the deforestation trajectory in Denmark typically began around 5500 years ago during the Neolithic Period and coincided with the first signs of arable cultivation recognised from cereal pollen. Its timing fits the

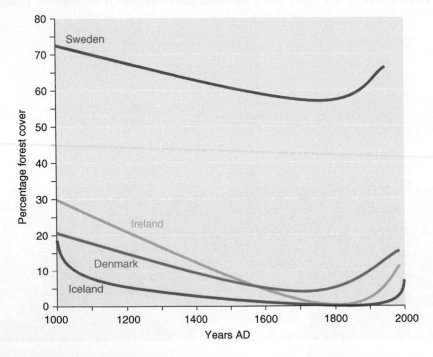

Figure 6.13 Estimated forest-cover trajectories since AD 1000. Source: Bradshaw, 2004. Reproduced by permission of Elsevier

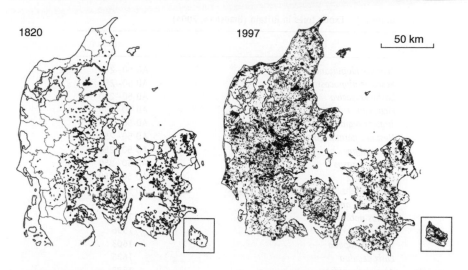

Figure 6.14 Danish forest cover in AD 1820 and 1997. Source: Møller, 1997. Reproduced by permission of Peter Friis Møller

pattern of spread of settled agriculture across Europe. Arable crops were cultivated on land cleared of forest, often by burning. However, grazing of livestock was probably a greater driving force for reduction of forest area through restriction of regeneration and winter browsing of young growth (Bradshaw and Mitchell, 1999). The introduction of sheep to Iceland, for example, was the driving force for rapid deforestation soon after human settlement (Figure 6.13).

The forest area in most European countries has been increasing during recent decades, both as a result of commercial planting and because of rural depopulation and the widespread collapse of traditional forms of land use, particularly extensive grazing (Agnoletti, 2006). Planting of exotic tree species on a small scale has been documented since Roman times and represents an escalation of human impact, adding to the deliberate transfers of fauna and flora associated with the spread of agriculture and the accidental transfers of weeds and pests. Genetic analyses of *Pinus pinea* (stone pine) populations in the Mediterranean region have revealed less interpopulation genetic diversity than in other *Pinus* species, probably as a consequence of widespread anthropogenic transfer of a few desirable phenotypes of this domesticated species with its tasty nuts (Vendramin *et al.*, 2008). The record of exotic tree introductions to Britain is well researched. Several species were introduced by the Romans, while a period of frequent introductions began about 500 years ago, continuing up to the present time, with frequent movements of species for horticultural and agricultural reasons (Table 6.3; Applebaum, 1972; White, 1997). Many exotic species neither regenerate naturally nor exchange genetic material with native species, but a small proportion of introduced species are invasive and pose conservation problems (Chapter 9). There is uncertainty over the native status of some woody species, such as *Arbutus unedo* (strawberry tree) in western Ireland, whose origin dates back at least 2000 years (Mitchell, 1993). *Acer pseudoplatanus* (sycamore) is a presumed early introduction to Britain, but given its highly successful invasive powers throughout much of northwestern Europe it may well have extended its range even without anthropogenic intervention (Bradshaw, 1995).

The increasing forest cover in many European countries can make it hard to detect the consequences of former human impacts. The socio-economic legacy of the Roman Empire is widely acknowledged throughout Europe, but its ecosystem impact may well have been

Table 6.3   Exotic trees in Britain (Bradshaw, 2004)

Exotic trees in Britain	Date of introduction
Acer pseudoplatanus	AD 50–250
Aesculus hippocastanum	AD 50–250
Castanea sativa	AD 50–250
Ficus carica	AD 50–250
Juglans regia	AD 50–250
Mespilus germanicum	AD 50–250
Morus nigra	AD 50–250
Prunus cerasus	AD 50–250
Quercus ilex	AD 50–250
Vitis vinifera	AD 50–250
Laburnum anagyroides	1560
Laurus nobilis	1562
Morus alba	1596
Tilia x europaea	1603
Larix decidua	1625
Acer platanoides	1682
Quercus suber	1699
Pinus nigra	1759
Alnus incana	1797

underestimated. One area of northeastern France was farmed under Roman rule for just 200 years between AD 20 and 250 and then abandoned (Dambrine *et al.*, 2007), yet the legacy of this brief period of land use can still be observed in the ground flora, with less woodland species and more plants characteristic of disturbed ground, despite almost 2000 years of subsequent continuous forest cover. Dambrine *et al.* (2007) suspect that the earlier period of cultivation altered the local soil structure and increased the soil phosphorus content in an effectively irreversible manner. It seems likely that similar alterations of soil properties are widespread around the world, but this has barely been investigated.

Not all of the recent reforestation of Europe is a result of new plantations, although these dominate in densely populated countries and regions; there are also large areas of secondary woody successions developing where traditional extensive grazing regimes have ceased to operate. Examples are large sections of the Spanish dehesas, mountain regions in northern Italy and the ecotone between mixed deciduous and boreal forest in Fennoscandia. Intensification of agriculture in all of these regions has reduced the pressure of domestic grazing in the forests, resulting in increasing forest cover through natural succession. The increase of forest cover within Europe, which gathered pace at the beginning of the 1900s, has been driven more by anthropogenic than natural processes, and both composition and structure differ from earlier forests on the same sites. The forest composition of southern Sweden changed from mixed deciduous to primarily coniferous during the last 1000 years, primarily driven by human activities, and the rate of change in forest composition was most rapid during the last 150 years (Figure 6.15; Lindbladh *et al.*, 2000). In northern Sweden, the most striking recent forest changes have been structural, with the loss of large individuals due to commercial forestry operations (Figure 6.16; Linder and Östlund, 1998).

The decrease and subsequent increase in forest cover that has taken place in several regions of Europe during recent centuries has an interesting parallel in forest dynamics of the northeastern USA. A 'complete transformation occurred in the regional and local landscape [of central New England] from AD 1770 to the present as the countryside was deforested, farmed and subsequently reforested' (Foster, 1992). European colonisation of New England led to forest clearance and agriculture, which was subsequently abandoned

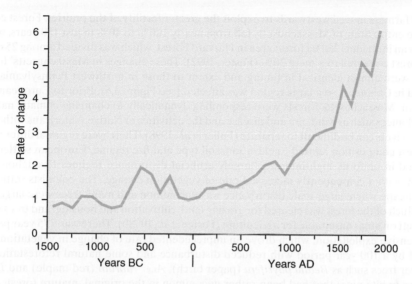

Figure 6.15 Mean values for rate-of-change analysis of 13 stand-scale pollen sites in southern Sweden. Rate of change is calculated from chord distance dissimilarities between samples within each site per 100 years (Lindbladh *et al.*, 2000). Reproduced by permission of Elsevier

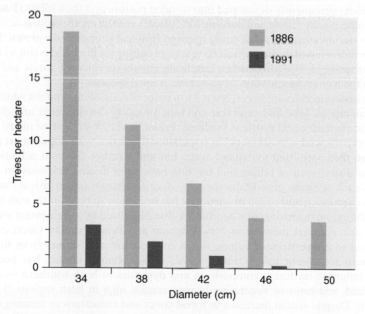

Figure 6.16 Diameter distribution of trees from Orsa *besparingsskog*, Sweden in AD 1886 and 1991. Source: Linder and Östlund, 1998. Reproduced by permission of Elsevier

as the farmers moved westwards to exploit the greater fertility of the prairies. Forest cover for the entire State of Massachusetts fell from nearly 100% to 40% in just 150 years, while within an individual 380 ha forest area in Harvard Forest, which was divided among 25 owners, forest cover fell to a mere 20% (Foster, 1992). These changes in Massachusetts' forest cover were almost identical in timing and extent to those in northwest Pennsylvania discussed in Chapter 3, so a large region was affected (see Figure 3.6). Prior to European settlement, Massachusetts forests were responding dynamically to changing climate, natural disturbances such as wind, fire and disease and the activities of Native Americans, although these drivers can be difficult to separate (Fuller *et al.*, 1998). There were regional differences in forest composition related to elevation, soil type and fire regime. European settlement resulted in uniform, high-intensity, largely artificial disturbance regimes throughout the region, with a consequently increased rate of vegetation change. The colonists settled in a landscape where large-scale disturbance was uncommon and many trees were large and old. Much of the forest was cleared for grazing land, cultivation and housing, and trees were only left on areas unsuitable for agriculture (Foster *et al.*, 1998b). There was a 75 year period of intensive exploitation, which drove an unprecedented rate of change in vegetation, followed by a 100 year period with reduced disturbance and some natural reforestation by pioneer trees such as *Betula payrifera* (paper birch), *Acer rubrum* (red maple) and *Pinus strobus* (white pine) that had been rather uncommon in the original, mature forest. During the period of maximum exploitation a homogenisation of vegetation took place, with a loss of the previous broad-scale vegetation patterns. Even as the forests grew back, disturbance was still frequent and the rate of vegetation change remained rapid. Agricultural land was often only gradually abandoned, with cultivated fields becoming converted to pasture, which subsequently developed into wooded pasture and then forest (Foster *et al.*, 1998). Periodic, patchy felling and burning took place, even during the reforestation, which selected for deciduous species that could resprout from cut stumps. Fully grown 'high' forest has only redeveloped during the last 60 years and cutting for firewood is still widespread. A major hurricane in 1938 also made a significant impact on Harvard Forest and has contributed to the high representation of early successional species.

Land use history in Harvard Forest has led to a range of ecosystem changes, which include drainage operations, lake and reservoir creation, property development and changes in atmospheric chemistry and nutrient loadings (Foster *et al.*, 1998). Native plants and animals have experienced a 'bottleneck' in population dynamics, with populations generally reduced and then permitted to expand again, but under rather different conditions from before. The disturbance of felling and burning have been disadvantageous for previous dominants such as *Fagus grandifolia* (beech), *Acer saccharum* (sugar maple) and *Tsuga canadensis* (hemlock), and beech in particular has been slow to recolonise, with its limited dispersal ability and the arrival of a beech-bark disease caused by a scale insect and fungus.

So despite a regional increase in New England and Pennsylvanian forest cover that has returned to pre-settlement values, forest composition now differs from that of the pre-European settlement forest (Fuller *et al.*, 1998). Human impact has become the dominant influence on forest composition and dynamics in both southern Sweden and New England, and rates of vegetation change remain high in both regions (Lindbladh *et al.*, 2000). Despite recent increases in forest cover and reductions in farming activities, there is no evidence of a return to forest types that predated the periods of agricultural intervention in either region. Climatic and soil conditions have changed, seed sources are different from before and there is still human influence on these forests. Foster *et al.* (1998) wrote that in New England 'broad-scale similarities in land use have overwhelmed regional environmental control' and these words apply equally well to the situation in the county of Småland, southeast Sweden. These studies show how historical land use has altered vegetation–environment relationships across large geographic regions. Historical data deserve to be considered in all contemporary studies of global change due to the long-lived legacies of past land use. This is likely to apply to many regions of the world, even though current knowledge of the past is patchy.

6.5 The tropics

Exploration of these legacies in tropical regions has not developed so rapidly and there is far less palaeoecological research to examine. However, there is growing evidence for early human impact on several tropical ecosystems, even if the areal extent of this impact is hard to assess. Upland valleys in Papua New Guinea support grasslands that are now maintained by frequent burning. Palaeoecological data from Kuk Swamp (1560 m above sea level) show that grasslands from cool, pre-Holocene times did not become reforested at the opening of the Holocene, in contrast to other upland valleys in the region (Denham *et al.*, 2003). Instead, periodic burning maintained grasses, ferns and subcanopy taxa that included the group to which bananas belong, Musaceae sect. *Eumusa*. The palaeoecological and archaeological evidence is more convincing for an intensification of human impact and banana cultivation beginning about 7000 years ago, but the authors were sufficiently confident with their older data to claim Kuk Swamp as a 'primary centre of agricultural development and plant domestication' (Denham *et al.*, 2003). Banana cultivation has also been detected in Cameroon dating from 2500 years ago and the plants must have originally been moved there by people from their native range in South East Asia (Mindzie *et al.*, 2001). Investigations at Nkang, Cameroon identified the earliest known bones of domestic sheep and goats from Central Africa, together with banana, from a forest environment with clearings. The landscape still retains a similar structure today (Mbida *et al.*, 2000). The dark earth soils (terra preta) of anthropogenic origin that have been estimated to cover 50 000 ha of central Amazonia have comparable ages, beginning to appear around 2500 years ago (Willis *et al.*, 2004). These soils are enriched with organic and inorganic waste of a pre-Colombian civilisation and surely indicate large populations that converted significant forest areas close to rivers into managed gardens (Glaser and Birk, 2012). Given the increase in charcoal from a marine core near Papua New Guinea 50 000 years ago (see Figure 6.2; Thevenon *et al.*, 2004), it would seem highly probable that future research will provide even more evidence for early human impact on tropical ecosystems. As tropical systems are so productive and rich in species, earlier impacts have left legacies that are harder to detect than in the relatively species-poor temperate and boreal biomes, where human impact has led to proportionately larger species turnovers and even changes in vegetation type. Current research into the long-term history of African savannahs may well provide evidence for one of the earliest large-scale changes in vegetation driven by manipulation of fire regimes by people (Verschuren, personal communication).

Deforestation trajectories in tropical regions appear to 'lag' behind those from temperate and boreal countries, as most tropical countries are still losing forest area and have not yet 'turned the corner' to the reforestation seen in the temperate and boreal zones of Europe (Figure 6.13). Why is this, when people have lived longer in the tropics? The answer is probably a combination of ecological and socio-economic factors. Trees grow faster in most tropical regions, so more timber has to be removed per year in order to remove forest cover. However, the most rapid reductions in forest cover in temperate regions have been caused in the past by the use of forests for grazing of domestic animals such as goats, sheep and cattle. The development of agriculture in tropical regions has taken a different course and only during the last century has deforestation gathered pace, through a combination of forestry operations, agriculture and other social factors. One might speculate that the tropical deforestation trajectories will follow similar courses to those from temperate and boreal regions and that forest cover will begin to increase once a value has been reached that threatens the viability of critical socio-economic and ecological systems. Could we learn from the past experience of the deforestation of Denmark? Unfortunately, it seems that every nation must discover for itself the loss of ecosystem services linked to the removal of the majority of its forest cover. Few environmental historians have any influence with national decision makers.

Summary: Global deforestation history

- The history of the last 5000 years of European forest cover is primarily one of deforestation, which has been largely driven by the course of agricultural development. However, European forest area has been increasing in recent decades, both as a result of commercial planting and because of rural depopulation and the widespread collapse of traditional forms of land use.
- European colonisation of northwestern USA led to forest clearance for agriculture, which was subsequently abandoned as farmers moved westwards to exploit the greater fertility of the prairies. These ecosystem dynamics took place in just 250 years.
- Despite scattered evidence for early human impact on several tropical ecosystems, deforestation trajectories in tropical regions tend to lag behind those in temperate and boreal countries, as most tropical countries are still losing forest area and have not yet turned the corner to the reforestation seen in the temperate and boreal zones of Europe.

6.6   Spatial upscaling of the timing and ecosystem consequences of human impact

Tracing national deforestation trajectories through time is one way of upscaling the time course of human impact on ecosystems, because in many countries deforestation has acted as a crude indicator of the scale of agricultural impact. Compiling and upscaling individual case studies is another approach to the important assessment of the global impact of human activities and supports modelling activities that show where we are taking the planet. Both archaeology and palaeoecology have traditionally been 'site-based' disciplines and past upscaling activities have been rather speculative. However, with the increase in quantity and quality of data in recent decades, we can now upscale with more confidence.

Björn Berglund led the vast interdisciplinary Ystad Project in southern Sweden from 1982 to 1990, which studied the landscape transformation of a 30 000 ha region from a wooded landscape into a thoroughly exploited agricultural zone (Berglund, 1991). He then placed the Ystad region in the broader context of northwest Europe by comparing pollen diagrams in a transect from Ireland to Estonia (Berglund, 2000). He identified two main phases of agricultural expansion throughout this region that led to pulses of deforestation and soil erosion centred around 1200 and 2900 years ago (Figure 6.17). The archaeologists and historians in the Ystad Project had argued for social changes, such as increases in population, improved social organisation and technical and economic development, as the main drivers of the observed phases of agricultural expansion. However, when Ystad was compared with a larger European area, Berglund (2000) was struck by the synchronicity of continental-scale changes in agricultural development, which could be linked with both improvements and deteriorations in the suitability of regional climate for agricultural production. Upscaling can certainly alter initial hypotheses about which factors might exert the dominating influence on landscape change. Climatic factors have often been judged to have greater explanatory power for vegetation dynamics at continental rather than local scales,

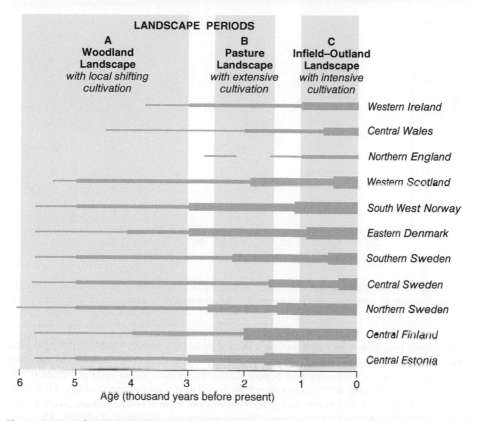

Figure 6.17 Deforestation patterns and development of the cultural landscape in northwest Europe (Berglund, 2000). Source: Berglund, 2000. Courtesy of PAGES, Bern; http://www.pages-igbp.org/

but the use of data–model comparison provides logical methods of evaluating competing hypotheses (Miller *et al.*, 2008).

The Australian drylands cover about 5 million km^2 and have a long and intriguing relationship with people. Smith *et al.* (2008) compiled 908 radiocarbon dates from 286 archaeological sites within this region and used these data as a proxy for human population size (Figure 6.18). This approach makes several assumptions and incorporates a potential bias towards better preserved, younger archaeological sites, but nevertheless is a novel approach to the estimation of past population dynamics, which had previously rested on insecure foundations for much of the Holocene (Boyle *et al.*, 2011). The Australian data indicated a widespread human movement into the arid zone 35 000–40 000 years ago, with major population increases at 19 000, 8000 and 1500 years ago (Figure 6.18). Smith *et al.* (2008) recognised six population events, occurring every 2000–3000 years, in which periods of relatively rapid population increase were followed by declines. Some of these population events could be loosely linked to known changes in past rainfall and temperature variability, but the relationships are complex and are likely to involve combinations of social and environmental factors.

A similar approach was adopted by Tallavaara and Seppä (2011) for a far smaller region comprising central and southern Finland and restricted to just the last 10 000 years. Using this more tightly constrained dataset, the Finns could demonstrate more convincing relationships between human population dynamics and environmental change. They found

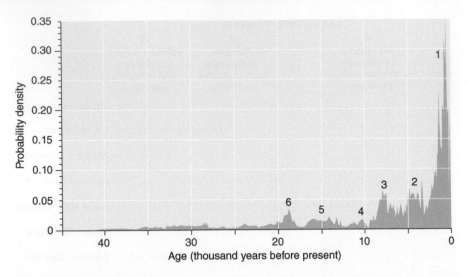

Figure 6.18 Probability plot for all ^{14}C ages from archaeological sites in Australian drylands (n = 908). Six population events are identified. Source: Smith *et al*, 2008. Reproduced by permission of SAGE Publications

a significant population increase during a period of high summer temperatures and high productivity of terrestrial, lacustrine and marine ecosystems that occurred between 7500 and 5700 years ago (Tallavaara and Seppä, 2011). There was an inferred population collapse between 5500 and 4000 years ago, which correlated with the onset of late-Holocene cooling and an associated major forest ecosystem shift from relatively species-rich mixed forest to a spruce-dominated coniferous boreal ecosystem (Seppä *et al*., 2009). This shift probably reinforced the negative effects of decreasing ecosystem productivity on food availability for the hunter-gatherer populations. Tallavaara and Seppä (2011) made the striking observation that following the intensification of agriculture 3500 years ago there was an apparent breakdown in these earlier links between human population size and the climate and environmental proxies. Even though agricultural systems are also sensitive to climatic change, they appear to provide more stable support for human populations than the natural resources that hunter-gatherers in Finland had previously exploited. There could be several reasons why agricultural societies appear to be less vulnerable to environmental change, and this is a promising area of research given concerns about food security during the current period of rapid climatic change.

Modelling past pollen–vegetation relationships using the REVEALS and LOVE models (Sugita, 2007a,2007b) that take into account pollen production and dispersal properties, has strengthened the theoretical foundation for presenting rather indigestible pollen data in an attractive, accessible format for a broad audience. Nielsen *et al*. (2012) have used REVEALS to reconstruct regional trends in forest cover as a series of Holocene maps covering Denmark and northern Germany and have speculated on the likely role of human impact through time (Figure 6.19). The maps share some features with full European maps produced by Kaplan *et al*. (2009), which also were generated by a model, but one based upon estimates of human population density, technology and the suitability of soils for agriculture; Kaplan's maps were therefore more indirect estimates of vegetation cover than Nielsen's pollen-based maps and did not include any natural variation in openness (Nielsen *et al*., 2012).

Nielsen's maps show that tree cover was not complete during the early Holocene and that the densest forest cover was modelled around 6700 years ago, just as the first signs

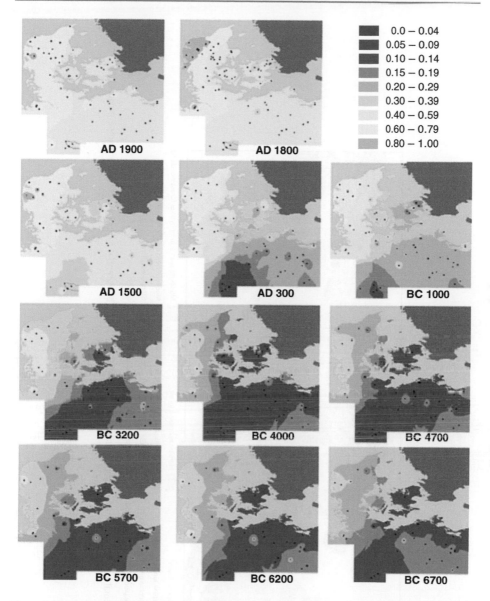

Figure 6.19 Reconstructed proportions of open, non-forest vegetation in Denmark and northern Germany, north-central Europe, at different dates. Source: Nielsen *et al*, 2012. Reproduced by permission of Elsevier

of cereal cultivation were detected. By 3000 years ago, deforestation driven by agricultural land use had created openings in the forest canopy that covered more than 20% of the land surface in the east and over 40% in the west of the study region, maintaining the natural east–west gradient in forest cover. Most of the southeastern part of the region was deforested between 500 and 1700 years ago, which was somewhat later than in Denmark. Openness was not purely dictated by agricultural history, as soil type and distance to the North Sea – a proxy for continentality – explained some of the differences between sites and regions (Nielsen *et al.*, 2012). Maps generated in this way can provide reliable

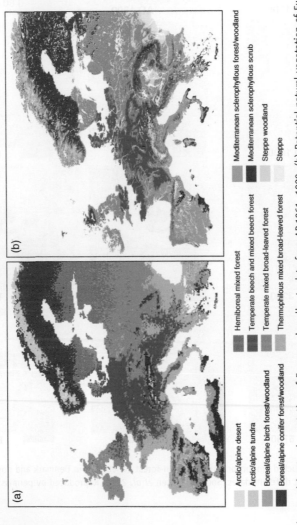

Figure 6.20 (a) Modelled potential natural vegetation of Europe using climate data from AD 1961–1980. (b) Potential natural vegetation of Europe produced by a panel of experts (Bohn *et al.*, 2003)

Arctic/alpine desert
Arctic/alpine tundra
Boreal/alpine birch forest/woodland
Boreal/alpine conifer forest/woodland
Hemiboreal mixed forest
Temperate beech and mixed beech forest
Temperate mixed broad-leaved forest
Thermophilous mixed broad-leaved forest
Mediterranean sclerophyllous forest/woodland
Mediterranean sclerophyllous scrub
Steppe woodland
Steppe

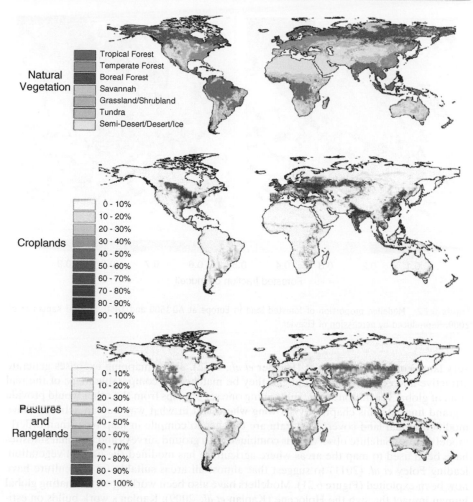

Natural Vegetation
- Tropical Forest
- Temperate Forest
- Boreal Forest
- Savannah
- Grassland/Shrubland
- Tundra
- Semi-Desert/Desert/Ice

Croplands
- 0 - 10%
- 10 - 20%
- 20 - 30%
- 30 - 40%
- 40 - 50%
- 50 - 60%
- 60 - 70%
- 70 - 80%
- 80 - 90%
- 90 - 100%

Pastures and Rangelands
- 0 - 10%
- 10 - 20%
- 20 - 30%
- 30 - 40%
- 40 - 50%
- 50 - 60%
- 60 - 70%
- 70 - 80%
- 80 - 90%
- 90 - 100%

Figure 6.21 Global extent of land cover change caused by humans. Potential natural vegetation (top panel) is compared with arable land (middle) and pastoral land (bottom). Source: Foley, *et al*, 2005. Reproduced by permission of the American Association for the Advancement of Science

summaries of human impact on forest ecosystems through the Holocene in areas where forest cover is the natural state of affairs and there are good sites for pollen analysis. For the many regions of the world where this is not the case, we must still rely on other, more qualitative methods for reconstructing the time course of human impact.

Plant ecologists are fond of the concept of 'potential vegetation', meaning the vegetation that would exist today if humans had never evolved. Climate-driven vegetation models are often used to generate global or continental potential vegetation maps – but rarely at finer spatial scales, as other, harder-to-model factors such as soil type, topography and disturbance might begin to be the dominating influences. The BIOME model and its successors (Prentice *et al.*, 1992) generated one of the most widely used global vegetation maps, and this mapping approach gives rather similar patterns in Europe to those generated by an expert group of plant ecologists based on practical experience (Figure 6.20; Bohn *et al.*, 2003). The ecologists' map treats large wetlands and alpine areas in a more realistic manner, as detailed modelling of these factors, while theoretically feasible, demands

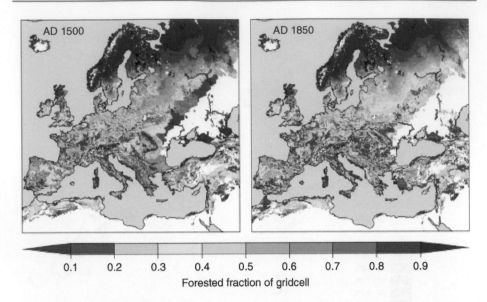

Figure 6.22   Modelled proportion of forested land in Europe at AD 1500 and 1850. Source: Kaplan *et al*, 2009. Reproduced by permission of Elsevier

very large computing resources (Hickler *et al.*, 2012). These mapping exercises generate attractive and colourful maps, but can they be matched by comparable maps of the real state of global vegetation? If so, subtracting one set of maps from the other would provide a grand finale to this chapter by showing where and in what way human activities have modified global land cover. Real data are tougher to compile in map form than orderly model output. Satellite observations combined with ground surveys and national statistics have been used to map the areas where agriculture has modified the natural vegetation, leading Foley *et al.* (2011) to suggest that almost all areas suitable for agriculture have now been exploited (Figure 6.21). Modellers have also been working on estimating global human impact through the Holocene (Kaplan *et al.*, 2009); Kaplan's work builds on estimates of human population size, over which there is considerable uncertainty, but their maps of the world at AD 1500 and 1850 showing the fractions of grid cells ($5 \times 5'$) still under natural vegetation are thought-provoking (Figure 6.22). The major patterns shown in the present-day, data-based maps of Foley *et al.* (2005) (Figure 6.21) are already in place, suggesting that considerable modification of the global land surface had already taken place prior to industrialisation (Kaplan *et al.*, 2009). This conclusion has relevance both for the long-term dynamics of the global carbon cycle (Chapter 7) and for current conservation and restoration efforts (Chapter 9). While there are as yet few appropriate datasets available by which to evaluate these simulations, current research synthesising the effects of global human impact through time is making good progress.

A timeline of global human impact on terrestrial ecosystems summarises its timing and scale (Figure 6.23). This is the type of speculative exercise that is fun to do on the back of an envelope and could only be published in a book, away from the demanding eyes of peer reviewers. Human impact began with the first accepted evidence for human mastery of fire. The greatest increases in impact are linked to the various – apparently independent – origins and rapid expansions of agriculture. There were also setbacks along the way,

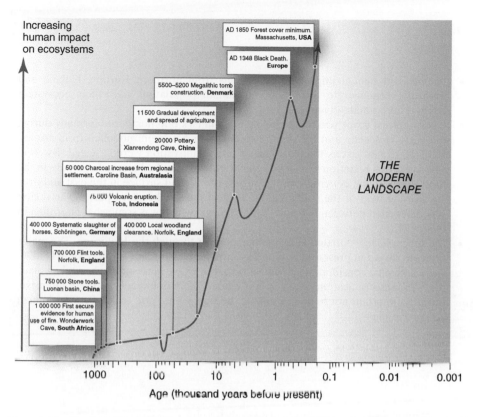

Figure 6.23 Timeline of human impact on terrestrial ecosystems based on events discussed in this chapter

brought on by large-scale natural disasters, sequences of crop failures, droughts, wars and diseases. The diagram includes population reductions and land abandonment following the enormous volcanic eruption at Lake Toba, Indonesia, the collapse of the megalithic building boom of northwest Europe and the worst outbreak of bubonic plague or 'black death'. The examples of human impact shown are a subjective selection from a large body of research, but they do illustrate the major trends and timing of human impact on ecosystems. Details of the last 200 years are not included on the diagram as their scale and rapidity would dwarf the earlier sequence of events. The message here is that pre-industrial modification of the surface of the Earth by people was considerable, and the use of fire in particular caused significant and sometimes irreversible ecosystem change. We shall enjoy revisiting this diagram in 10 years' time when we anticipate that both the timing and extent of human impact on terrestrial ecosystems will have been pushed even further back in time.

The history of global human impact on ecosystems during the last 200 years or so is outlined in Chapter 1 and covered in depth by numerous other books and papers. Our focus here has been on the longer-term development of human impact on ecosystems. Several authors have recognised a new phase of human–ecosystem interactions beginning around 1800 with the onset of the Anthropocene (Steffen *et al.*, 2007). This period is generally well reviewed elsewhere, while our survey of earlier human impact on ecosystems has not been so thoroughly covered.

Summary: Upscaling the consequences of past human impact

- Analyses of human impact on ecosystems from northern Europe, Finland and Australia demonstrate links between human population size, their ecosystem impacts and climate dynamics.
- Natural forest cover was not complete in Denmark and northern Germany during the early Holocene and the densest forest cover has been modelled from around 6700 years ago, just as the first signs of cereal cultivation were detected. By 3000 years ago, deforestation driven by agricultural land use had created openings in the forest canopy that covered more than 20% of the land surface in the east and over 40% in the west of the study region. Most of the southeastern part of the region was deforested between 500 and 1700 years ago, which was somewhat later than in Denmark.
- Modellers have estimated global human impact through the Holocene. The major patterns shown in present-day, data-based maps were already in place by AD 1500, suggesting that considerable modification of the global land surface had taken place prior to industrialisation. The last 200 years of human impact are more intensive, but it is important not to underestimate the significance and scale of earlier human ecosystem impact.

7

Millennial Ecosystem Dynamics and Their Relationship to Ecosystem Services: Past and Future

'The changes that have been made to ecosystems have contributed to substantial net gains in human well-being and economic development, but these gains have been achieved at growing costs in the form of the degradation of many ecosystem services, increased risks of nonlinear changes, and the exacerbation of poverty for some groups of people. These problems, unless addressed, will substantially diminish the benefits that future generations obtain from ecosystems.'

MEA (2005)

7.1   Introduction

Throughout their long history, humans have relied on the environment in which they live for their survival. In this environment it is not only physical or abiotic aspects such as climate that determine survival but also interactions with a range of other species, both plants and animals. Humans tend to classify and give structure to their surroundings, and the concept of 'ecosystems' is one such classification. Ecosystems can be described in many ways and at a multitude of scales, but they always contain a mixture of species that interact with each other and with aspects of the physical environment such as climate and soils. Tansley (1935) defined an ecosystem as 'the fundamental concept appropriate to the biome considered together with all the effective inorganic factors of its environment … In an ecosystem the organisms and the inorganic factors alike are components which are in relatively stable dynamic equilibrium'. Humans do not of course live remote from or somehow 'outside' ecosystems; rather, they form integral parts of many ecosystems and their presence is often significant within (and outside) them. They can live in some sort of dynamic equilibrium within an ecosystem, or they can alter it in positive or negative ways. Humans are also

Ecosystem Dynamics: From the Past to the Future, First Edition.
Richard H.W. Bradshaw and Martin T. Sykes.
© 2014 John Wiley & Sons, Ltd. Published 2014 by John Wiley & Sons, Ltd.
Companion Website: www.wiley.com/go/bradshaw/sykes/ecosystem

critically reliant on ecosystems, the diversity of species within them and the range of goods and services they provide for their survival. In recent times, significant effort has been put into classifying the services provided by ecosystems, particularly through the Millennium Ecosystem Assessment (MEA), which assessed the interactions between ecosystems, society and humans in general (MEA, 2005). It also explored the actions that would improve conservation of biodiversity and the sustainability of ecosystems and, by extension, their services.

The concept of 'ecosystem services' is much older than we might imagine, although it may not have been explicitly defined as such. As with many things, it was the ancient Greeks, and this time Plato (in 360 BC), who suggested that destruction of forests leads to soil erosion and drought and therefore to loss of service. Marsh (1864) was among the first to make a systematic assessment of human influence upon and interactions with the natural world from a conservationist's point of view and he suggested that the earth's natural resources are not unlimited, taking as an example the declining soil fertility in the Mediterranean. More recently, Ehrlich and Ehrlich (1981) became concerned about the consequences of the recent and rapid extinction of species. They described the potential effects on ecosystem services by analogy to removing the rivets from the wing of an aeroplane. Remove or lose one and the plane can still fly safely; removal of a second is probably not a problem either. But as rivet removal continues, at some point the wing is likely to drop off and the plane will be unable to fly. In a similar way, the loss of one species in an ecosystem may not seem to be a problem for humanity but as more species are lost, the ecosystem, its services and possibly the society dependent upon them become more likely to collapse.

Ecosystem services have been defined in a number of ways, depending partly on the author's approach. Daily (1997), for example, coming from an ecological standpoint, defined ecosystem services as 'the conditions and processes through which natural ecosystems and the species that make them up, sustain and fulfil human life'. A more economic viewpoint is expressed by Costanza et al. (1997): 'The services of ecological systems and the natural capital stocks that produce them are critical to the functioning of the Earth's life-support system. They contribute to human welfare, both directly and indirectly, and therefore represent part of the total economic value of the planet.'

Many basic services provided by ecosystems are relatively easy to quantify and indeed value in economic terms, including such vital services as food, wood and fresh water. Other services, possibly equally important, are more difficult to assess and quantify via any monetary valuation, including recreational, cultural, aesthetic and spiritual aspects, many of which can be said to make humans human. There has been a tendency to consider all ecosystem services purely in economic terms, and if this cannot be done to dismiss them as having no 'value' to society. To ignore the noneconomic values of such services can be viewed as at least myopic, if not downright foolish.

Gretchen Daily and the contributors to her seminal edited volume aimed to address the lack of understanding and appreciation by many levels in society of the total dependence of humanity on the services provided by ecosystems (Daily, 1997). Daily is also one of the leaders of the Natural Capital Project (www.naturalcapitalproject.org). Natural Capital aims to apply the economics of capital to ecosystem services and to develop tools that will quantify the value of nature, which are used by stakeholders and decision makers to assess the importance of natural capital and to explore the trade-offs between different possible choices in exploiting ecosystems. The project also assesses how investments in nature can generate social value. Similar concepts were used by Costanza et al. (1997) to estimate the economic value of 17 ecosystem services in 16 biomes. The result was an average of US$33 trillion a year, even though at the time global gross national product (GNP) was estimated at just US$18 trillion annually. This raises many issues, not least of which is the

Figure 7.1   Stockholm's National Urban Park (www.nationalstadsparken.se)

implication that if the actual value of services provided by ecosystems were to be included in commodity prices, those prices would be substantially increased. These numbers are huge and incomprehensible to ordinary human beings, but simpler estimations of the economic values of ecosystem services have been made.

At a completely different scale, Hougner *et al.* (2006) conducted an economic valuation of the seed dispersal service carried out by Eurasian jays (*Garrulus glandarius*) in the Stockholm National Urban Park, which contains one of the largest populations of giant oaks in Europe, along with many other rare species (Figure 7.1). As the world's first National Urban Park, protected by law, this is an important recreational area, and it has many visitors. It thus provides a direct cultural service to humanity. The Eurasian jay collects and stores between 4500 and 11 000 oak acorns per year as winter food. It chooses viable acorns and buries them along the forest edges, away from predation and at the ideal depth for germination. A survey showed there were 42 pairs of jays in the park. If their service were to be replaced in some way by human labour, the cost would be high. If acorns were planted, the replacement cost would be 1.5 million SEK (ca. £150 000 or €170 000), or 35 000 SEK (£3500 or €4000) per pair of jays. If the more reliable technique of planting saplings were used, the replacement cost would be 6.7 million SEK (£670 000 or €770 000), or 160 000 SEK (£16 000 or €18 400) per pair. This is a relatively simple example of a service and its valuation, but it fails to factor in the cost of maintaining the nearby dense coniferous forests where the jays build their nests and raise their young, safe from predators. This raises issues about the actual requirements of maintaining a service, as most do not operate in isolation and are dependent on other aspects such as the presence of surrounding ecosystems.

The ecosystem service concept firmly places people at its centre and is a formal recognition of the dependence of the human race on sustained ecosystem function. Ecosystems are sources of biophysical supplies that have been used in different ways by human populations through time. As these populations have increased in size, the demands made on ecosystems have increased, bringing us to a currently emerging crisis where our exploitation of several ecosystem services is clearly not sustainable.

We adopt a long-term approach to ecosystem services in this chapter. There are compelling reasons why this approach is useful, although it has rarely been used before. First, human societies have confronted numerous environmental crises of different types in the past. Admittedly, each new crisis presents its own challenges in a unique context, but successful responses to past crises place the weight of history behind the positions taken by environmental advocates, which can outweigh the arguments developed by political and corporate interests (Orr, 2003). Second, knowledge of the past indicates where we are headed, and palaeodata can assist with the development and validation of models that explore future scenarios in a more sophisticated manner.

In this chapter, we first define and introduce the major ecosystem services and then explore the current pressures on them, which are widely perceived to be unsustainable. We place our use of ecosystem services in a long-term temporal perspective in order to provide background to how we have arrived at the present state of affairs. Finally, we explore possible future scenarios.

Summary: Introduction

- Humans are reliant on ecosystems for their survival, due to the wide range of goods and services they provide.
- The value of any particular ecosystem service is currently assessed in monetary terms but this is perhaps a rather short-sighted approach.
- Many ecosystems services, such as food, wood and fresh water, are relatively easy to quantify and value financially. Others, such as recreational, cultural, aesthetic and spiritual services, are more difficult to value in this way.

7.2   MEA classification

Ecosystem services were first defined by the MEA (2005). The MEA's definition was subsequently enlarged as part of the EU-funded RUBICODE project (www.rubicode.net) to 'benefits that humans obtain from ecosystems that support, directly or indirectly, their survival and quality of life. These include provisioning, regulating and cultural services that directly affect people, and supporting services needed to maintain the direct services. They are a subset of ecosystem processes, which include roles that are not easily definable in terms of human needs.'

7.2.1   Provisioning services

These are perhaps the easiest to identify, and in many cases to quantify. They are simply the products that can be obtained from ecosystems, such as food, fibre, fuel, fresh water, natural medicines and pharmaceuticals. One of the most important of these is food, which is mainly provided by agricultural, grassland, mountain, river and lake ecosystems. In Europe, much food provision is through intensive agriculture, although extensive and more traditional agriculture are still practised over large areas as well, especially in seminatural grasslands and uplands (Harrison *et al.*, 2010). Fibre is also provided by agro-ecosystems, including wool, cotton, silk, flax and hemp. Forests provide timber for building and fuel. Fresh water is often also provided by forest ecosystems, and it is important that it be drained through a florally diverse ground layer. Such ecosystems contribute directly to provisioning from rivers and lakes. Natural medicines and biochemicals are provided by a number of ecosystems. The maintenance of high levels of biodiversity is important for future sources of as yet undiscovered treatments.

7.2.2 Regulating services

These are the benefits obtained by an ecosystem through some sort of regulation. They include such things as climate regulation, water regulation, erosion regulation and pollination. Pollination is the most frequently cited service and is one for which there has been substantial research. It is in serious decline worldwide, both from wild bees and from managed honey bees, which is having detrimental consequences for many agricultural crops. Holzschuh *et al.* (2012) explored which bees were most important as pollinators in a study using sweet cherries, which are highly dependent on insects for pollination. Honey bees visited cherry flowers more often than wild bees and were twice as abundant. But fruit set was significantly improved through the visits of wild bees, partly due to their more efficient pollination. Wild bee visitation was related to the biodiversity of the surroundings and was more frequent where the landscape was composed of highly diverse seminatural habitats.

Of the main world crops that are consumed directly by humans, 70% are insect pollinated (Klein *et al.*, 2007). The economic value of pollination globally for crops that rely on pollination, including vegetables, fruits, edible oil crops, nuts and spices, is €153 million, making up 9.5% of the value of human food agriculture (Gallai *et al.*, 2009).

The regulation of climate is important. Forests, for example, play an important part in the global carbon cycle, as they sequester atmospheric CO_2 and store it as long-term carbon, both in the woody biomass and in forest soils, which has a direct effect on climate. Other ecosystems also store carbon, including wetlands and peatlands, as well as seminatural grasslands, which may have significant carbon storage, especially belowground. Other examples of regulatory service include the seed dispersal by jays described in Section 7.1 and pest regulation through habitat management, which is an important service in agricultural ecosystems (Landis *et al.*, 2000). Erosion regulation is important in mountain areas and mountain forests are often used to provide this service.

7.2.3 Cultural services

Cultural services are the nonmaterial benefits people obtain from ecosystems, such as recreation and fitness, aesthetic appreciation, cultural heritage and spiritual and religious experiences. They never sit easily with the other, more science-based services, even though for many people they are the most directly appreciated. We explore these services in Chapter 8.

7.2.4 Supporting services

These are often not considered in the same light as other services. They can be classified as those services that are required to support indirectly or in the long term the production of other ecosystem services and they have an indirect or long-term impact on people and society. They include soil formation, photosynthesis and primary production. Soil fauna, for example, provide a supporting service via nutrient cycling and soil organic matter decomposition, which benefits local farmers through increased crop production and the wider population through the availability of food and fibre (Postma-Blaauw *et al.*, 2006).

Another example is the provision or restoration of habitat, for example for an endangered species. *Maculina* species of butterfly are obligate ant parasites that have very specialised requirements with regard to host plants. The large blue (*Maculina arion*) lays its eggs in wild thyme flowers, which are usually found in dry grasslands. It also

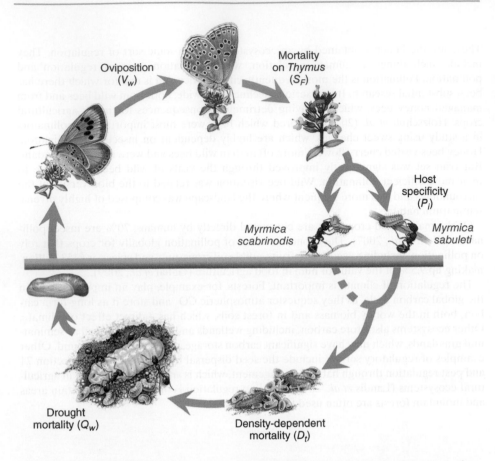

Oviposition (V_w)

Mortality on *Thymus* (S_F)

Host specificity (P_i)

Myrmica scabrinodis

Myrmica sabuleti

Drought mortality (Q_w)

Density-dependent mortality (D_t)

Figure 7.2   Life cycle of the butterfly *Maculinea arion* (large blue). Adult butterflies lay their eggs on thyme species when they are flowering, from June to July. The resultant larvae feed on the flowers for 3 weeks. *Myrmica* ant workers find these larvae and adopt them into their nests underground, where they eat most of the ant brood. After 10 months the larvae pupate while still in the nest, and they emerge as adults 2–3 weeks later. Reproduced by kind permission of Jeremy A Thomas OBE and Richard Lewington

requires that the ant genus *Myrmica* be in the same habitat so that it can be parasitised (Figure 7.2). Thomas *et al.* (2009) describe successful measures taken in the UK to conserve this threatened species, which involved recognising that the host ant species had a very narrow niche and that small changes in grazing and therefore the structure of the grassland had caused it to be outcompeted by other ants that were not suitable hosts. Once appropriate management regimes had been established, the host ant responded rapidly. Successful releases of *Maculina* collected from a Swedish population on Öland then followed.

Monetary valuation of such habitats is difficult and somewhat arbitrary, and anyway may not be desirable. There is considerable pressure on this type of grassland habitat, however, from climate and land use change and from habitat fragmentation. Conservation should be focused on providing large areas of suitable grasslands and heathlands that will be more resilient to environmental changes.

> ## Summary: Millennium Ecosystem Assessment Classification
>
> - Ecosystem services include provisioning, regulating and cultural services, which directly affect people, and supporting services, which impact on the other services and thus indirectly on society.
> - Provisioning services are the products that can be obtained from ecosystems, such as food, fibre, fuel, fresh water, natural medicines and pharmaceuticals.
> - Regulating services are the benefits obtained through regulation, such as climate regulation, water regulation, erosion regulation and pollination.
> - Cultural services are the nonmaterial benefits people obtain from ecosystems, such as recreation and fitness, aesthetic appreciation, cultural heritage and spiritual and religious experiences.
> - Supporting services are those services that are required to support indirectly or in the long term the production of other ecosystem services, such as soil formation, photosynthesis and primary production.

7.3 The current crisis in ecosystem services

As we saw in Chapter 1, the global economy is almost five times the size it was half a century ago and such a rapid increase has no historical precedent (Jackson, 2009). The associated increase in use of finite natural resources and management of increased land areas has led to rapid conversion of terrestrial biomes, with loss of species and consequent modification of ecosystem services (Figure 7.3). There is a general consensus that we are facing an unpleasant cocktail of disrupted and disappearing ecosystem services, with growing competition between individual services as terrestrial ecosystems are more fully exploited. Increasing concern about human impact on the environment gathered pace with the publication of Rachel Carson's *Silent Spring* in 1962, which contributed to the growth of 'environmentalism' as a political movement and the rise of green parties in several countries (Carson, 1962; Orr, 2003). Some earlier regional crises have been averted by government regulation; for example, the ability of the atmosphere to cleanse itself of pollutants has declined since pre-industrial times, yet successful emissions controls in several European countries have alleviated the most pressing problems of air pollution in many regions. However, examples of ecosystem services in trouble are now rapidly appearing as a result of increasing pressures and competition for land. Most of the earth's surface that is suitable for arable agriculture is now exploited and competition for its use is becoming stiff, with significant social consequences. For example, transnational land acquisitions, coarsely known as 'land grabs', are an active policy of several governments and corporations in developed countries. These land acquisitions are driven by concern over how to meet future food and energy requirements in the grabbing countries. The targets are developing or sparsely populated countries, and the majority of cultivated land in countries such as Gabon, Liberia and the Philippines has already changed ownership in this way during recent years (Rulli *et al.*, 2013).

On top of competition for land, there is also developing competition for the use of foodstuffs. Maize, together with rice and wheat, provides over 30% of the essential daily food to more than 4.5 billion people in 94 developing countries and is cultivated on nearly 100 million hectares of land (Shiferaw *et al.*, 2011). The nature of demand for maize has changed during recent years, with a significant increase in its use as animal feed, driven

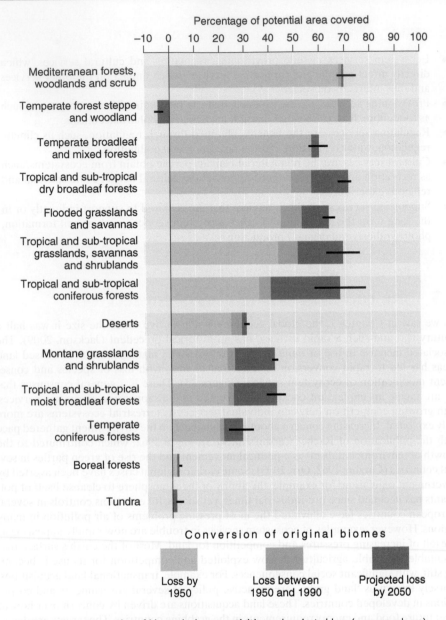

Percentage of potential area covered

Conversion of original biomes

| Loss by 1950 | Loss between 1950 and 1990 | Projected loss by 2050 |

Figure 7.3   Conversion of biomes by human activities and projected loss (www.maweb.org)

by increasing affluence in China, India and other countries and a consequent demand for luxury foods such as milk, eggs and meat. The rapidly developing bioethanol programmes in the USA and elsewhere are also based on maize, and these multiple, growing demands have driven maize prices up by 48% since 2008 (Figure 7.4). Such price increases have been devastating for poor people in many parts of the world and have contributed to widespread food riots. The demand for maize is forecast to double between now and 2050 (Rosegrant *et al.*, 2013), so the present situation is set to deteriorate further unless there are changes in policy.

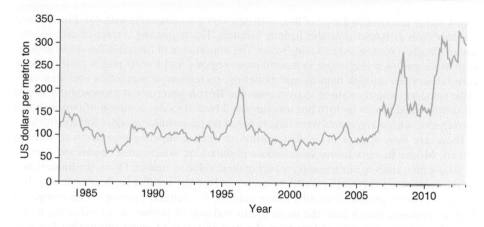

Figure 7.4   Maize price development during the last 30 years (USDA Market News)

The use of maize in biofuel is a controversial topic and needs to be viewed in a broader ecosystem services perspective. Reduced carbon emissions from the transport sector are used as a major justification for the use of biofuels, and their production (whether from willows, maize, sugar beet, oilseed rape or palm oil) has become an important provisioning service. Biofuels are controversial not only because the more agricultural types such as maize, sugar beet and oilseed rape cause loss of food production capacity but also because of their true contribution to the reduction of greenhouse gases. Biofuel is of course not a fossil fuel, but energy is still expended during its production process. According to Tilman *et al.* (2009), the resultant 'food, energy, and environment trilemma' needs some common-sense guidelines if biofuel production is to achieve a satisfactory resolution. These authors suggest that the biofuel industry should be based on the use of (1) perennial plants grown on degraded, abandoned agricultural land, (2) waste from forestry and agriculture and (3) municipal and industrial waste, in order to minimise conflicts between competing ecosystem services. The current tendency to clear forest for biofuel production can lead to increased greenhouse gas emissions, biodiversity loss and higher food prices, with undesirable consequences (Tilman *et al.*, 2009).

7.3.1   How did we get here? A palaeo perspective

The long-term development of ecosystem services is inextricably linked to human population size, both global and local, and its relationships with changing ecosystems. We saw in Chapter 3 the uncertainty that exists about the size of human populations during the Holocene, but certain relationships between population size and ecosystem services can be explored. As human populations have grown and urbanised, they have made different demands on ecosystems for services. Forest ecosystems provided basic provisions for Mesolithic peoples in the form of game, fruits, nuts, fungi and honey. Following the European development of agriculture, forest products became of less importance for urbanised societies and forest ecosystems were either cleared to create arable land or managed as rough grazing for domestic animals. Forests had always provided timber for construction and fuel, but with the development of the forest industry, to supply first building materials for industrial cities and subsequently paper, their provisioning role changed again. Domestic animals were removed from forests in Scandinavia, often by national law, and the primary economic value lay in the timber. Today the ecosystem services of the forest sector are again in a state of flux, with new services competing with old. Provisioning services such as timber now also include biofuels. The regulating service of carbon storage in forests is recognised as having value. The importance of forests in flood and erosion regulation guides management in mountainous regions, and forests play a central role in cultural services through hunting and gathering, recreation, conservation and their symbolic value as close-to-'natural' ecosystems. The British government proposed the sale of all state-owned forests in 2010 but was surprised by a vigorous coalition of opponents to the scheme, whose arguments were largely based on the cultural services forests provide.

There are now sources of conflict arising between competing ecosystem services in forests. Moose hunters favour large moose populations, which cause significant damage to young pine trees in Scandinavia, reducing their value as timber. Often the hunters and forest owners are the same people, so each resolves the conflict in their own way. But in areas close to groups of wealthy urban hunters, the cultural hunting values can give a higher economic return than the more traditional sale of timber. It is fascinating to see how the provisioning role of forests in the past in terms of game production has now become an important cultural service.

7.3.2 Provisioning services in the past

The major transition in provisioning services was the switch from hunting and gathering to agriculture, which took place independently on at least three continents at various times after the last ice age. Much has been written about this transition and its consequences, but one of the most intriguing and relevant issues for contemporary agriculture is the potential influence of climatic change on the origin, spread and sustainability of agricultural systems. Hunting and gathering is energetically far more efficient than primitive agriculture, which provides low yields, so a stimulus is needed to drive the transition to agriculture. This could be increased population pressure leading to degraded resources or external agencies such as climatic change. Our thinking about the origins of agriculture is in a state of almost constant flux. There appear to be at least 10 different places around the world where agriculture was independently developed, and the antiquity of domestication is being pushed back in time with new discoveries, but as yet no evidence of domestication predates the opening of the Holocene 11 500 years ago (Table 7.1; Price and Bar-Yosef, 2011). The first stable village sites appeared in China about this time, but people still primarily relied on hunting, fishing and gathering. Although millet was domesticated about 10 000 years ago, several millennia passed before domesticated plants, including millet and rice, made significant contributions to diet and agricultural systems with cultivated fields were put in place (Cohen, 2011).

Our knowledge of the origins of agriculture in the Fertile Crescent of the Near East has rapidly developed in recent years, based on the study of new sites, increased dating accuracy and analyses of ancient DNA. It is now believed that plant and animal domestication both occurred at about the same time, at least 11 500 years ago, coincident with the rapid climatic changes of the opening of the Holocene (Zeder, 2011). There seems to have been a long transition period for the next 4000 years, during which hunting and gathering were combined with the experimental use of domesticated plants and animals. Zeder (2011) describes a gradual decrease in the use of a wide range of native plants from the Euphrates flood plain, with increasing exploitation of the wild ancestors of early crops such as barley, emmer wheat, lentils, chickpeas and beans. People actively modified local plant communities and began to cultivate wild cereals and pulses alongside traditional hunting and gathering activities. Sheep and gazelle remains are found together, but gazelle were never domesticated, as they probably ran too fast. Wild einkorn wheat would not have responded well to the increasing heat and aridity of the early Holocene and as a chosen target for domestication would have needed irrigation and horticultural attention if it were to survive.

Table 7.1 Approximate dates for the appearance of domesticated species in various parts of the world

Location	Species	Date of appearance (years before present)
Southwest Asia	Plants	11 500
	Animals	10 500
China	Millet	10 000
	Rice	>7000
Mexico	Corn	5000
South America	Plants	10 000
	Animals	6000
New Guinea	Plants	>7000
South Asia	Plants	5000
	Animals	8000
Africa	Plants	5000
	Animals	5000
Eastern North America	Plants	5000

Source: Prince and Bar-Yosef, 2011. Reproduced with permission of University of Chicago press

By at least 10 500 years ago, key elements of the developing agricultural system had been shipped 160 km over to Cyprus, including einkorn, emmer, barley and morphologically wild pigs, sheep, goats and cattle, together with fallow deer and fox (Vigne *et al*, 2011). Zeder believes that these early Cypriots took with them the range of useful resources with which they were familiar; a combination of managed and wild populations of plants and animals. The earliest known evidence for human colonisation of Cyprus is of hunter-gatherers, who arrived about 2000 years earlier, probably also from the Levantine mainland to the east (Vigne *et al.*, 2011).

As in Cyprus, the transition to agriculture was probably more abrupt in regions where the technology was imported. Piles of discarded shells, charcoal and simple stone tools, known as 'kitchen middens', from Mesolithic beach parties can be found on beaches in northwestern Europe, including Denmark and the British Isles (Figure 7.5). The Danish middens provide evidence of diverse shellfish menus, including abundant oysters, prior to the development of agriculture around 6000 years ago. During this period there was a 3 m tidal range in the fjords of northern Denmark. However, coincident with the first evidence for agriculture in Denmark the tides virtually disappeared and the abundance and diversity of shellfish resources were greatly reduced. The disappearance of the tides, which are still absent from the region today, was caused by the interaction of two tidal currents in response to a change in land–sea geography that was still adjusting to the disappearance of glacial ice (Petersen *et al.*, 2005). With the loss of the edible shellfish, the beach parties were over and it is tempting to speculate that this contributed to the switch to agriculture, which was initially a more labour-intensive and uncertain way of life (Petersen *et al.*, 2005). The agricultural system that came to Denmark was based on that which had spread across Europe from the Near East, comprising cereal cultivation and husbandry of cattle, pigs, sheep and goats (Madsen, 1982).

The origins of agriculture represent an intensification in provisioning services that was a significant step towards the current situation. But along the way there were periodic crises in food supply that exposed the vulnerability of earlier societies. Chinese written history is long, and it describes how the Zhou tribes and their capital cities moved five times along the Yellow River away from semi-arid land between 3900 and 2300 years ago as a result of drought and subsequent famines, plagues and invasions by nomadic tribes

Figure 7.5   A Mesolithic kitchen midden on a beach in Sligo, Ireland being examined by two eminent Quaternary scientists: Professor Pete Coxon (left) and Professor Bob Devoy (right)

(Huang and Su, 2009). The movements of peoples as a consequence of failing provisioning services can be traced far back in time. There are examples known from sedentary population centres from several parts of the world, and the decline of the Classic Maya civilisation on the Yucatan Peninsula in Mexico has been studied in detail. The lowland Maya population had reached 4 million people by AD 800 but fell to a few hundred thousand during the following 150 years (Medina-Elizalde *et al.*, 2010). A high-resolution $\delta^{18}O$ record from a stalagmite on the peninsula shows eight severe droughts of 3–18 years' duration during this 150 year period, with reductions of 35–50% in estimated annual rainfall (Figure 7.6). Each drought must have increased rivalry for resources and caused famine, disease and civil unrest, while the intervals between them permitted partial recovery. Medina-Elizalde *et al.* (2010) speculate that these droughts contributed to increased warfare and struggle for resources, which led to the eventual collapse of the civilisation. A similar reduction in rainfall is forecast for later this century throughout Central America.

There has been little flexibility through time in human demands for food and water, just a continual increase in demand that matches population growth. However, our demand

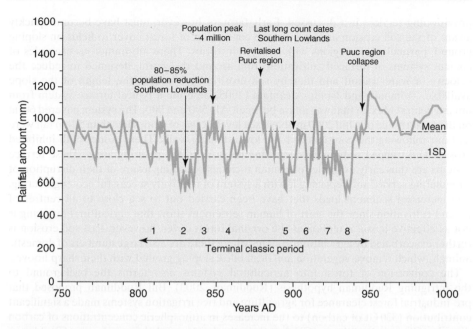

Figure 7.6 Annual precipitation reconstructed from an oxygen isotope stalagmite record from the Yucatan Peninsula, Mexico across a 250 year period. Eight droughts were identified during a 150 year period, which correlate with demographic events. Source: Medina-Elizalde *et al*, 2010. Reproduced by permission of Elsevier

for fuel from ecosystems has changed significantly through time. Since the first use of fire by people, wood has been the major fuel source. It has been converted into charcoal for ease of transport, storage and combustion efficiency for at least several millennia. With the onset and spread of agriculture, forest clearance became so widespread that wood for fuel and construction became a rare resource in more heavily populated countries. Forest cover in England had reduced from an estimated 90% in the early Holocene to just 15% by AD 1086, and there are similar deforestation trajectories for other European countries (see Figure 6.13). With timber rare and thus reserved for construction, including shipbuilding, alternative energy sources became economically viable. Peat was a second-best choice, despite the hard work needed to dry and transport it. There is evidence for peat cutting in England from over 2000 years ago. It was a major industry in the county of Norfolk between AD 1100 and 1300, and its exploitation altered the shape and hydrology of the Norfolk Broads wetland system. The Norfolk peat excavations were flooded and largely abandoned by AD 1500, but peat cutting continued over much of the landscape of Ireland and upland Britain. Rotherham (2008) has estimated that more peat had been cut in the South Pennines than in Norfolk by AD 1500 (c. 34 million cubic metres) and exploitation continued there until coal largely replaced peat as a major urban fuel source in the late AD 1700s. The shift from wood to peat to coal and oil is one from a renewable resource to partially fossilised material (peat) to fossil fuels, which are only replenished, if at all, over geological timescales. As these resources are now becoming scarce, there is renewed interest in sustainable biofuels, including forest products.

7.3.3 Regulating services in the past

Few changes in regulating services in the past are likely to have had significant impact on earlier human populations, as their attention was probably more focused on the provisioning services just discussed. Early farmers, however, must have become quickly aware of the soil erosion arising from the conversion of forest cover to fields on sloping ground, particularly in regions with heavy rainstorms. There are numerous examples of ancient systems of terraced cultivation from around the world, designed to reduce the velocity of water runoff and thereby soil erosion by breaking the length of the slope available. Trombold and Israde-Alcantara (2005) describe a typical terrace system from upland central Mexico that was in use between AD 500 and 900. The system covered 60 ha on the eastern side of a hill 2000 m above sea level. Rows of stones less than 50 cm tall were laid out following the contours to form low retaining walls. These stones both inhibited erosion and trapped sediments, deepening the soils in the planting areas. These terrace systems are thus early examples of human societies becoming aware of their disruption of a regulating service and replacing it with a system of relatively successful eco-engineering. The increased sediment loads that have been carried out to sea close to all centres of upland cultivation since the start of human settlement show that agricultural terracing is not as effective in soil retention as the original forest cover, however. Past soil erosion is further exacerbated by agricultural systems where there are also large numbers of domestic animals, which remove vegetation and destabilise sloping ground with their sharp hooves.

The conversion of forest into agricultural systems also forms the background to the intriguing Ruddiman hypothesis (Ruddiman, 2003). Bill Ruddiman proposed that pre-industrial forest clearance for agriculture and rice irrigation systems made a significant contribution (330 Gt of carbon) to the increases in atmospheric concentrations of carbon dioxide (CO_2) and methane (CH_4) seen in the last several thousand years (Ruddiman et al., 2011b). Consequently, early human activities influenced the global carbon cycle and its associated climate regulation service, which had previously been controlled by marine and terrestrial ecosystem dynamics during the glacial–interglacial cycles of the Quaternary. Indeed, human activities have potentially contributed to the delay or even cancellation of the next ice age. This hypothesis has provoked lively discussion, which in the best traditions of science has stimulated new research into the global carbon cycle and specifically the scale of influence of past civilisations on ecosystem conversion. Ruddiman's hypothesis still enjoys considerable support as we write. The release of CO_2 from biomass and soils following deforestation in the past is indisputable and the debate is simply about the scale and timing of this release.

Ruddiman also considered human impact on CH_4 release, which is a far more potent greenhouse gas than CO_2. Methane is naturally released from wetland ecosystems, but also from rice paddy fields and cattle (both front and rear ends). The observed downward trends in methane emissions for the previous six interglacials, which might have been driven by increasing desiccation of mid-latitude wetlands during each interglacial period, strongly suggest that the anomalously increasing trend of methane emissions during the last 5000 years has an anthropogenic origin (Figure 7.7; Ruddiman et al., 2011b). Natural methane emissions from northern tropical and boreal wetlands, which are two of the largest planetary sources, were actually falling during this time, although emissions from the south Amazon basin were weakly increasing. The major increases in methane supply are therefore thought to originate from Asian rice paddy fields and the increasing global populations of domestic animals.

The causes of increases in atmospheric CO_2 during the last 7000 years are more controversial. They are potentially influenced by the enormous stores of carbon held in the oceans. Coral reef construction is thought to have accelerated after global sea levels more or less stabilised at high levels around 7000 years ago. Building reefs made of $CaCO_3$ and $MgCO_3$ extracted carbonate ions from the ocean, which consequently became saturated with CO_2, with the excess perhaps escaping into the atmosphere (Ruddiman et al., 2011b). This coral reef hypothesis proposes that reef building has been a major contributor of atmospheric CO_2, although this is a hard process to quantify (Ridgwell et al., 2003). Another process

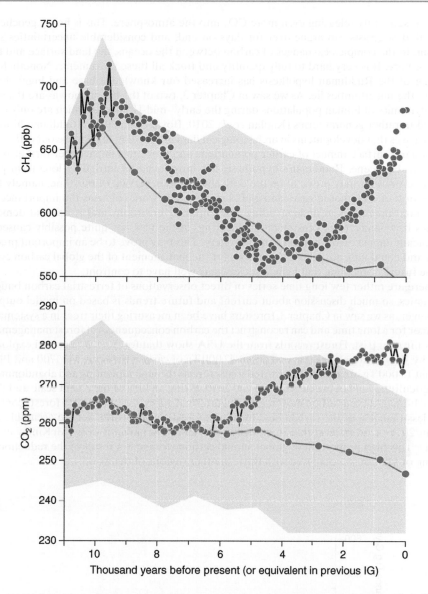

Figure 7.7   Comparison of Holocene (red) and mean values from six previous interglacial (IG) periods (blue) for CH_4 and CO_2 records from the Antarctic ice core Dome C. Light-blue shading shows one standard deviation around the previous interglacial means. Source: Ruddiman *et al*, 2011. Reproduced by permission of SAGE Publications

that can release CO_2 into the atmosphere is delayed ocean carbonate compensation, with the rapid early Holocene transfer of carbon into terrestrial ecosystems thought to have moved ocean carbonate chemistry out of equilibrium. Organic matter accumulates following the glaciation and the carbon thus moves on to land as it is fixed by vegetation . When the land is fully revegetated and terrestrial biomass stops increasing, the rate of increase of terrestrial carbon storage slows right down to form an equilibrium. Consequently, more of the atmospheric CO_2 becomes dissolved in the oceans, increasing acidity, dissolving $CaCO_3$

and subsequently releasing even more CO_2 into the atmosphere. This is heavy geochemistry that scientists can argue over for days on end, and considerable uncertainties still remain in the complex exchanges of carbon between the oceans, the land surface and the atmosphere. It is very hard to fully quantify and track all these movements. Nonetheless, testing of the Ruddiman hypothesis has increased our knowledge base and highlighted where the uncertainties lie. As we saw in Chapter 3, two of the big unknowns are the size and dynamics of human populations during the early–mid-Holocene, which are only estimated in rather general terms (Kaplan *et al.*, 2010; Boyle *et al.*, 2011). Ruddiman himself emphasises that developments in archaeological data are tending to increase our estimates of the land-surface impact of earlier populations and move this impact ever further back into the Holocene. Ruddiman's hypothesis has introduced an intriguing long-term perspective on what may prove to be the critical ecosystem service of our time, namely the regulation of greenhouse gases and global climate. His work confirms the importance of millennial timescales in understanding contemporary environmental issues and demonstrates how inadvertent ecosystem 'engineering' in the past has quite possibly caused a significant impact on the current climate system. This may prove to be an important precedent and could indicate potential methods for the management of the global carbon cycle in the future, something that human society may well have to confront.

There are rather few long time series of direct observations of terrestrial carbon budget dynamics, so much discussion about current and future trends is based on model output. However, as we saw in Chapter 3, foresters have been measuring their trees in a systematic manner for a long time and can reconstruct the carbon consequences of forest management back into the past. Forest records from the USA show that forest clearance and exploitation by European settlers released about 42 000 Tg of carbon between AD 1700 and 1935. About 15 000 Tg of carbon was restored to the forests through replanting and abandonment of agricultural land between AD 1935 and 2010, giving an idea of the recent scale and timing of human-driven carbon movements (Figure 7.8; Birdsey *et al.*, 2006). US forests contain just less than 4% of the carbon held by all the forests in the world, yet are accumulating about 20% of the carbon that is annually sequestered in global forests (McKinley *et al.*, 2011). This high proportion of global annual carbon storage is a result of the reduction in felling of US forests compared with those in other regions of the world.

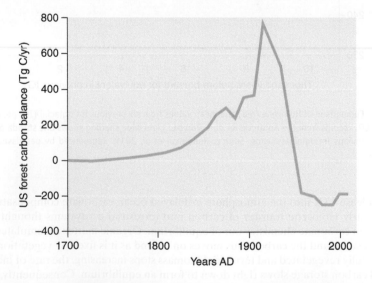

Figure 7.8 Recent forest carbon emission and sequestration in the USA. Source: Birdsey *et al* 2006. Crop Science Society of America

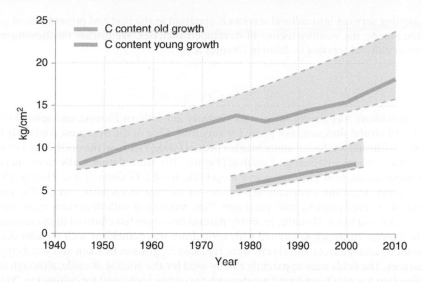

Figure 7.9 Change in aboveground carbon stock for old-growth stands (1945–2010) and young-growth stands (1977–2002) from Lady Park Wood, UK. Dashed lines: 95% confidence limits for a typical transect from a linear mixed model

Permanent forest plots and transects have been monitored in several countries and can be used to reconstruct the carbon dynamics of individual woodlands. The vegetation in Lady Park Wood, a temperate deciduous forest in the Wye Valley, UK, has been periodically recorded since it became an unmanaged nature reserve in 1944. Conversion of tree measurements into kilograms of carbon per hectare shows a 73% increase in sequestered carbon in the old-growth sections of the forest over a 57 year period (Figure 7.9). Sections of the forest had been clear-felled in 1943 and the carbon stored in biomass rose from 0 to 9 kg/ha in 57 years. We used the dynamic vegetation model LPJ-GUESS (see Chapter 2) to estimate how much of this increased storage could be ascribed to climate change and successional processes. While the regrowth of the clear-felled areas closely tracked the model forecast, the old-growth model forecast was for relatively constant carbon storage of about 14 kg/ha. The observed increase in carbon storage in the old-growth sections was most likely a consequence of the relaxation of management, both grazing and selective felling, that occurred prior to 1944. The development of Lady Park Wood during recent decades is typical of that of many unmanaged temperate woodland areas in Europe. These were actively exploited until the mid-1900s but now temperate woodland is the only global biome forecast to actually increase in area in the near future by the MEA, as such woodlands are rarely used for grazing today and the commercial demand for timber products in Europe has shifted to plantations (Figure 7.3). The recent history of European temperate woodlands has included a significant withdrawal of atmospheric CO_2 and passive woodland management is one of the most effective ways in which people can influence important regulating services such as carbon sequestration.

7.3.4 Cultural services in the past

Trees and forested ecosystems have held a special place in the culture, belief and mysticism of human societies throughout time. As we shall see in Chapters 8 and 9, nature conservation policy is not purely driven by scientific values but also draws on the widespread regard for cultural ecosystem values keenly felt by many societies. The development of earlier provisioning services into cultural services is apparent in the past and present role of game hunting among the wealthy sectors of developed societies. We discuss this development and other cultural services in detail in Chapter 8.

7.3.5   Supporting services in the past

Early agriculture was a risky activity, particularly in northern Europe, and several of the initial agricultural sites were not in use for very long. One of the best preserved early field systems in Europe was found under blanket peat at Céide Fields on the north Mayo coast in Ireland by archaeologist Seamus Caulfield (Figure 7.10). Buried stone walls surround large, rectangular fields, oval enclosures and megalithic tombs. O'Connell and Molloy (2001) have reconstructed the environmental history of the site in exquisite detail. The region was forest-covered prior to 5840 years ago. Pine was mixed with deciduous trees, including oak, elm and hazel. Periodic, probably natural fires have left charcoal in the sediments. The first forest clearance associated with agriculture was coincident with the elm decline (Chapter 5) and the trees and shrubs were replaced by grasses, bracken and weeds typical of pastures. The fields were apparently chiefly used for the grazing of cattle, although some cereal pollen has also been found nearby with two stone tools used for cultivation. Yet the fields were only in use for about 500 years before a heathy woodland cover had developed at the site by 5100 years ago. Abandonment occurred over a 50 year period. Blanket peat was developing elsewhere in the area at the time, which may have contributed to site abandonment, but the local vegetation record showed an increase of heather (*Calluna vulgaris*)

Figure 7.10   Céide Fields, County Mayo, Ireland, the oldest known stone-walled field system in the world, now buried under blanket peat. © National Monuments Service, Dept of Arts, Heritage and the Gaeltacht.

and crowberry (*Empetrum nigrum*), which grow on dry heathland. These plants indicate reduced soil fertility and often appear following overgrazing, so the short occupation of the Céide Fields might have been caused by failure of supporting services following heavy site exploitation and insufficient replacement of soil nutrients. Short periods of arable farming were common elsewhere in Ireland as well, and at nearby Carrownaglogh, County Mayo a 250 year farming episode was terminated by waterlogging, as here blanket peat did inundate the site around 800 BC (O'Connell, 1986). O'Connell believes that at Carrownaglogh agriculture was 'pursued to the point of soil exhaustion', which initiated podzolisation of the soil and the formation of an iron pan, which contributed to waterlogging, peat formation and eventual site abandonment.

Maintaining site fertility prior to systematic management of manure and fertilisers was a problem in boreal Scandinavia as well, where a particular type of slash-and-burn agriculture was practised for much of the last 2000 years. Långrumpskogen is now a swamp-forest nature reserve in northern Sweden famous for its insect fauna, which includes so-called 'indicators' of long-term forest continuity. Segerström *et al.* (1994) collected just 40 cm of sediment from the site and noticed charcoal at the base of the core. Back in the laboratory, they were able to reconstruct the course of events in the wet forest. In about AD 1500 a small area of wet spruce forest was burnt. The trees had probably been ring-barked earlier, in order to kill them and provide dry fuel. A series of small-scale rye (*Secale cereale*) cultivations were made over the next 20 years or so, with some cereal production each year in small, burnt clearings within the forest. At each site, cultivation was abandoned when the size of the harvest was reduced owing to shortage of nutrients, which was probably after just two or three seasons. Rye pollen was found in several basal peat samples, collected from transects across the site (Figure 7.11). Following rye cultivation, the whole site was used as a hay meadow, as shown by peak values of grass (Poaceae) pollen, and for a while the entire swampy basin was almost totally deforested. Grazing indicators in the pollen record such as sorrel (*Rumex*) and buttercup (*Ranunculus*) indicate that cattle were then let on to the site during the summers for maybe 300 years or so. The present forest cover is of relatively recent origin. This reconstruction of a Scandinavian slash-and-burn episode is probably typical for a type of agricultural activity that spread from the east into Sweden. It was widely practised in Finland and Sweden for maybe as long as 2000 years in places, but ceased completely during the mid-1900s. During the 1990s, an old-timer described to me (R.B.) how he had burnt forest for short-term rye cultivation during the 1930s, but said he was the last practitioner of this technique that he knew of, as it was illegal at that time. Shifting cultivation like this has been practised in one form or another throughout the world, and prior to the organised use of fertilisers it was the common method for maximising crop production when there was plenty of land available. Soil fertility is a key supporting ecosystem service that requires management by replacement of extracted nutrients. Today, artificial fertilisers and manure are frequently used, but in former wooded landscapes, wood ash was employed, followed by very long fallow periods to permit full forest recovery.

Genetic resources are an integral part of supporting ecosystem services. Rich genetic variation was a major factor exploited in the domestication of plants and animals, to accentuate desirable traits such as seed production or docility. Ancient DNA analyses are yielding new insights into how genetic variation has altered within species. The most ancient genetic detail, apart from in humans, is available for horse, reindeer, bison, musk ox and the extinct woolly rhinoceros and mammoth. The distribution ranges of all extant species are smaller today than they were 50 000 years ago, with the greatest reductions for reindeer and bison (Lorenzen *et al.*, 2011). Climate change appears to have been the major driver of population change over the last 50 000 years and can largely explain the extinction of Eurasian musk ox and woolly rhinoceros. Genetic diversity has reduced in conjunction with range changes and was at minimum values for bison during the Holocene timeframe for domestication (see Figure 4.18). Today we have considerably reduced the natural ranges

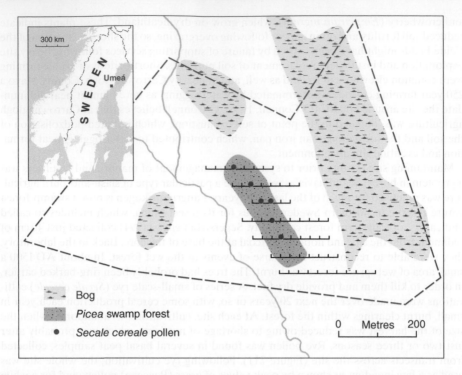

Figure 7.11   Långrumpskogen nature reserve, Sweden. The transects show the locations of the peat–soil interface samples and the filled circles mark the samples that contained *Secale* pollen. Source: Segerström *et al*, 1994. Reproduced by permission of Elsevier

and genetic diversity of most important agricultural species, but there was also signifi-cant 'natural' genetic loss associated with rapid climatic change during the last ice age. There is a growing awareness of the importance of a broad genetic base for the mainte-nance of food security during a period of rapid climate change, as shown by initiatives such as the creation of cereal seed banks (Sachs, 2009) and the Rare Breeds Survival Trust (www.rbst.org.uk).

Summary: Ecosystem services and the past

- The four major categories of ecosystem service are provisioning, regulating, cultural and supporting. Increasing global population is imposing unsustainable demands on ecosystem services, such that several are reaching crisis points. Transnational land grabbing and rapidly increasing food prices are indicative of intense competition for provisioning services.
- The multiple origins of agriculture and abandonment of hunter-gathering lifestyles represent a major past transition in the development of provisioning services.
- Soil erosion and release of carbon from forest ecosystems following tree felling are examples of how human activities have disrupted regulating services in the past. Management systems have subsequently been developed to protect them.
- Past loss of soil fertility and genetic diversity of domesticated species have demon-strated the importance of supporting ecosystem services.

7.4 Ecosystem services and the future

Modern society is highly complex but is still absolutely dependent on the proper functioning of ecosystems for its survival. Projecting into an uncertain world the possible interactions between natural systems and societal systems is difficult. Monitoring these interactions and likely changes is important, and the status of the biodiversity in an ecosystem is a major measure of its health. One means of monitoring is through the development and use of indicators of biodiversity; once developed, these can be followed through time as they change, usually due to loss of species. They aim both to reflect the state of the environment and also to monitor the responses to any changes as a result of policy. Such an indicator could simply be the number of particular species in a disturbed or polluted area as compared to the number in a more pristine area (EEA, 1999). They are used to supply information about a particular environmental problem, support the development of policy, monitor the responses or effects with regard to the selected policy and raise public awareness. A wide variety of indicators have been developed. Feld *et al.* (2009) examined 531 different indicators, mostly associated with biodiversity and ecosystem services, across a range of ecosystem types. Normally, biological indicators are applied at regional and fine scales, and rarely across ecosystem types. Abiotic indicators are usually used at broader scales, from the region to the continent. Indicators of biodiversity are usually associated with species richness but the important aspects of biodiversity, such as function, structure and genetic aspects, are poorly addressed. In addition, they only fulfil in part the current need to better understand the relationships between humans and nature in a modern and increasingly complex world.

This complexity must be addressed more fully. Berkes and Folke (1998) expressed the view that the ecosystem concept based on a purely scientific approach is deficient as it lacks the dynamics of social and ecological interactions and does not explore the idea of *humans-in-nature*. The concept of the socio-ecological system (SES) has been developed in recent years to address some of these shortcomings. Gallopin (1991) defined an SES 'as any system composed by a societal (or human) subsystem and an ecological (or biophysical) subsystem'. He suggested that the SES concept is an appropriate tool for use in sustainability research and can be used at any scale from the local to the global. It is important to recognise that these systems are reliant on each other and are 'nondecomposable' into their constituent parts (Gallopin 2006).

Such conceptual frameworks are an approach, rather than a model or a theory. They are designed to provide a framework within which exploration of the characteristics of an ecosystem can be made in terms of aspects of society. Their aim is to focus on key outcomes that will lead to sustainable ecosystems (Berkes and Folke, 1998). There are many frameworks that can be used in such analyses, although not all are relevant to our approach to ecosystem services. The MEA contains a well-known framework that links biodiversity, ecosystem services and human well-being (Figure 7.12).

Rounsevell *et al.* (2010) devised another approach by integrating the DPSIR (Driving Forces-Pressures-State-Impacts-Responses) framework within an SES and thereby connecting the outcomes directly to those parts of society that benefit from the service. The DPSIR approach describes relationships between the environment and human society (Figure 7.13) and is an important tool in environmental analysis (EEA, 1999).

The DPSIR provides a structure within which different indicators can be evaluated in terms of environmental policy and policy responses. There are five elements involved: drivers, pressures, states, impacts and responses. *Drivers* are the underlying factors that cause environmental changes, such as increasing atmospheric greenhouse gases. These lead to *pressures*, which are the actual variables that influence the environment, such as changes in temperature and precipitation. *States* are those elements of an environment that will be influenced by these changes, such as a specific crop production, species presence/absence

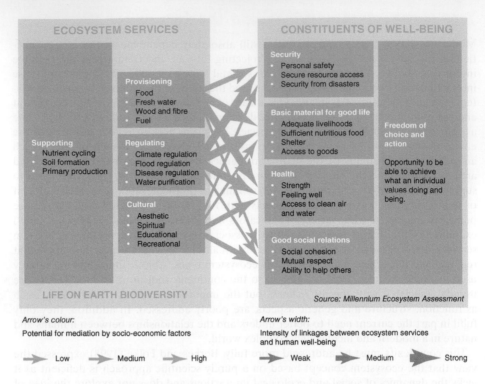

Figure 7.12   Framework linking biodiversity, ecosystem services and human well-being used in the MEA

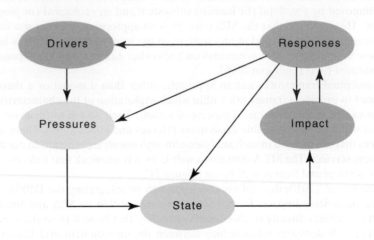

Figure 7.13   The DPSIR framework, which can be used to address environmental issues (EEA, 1999). Source: http://www.eea.europa.eu/publications/TEC25. Reproduced by permission of the European Environment Agency

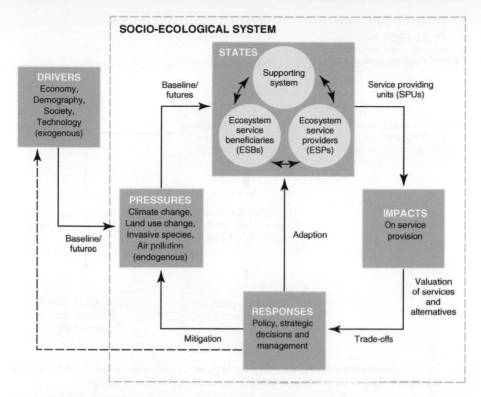

Figure 7.14 Framework for Ecosystem Service Provision (FESP), using a modified DPSIR structure Source: Rounsevell *et al*, 2010. Reproduced by permission of Springer

or distributions, while the *impacts* in this case would refer to the food insecurity or biodiversity loss that results. *Responses* include how society will respond to the impacts, for example by changing relevant policies.

Rounsevell *et al*. (2010) took the DPSIR idea further by integrating it into a new framework, the FESP (Framework for Ecosystem Service Provision), which can be used to assess the impacts of environmental change on the provision of ecosystem services and the implications for policy and management decisions (Figure 7.14).

This approach has the advantage that it allows comparisons between different and competing ecosystem services and clearly shows that a service depends on both attributes of the people receiving the benefits and the biological attributes of the service provider. A simple stepwise implementation using the jay example given in Section 7.1 describes this approach (Figure 7.15; Rounsevell *et al*., 2010).

- *Step 1*: This step relates to the ecosystem service beneficiaries (ESBs); that is, the visitors (both local and tourists) to the park.
- *Step 2*: The service (provided to ESBs) is cultural: the park is used for recreation.
- *Step 3*: The ecosystem service provider (ESP) is the oak forest and not the jays – the jays provide a service to the oaks.
- *Step 4*: Drivers such as global change may affect both the oaks and the jays, possibly through changes in temperature, precipitation or disease prevalence.
- *Step 5*: A census of the number of jays required to maintain the oak forest is carried out.

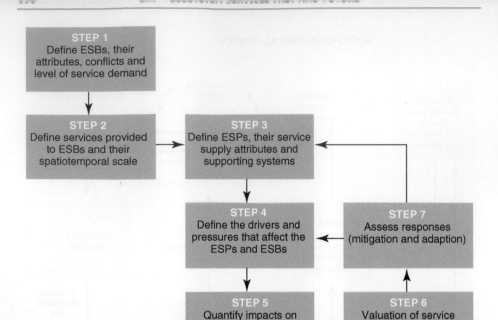

Figure 7.15 Implementation of the approach described in Figure 7.14, using the jay and Stockholm National Urban Park example. Source: Rounsevell *et al*, 2010. Reproduced by permission of Springer

- *Step 6*: Alternative ways of providing the service are valued, such as seeding or planting of the oaks.
- *Step 7*: The possible policy responses required to mitigate or adapt to the service are assessed. As the jays are clearly the most cost-effective way of providing the service, conservation strategies aimed at maintaining them may be required (Rounsevell *et al.*, 2010).

Such conceptual frameworks can provide the structure within models of the future, which can be used to explore a range of possible outcomes for a range of different ecosystem services. They can also form structures within which assessments of the sustainability of ecosystem services can be made. However, they are in many ways a work in progress.

Summary: Ecosystem services and the future

- Indicators are commonly used to assess biodiversity and to follow the rate of biodiversity change – usually loss – through time. They aim both to reflect the state of the environment and to monitor the responses to any changes as a result of policy.
- A socio-ecological system (SES) combines both societal (or human) and ecological (or biophysical) subsystems, e.g. the MEA framework.
- Such conceptual frameworks can provide the structure within models of the future, which can be used to explore a range of possible outcomes for a range of different ecosystem services.

7.5 Relating the maintenance of biodiversity to ecosystem service provision

The innovative European project RUBICODE (Rationalising Biodiversity Conservation in Dynamic Ecosystems; www.rubicode.net) examined the links between biodiversity and the provision of ecosystem services by evaluating those aspects of biodiversity that are specifically relevant to society (Harrison *et al.*, 2010). It aimed to increase understanding of the value of these services and thus the potential cost to society if they are lost. In order to maintain these services, it is important to know something about the service providers, namely the plants, animals and other organisms, as well as the dynamics of the ecosystems in which they live and therefore their influence on provisioning ecosystem services (Harrison *et al.*, 2010). Harrison *et al.* (2010) summarised the trends of various ecosystem services throughout Europe since 1950, as well as their current statuses, showing, perhaps not unexpectedly, that managed ecosystems are much more important than seminatural systems for service provision of such items as food, fuel, timber and fresh water. However, seminatural systems are more important for services such as genetic resources. Cultural services as a whole are generally important in all ecosystems, although spiritual and religious services are not always viewed as being significant. The situation is more complex for regulating services, with forest and mountain ecosystems being very important for such services, while agro-ecosystems are often regarded as negative for regulation. Importantly, Harrison *et al.* (2010) point out that most ecosystems are multifunctional, often contributing a range of different services. Many services have been degraded over recent years, at least in Europe.

It must be remembered that it is important to relate the maintenance of biodiversity and the provision of ecosystem services to what policy makers can or may wish to do. The RUBICODE project identified, with help from a wide range of stakeholders, a number of important research priorities, including: the quantification of biodiversity in ecosystem service provision, which includes as many organisms as possible, for example uncharismatic groups such as invertebrates and fungi; development of integrated assessment of ecosystem services at a range of scales, both in time and space; identification of thresholds at which ecosystem service delivery changes, perhaps irreversibly; quantification of the effects of socio-economic and environmental drivers on ecosystems; improvement of the classification of ecosystem services and values; and development of indicators for ecosystem services (Anton *et al.*, 2010).

7.6 Scenarios of possible futures: some different approaches

'Beyond a general knowledge of the overall importance of ecosystem services, we have only a few concrete examples of the value of these ecosystem services. Even more rudimentary is our understanding of how these ecosystem services will change due to climate change, how these changes will affect people and the economy, and how the economy will respond to these changes. The problem isn't just uncertainty about the future, but our general lack of understanding of how to deal with risk and make decision under this uncertainty.'

Shaw *et al.* (2011)

As we have discussed earlier, the maintenance of ecosystem services is important for the sustainability of human society and well-being. The many and varied stresses placed by humans on ecosystems and their services in the past and at present are in reality relatively minor in comparison to those likely to occur in the future, as increased population pressure and global change in all its aspects intensify. If humanity wishes to survive and flourish into the future then it is necessary to develop some strategies that can respond to these changes. Precognition would be ideal, or failing that a good guess. However, Rounsevell and Metzger (2010) provide a fitting quote from Herbert Kahn, a so-called 'futurist': 'The most likely future isn't'. Futurists are interdisciplinary scientists and social scientists who try to predict the future using systems approaches. It is not possible to predict into the future with any reliability due to lack of information, limited experience and flawed thinking. One can explore possible futures, but must be aware that none are likely to come into existence. Scenarios are one way in which the future can be explored. These have been described as 'plausible, provocative and relevant stories about how the future might unfold. They can be told in both words and numbers. Scenarios are not forecasts, projections, predictions, or recommendations, though model projections may be used to quantify some aspects of the scenarios' (MEA, 2005). In a similar vein, an Intergovernmental Panel on Climate Change (IPCC) special report on future greenhouse gas emissions (Special Report on Emission Scenarios, SRES) described scenarios as 'alternative images of how the future might unfold … an appropriate tool with which to analyse the driving forces that may influence future emission outcomes and to assess the associated uncertainties' (IPCC, 2000).

The concept of scenarios is not new. As we have seen elsewhere in this book, the idea can be traced back to the ancient Greeks, in this case to Plato's *Republic*. More formal techniques for scenario building were developed by Prussian military strategists in the nineteenth century. The military aspect continued with modern-day techniques in the Cold War period after the Second World War. Further developments occurred as a result of social policy planning and from needs within the business community (Rounsevell and Metzger, 2010).

The first global environmental scenarios using mathematical models were developed by Mesarovic and Pestel (1974) and the Club of Rome as environmental concerns about the state of the planet and humanity's reliance on well-functioning ecosystems and their services became more and more pressing. Models such as IMAGE (Chapter 2) were part of this development. However, numerical models were limited in their use as they were restricted to well-known systems and short periods of time. On the other hand, more qualitative or 'narrative' approaches to describing alternative futures, while useful in some ways, were not very scientific in their approach. It was the Global Scenario Group (www.gtinitiative.org/gsg), created in 1995, that combined both quantitative methods and qualitative storyline approaches to create much more detailed scenarios. These developments have since been used in many studies at a variety of scales, from global to local (e.g. Raskin *et al.*, 1998).

Three types of scenario are described by the Global Scenario Group, namely *conventional worlds, barbarisation* and *great transitions*. Conventional worlds are basically the continued evolution of the current system without any major transformations or discontinuities. Barbarisation, on the other hand, explores the possibility of the deterioration or even collapse of the social and economic order (think *Mad Max*). Great transitions involve fundamental changes in socio-economic systems and values, the development of sustainability and a transition to a more balanced society, with regard to both ecosystems and socio-economic aspects. This combined approach has become the basis for all major later assessments, such as those done by the IPCC and MEA.

Over the intervening years, many different scenarios have been developed, although in environmental science three different types tend to dominate: *exploratory, normative* and *business-as-usual* (Rounsevell and Metzger, 2010). The IPCC emission scenarios, which we will discuss in the next subsection, are a typical example of exploratory scenario storylines. Basically, they are plausible storylines about different socio-economic developments that can be compared over the next century. There is potential for very different outcomes.

Normative scenarios relate to 'desired futures', so that the storyline describes events and relationships that lead from the current situation to a wished-for future. Policy scenarios such as the Convention on Biological Diversity or European Union goals on renewable energy are good examples. Business-as-usual scenarios are used in policy analysis over the short term, where policy effects are likely to dominate.

7.6.1 IPCC Special Report on Emission Scenarios

The level of greenhouse gases in the atmosphere directly affects climate and thus ecosystems and ecosystem services. The future level of greenhouse gas emissions is highly uncertain and is dependent upon a number of driving forces, including human population levels, socio-economic developments and technological advances. Scenarios are an important tool for exploration of possible futures as a result of these changes. The IPCC SRES comprises a set of exploratory scenarios, developed in different versions between 1990 and 2000 (Girod et al., 2009). Each series is significantly different in terms of both methods and assumptions, as understanding of emissions and impacts has increased through the years. The most widespread and most frequently cited is the latest series (IPCC, 2000), and it is this series we will describe here. SRES scenarios are used in modelling exercises projecting into the future. Ideally, more than one storyline and family of scenarios will be used, remembering that there is no single 'best-guess' scenario. Even though there are many uncertainties in these scenarios and in the models that use them, the authors of SRES (IPCC, 2000) suggest they 'may provide policymakers with a long-term context for near-term analysis'. Four different qualitative storylines describe different demographic, socio-economic, technological and environmental pathways. Climate policies were excluded from the SRES terms of reference. Families of scenarios were developed for each storyline by using different models with similar assumptions, giving a total of 40 different scenarios over all storylines. All scenarios were viewed as equally valid, with none more or less likely than the others.

- The *A1 storyline* and scenarios (Figure 7.16) are global, convergent and economic in outlook. Rapid economic development continues and as a result the differences between rich and poor countries disappear. Solutions are market- and consumption-based, with high rates of investment in education, technology and national and international institutions. There is a rapid introduction of efficient technologies. Global population peaks in the middle of the twenty-first century and then declines towards its end. The A1 storyline is unusual in that it is further divided based on energy sources, so that A1FI is carbon-intensive, using up all the fossil fuel as quickly as possible, A1T relies on the development of non-fossil and renewable energies and A1B is somewhere in the middle, using both non-fossil and fossil fuels. In general in this family of scenarios the environment is valued, or at least the amenities that are provided by it. There is a focus away from the conservation of ecosystems and their services toward a more active management and an emphasis on ecological resilience.
- The *A2 storyline* and scenarios (Figure 7.16) are regional in outlook and describe a heterogeneous world formed from different economic regions. The important point is self-reliance with regard to resources. There is less trade, slower capital and stock turnover and slower technological change. There are fewer interactions generally among different regions. As a result, economic growth is uneven and an income gap remains between rich and poor societies. There is also less mobility of people and ideas. Population growth rates are about the same as at present or even increase, and the projected global population is the largest of all the storylines, at around 15 billion by 2100. Some regions are more welfare-orientated and therefore have less income inequality, while other regions have less government but greater inequality.

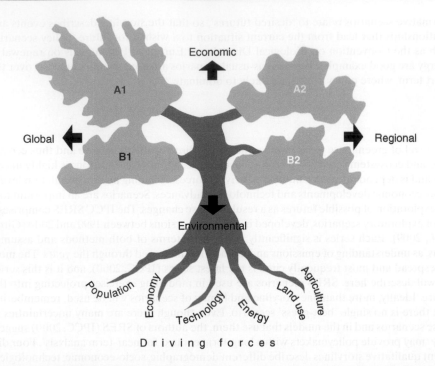

Driving forces

Figure 7.16   The IPCC SRES scenarios. These are divided into four main scenario groups (Global, Regional, Environmental and Economic). The driving forces are greenhouse gas emissions. Source: IPCC, 2000. Reproduced by permission of Cambridge University Press

There are regional and local responses, but not global, to environmental issues, and as a result there are inconsistent outcomes across regions. Given the population size, the provisioning of food is of greater importance than in other storylines. Agricultural production is one of the areas where innovation and research is important.

- The *B1 storyline* (Figure 7.16) is similar to A1 in that it is a convergent global world with similar trends in population growth and decline. There is also a high level of economic activity. However, in this family the activity is different, with a move towards a service and information economy and the development of technologies that are clean and resource-efficient. Environmental and technological change and social aspects of society are all important. Technologies that use post-fossil fuels are developed and applied. Some of the scenarios within this storyline reflect the earlier *great transition scenario*, with its emphasis on sustainability. Environmental quality is important in this storyline and greenhouse gas emissions are lower than in the others. The negative environmental effects of rapid development are addressed and resolved either locally, nationally or globally. The effects of air pollution from acid rain are eliminated, for example. Land use is managed in an appropriate way with regard to the environment. Cities are compact and public-friendly, suburbia is tightly controlled. Low-impact agriculture is practised and there are large areas of wilderness. Food prices are high and meat consumption low, with a reduced impact on ecosystem services.

- The *B2 storyline* (Figure 7.16) is both environmental and regional. It also emphasises local solutions with regard to economic, social and environmental sustainability. Global population increases, but not as fast as in A2. Human welfare, equality and the environment are important. Economic development is moderate and there are fewer and more uneven technological changes than in either A1 or B1. International institutions decline in importance as decision-making becomes more local or regional. However, environmental protection at the global level remains important, although policies are likely to be more successful at local and regional scales. Education levels are high and community initiatives and social innovation are therefore also high. Land use management is well integrated and urban infrastructure is innovative, and as a result there is less dependence on private transport and less urban sprawl. Local food production and self-reliance are important, and at the same time less meat is eaten. Environmental policies are carried out regionally, leading to fewer emissions, less pollution and less acidification.

7.6.2 MEA scenarios

Most of the scenarios described so far have strong social and economic elements but are relatively little concerned with the ecology and dynamics of ecosystems. They care about environmental change – for example, the IPCC emission scenarios take into account a number of environmental issues, including feedbacks between climate, land use and emissions – but less about the complex interactions within ecosystems. Ecosystems, through their ecology and variety of feedbacks, are intimately linked to the many services needed by humanity. Improving one ecosystem service, such as food production, can lead to feedbacks with regard to others, such as changed rainfall patterns, increased soil erosion and pests, new diseases and reduced water supply and quality (MEA, 2005). The global MEA scenarios are concerned with this ecological complexity, but in addition take note of socio-economic change. They are based on plausible changes to ecosystems and their services between now and the year 2050, and the effects these changes might have on humans and society. Four scenarios were devised by the MEA that aim to address these points: *Global Orchestration, Order from Strength, Adapting Mosaic* and *Technogarden*. A brief summary of each follows, but much more detail is available in the MEA (2005).

- The *Global Orchestration* scenario (Table 7.2) is based on an interacting and cooperative global society. It includes strategies that promote a more equable distribution of wealth and access to improved education, health care and infrastructure. It is an individual-focused scenario in which free markets and trade liberalisation are important. Increased levels of personal income lead to increased demand for certain ecosystem services, such as meat, fish and vegetables. As a result, there is a decline in other services such as those from forests as forest is removed and crops are grown instead. It is interesting to note however that this scenario seems not to be concerned with the issue of biofuels, perhaps because their profile was not so high when the MEA was published. In addition, because of the global economic expansion, some essential ecosystem services decline, such as the availability of drinking water, which affects poor communities more than others. The MEA considers that the key challenge to ecosystem services in this scenario is abrupt and unpredictable changes in the ecosystems.
- In the *Order from Strength* scenario (Table 7.2), the keywords are 'regional' and 'fragmented'. Countries are very much focused inwards, concerned with their own protection and interests above anything else. Trade is severely restricted, security is enhanced and there are many more restrictions on the movement of people and information. Strategic businesses such as energy and water suppliers are under much greater government control. Global inequality increases and national environmental policies are all about securing natural resources. Such global agreements as those on climate change, fisheries and the like are only weakly adhered to by individual countries, leading to global environmental decline. There is an increasing gap between rich and poor countries and between the rich and poor within countries. Ecosystem services are more

Table 7.2   The main characteristics of the MEA scenarios (MEA, 2005)

Scenario name	Dominant approach for sustainability	Economic approach	Social policy foci	Dominant social organisations
Global orchestration	Sustainable development, economic growth, public goods	Fair trade (reduction of tariff boundaries), with enhancement of global public goods	Improved world, global public health, global education	Transnational companies, global NGOs and mulitlateral organisations
Order from strength	Reserves, parks, national-level policies, conservation	Regional trade blocs, mercantilism	Security and protection	Multinational companies
Adapting mosaic	Local–regional co-management, common-property institutions	Integration of local rules regulating trade, local nonmarket rights	Local communities linked to global communities, local equity important	Co-operatives, global organisations
Technogarden	Green technology, eco-efficiency, tradeable ecological property rights	Global reduction of tariff boundaries, fairly free movement of goods, capital and people, global markets in ecological property	Technical expertise valued, following opportunity, competition, openness	Transnational professional associations, NGOs

vulnerable in this scenario, with shortages of food and water in poorer regions leading to malnutrition in some areas.

- The *Adapting Mosaic* scenario (Table 7.2) is local and regional in nature, with a strong emphasis on the local management of SESs. Local institutions are strengthened and investment is made in improving knowledge about ecosystems and their functioning and management. Communications technology improves and barriers to the spread of information are thus reduced. National governments are decentralised with the idea that local management will develop its own adaptive strategies as needed. However, the local focus means that global problems such as climate change, fisheries and pollution are ignored, which may lead to major environmental problems. Ecosystem services are viewed as important in this scenario.
- The *Technogarden* scenario (Table 7.2) is global in outlook, based upon technology and 'natural capitalism'. Ecosystems are managed to provide optimum services for society, and technology and market-orientated approaches aim to solve issues with the environment. Unexpected breakdowns to delivery of these services can occur as a result of excessive control and management of the system, however. Consumers are required to pay for the services they receive from ecosystems. Society becomes better educated with regard to the value of ecosystem services and green technology in general. Farmers are encouraged to diversify into other ecosystem services beyond food. Markets for ecosystem services increase and the reliable provision of these services helps poorer countries to become richer. This increased reliance on technology does however bring some dangers, as any failures can have far-reaching consequences. Nature in which humans are not in some way involved becomes rarer as the 'gardening of nature' continues.

7.6.3 ALARM scenarios

A third group of scenarios was developed as part of the EU project ALARM (Assessing LArge-scale Risks for biodiversity with tested Methods; www.alarmproject.net), aimed at exploring the risks to biodiversity within Europe and the policy options that might mitigate them. The ALARM scenarios were based on the underlying drivers for environmental change described within the IPCC SRES scenarios (IPCC, 2000) and in particular reflected the range of climate-model and emission uncertainties in these scenarios (Fronzek *et al.*, 2012). Increases in mean annual temperature of 3–6 °C in Europe by the end of the twenty-first century over those at the end of the last century were projected, with annual precipitation increasing in northern Europe and decreasing in the south. In total, six scenarios were devised; three were integrative in nature and were developed using an econometric model (GINFORS, Global Interindustry Forecasting System; Stocker *et al.*, 2012) and three were developed as 'shock' scenarios (Spangenberg *et al.*, 2012).

The three integrative scenarios were called *Business as Might be Usual* (BAMBU), *Growth Applied Strategy* (GRAS) and *Sustainable European Development Goal* (SEDG) (Figure 7.17):

- In the *BAMBU* scenario, the policy decisions already made in Europe are enforced. It is not a business-as-usual scenario based on extrapolating past trends but rather relates to both recent decisions and those that are about to be implemented. No additional

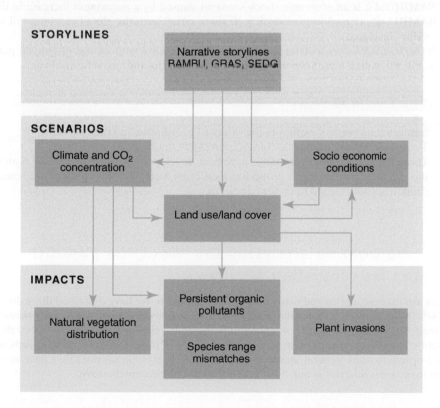

Figure 7.17   The ALARM storylines and scenarios, based on Figure 1 of the editorial introducing the special issue of *Global Ecology and Biogeography* about ALARM scenarios (Settele *et al.*, 2012)

measures are taken. It is a mixed bag of liberal markets and sustainable socioenviron-mental policies. This scenario fits closest to the SRES A2 storyline.

- *GRAS* is liberal, focused on growth and policy-driven. It is based on the idea of a world dominated by the market, free trade and globalisation. It is about deregulation and economic growth. Environmental policy is concerned with repairing damage and taking limited preventive actions based on cost–benefit analysis. This scenario corresponds to the SRES A1FI storyline and its assumptions.
- *SEDG* is about sustainability of society through integrated social, environmental and economic policies. It concerns competitive economy, gender equality, international cooperation and a healthy environment. It is precautionary in approach, trying to reduce uncertainty and avoid future as-yet-unknown problems. This scenario is closest to the SRES B1 storyline but differs significantly with regard to the level of greenhouse gas emissions.

The three 'shock, wild-card or hazard-driven' scenarios were *Cooling Under Thermohaline Collapse* (GRAS-CUT), *Shock in Energy Price Level* (BAMBU-SEL) and *Contagious Natural Epidemic* (BAMBU-CANE):

- *GRAS-CUT* is an environmental shock scenario, describing the collapse of the North Atlantic ocean circulation and the loss or reduction of the Gulf Stream, which is so important for western Europe's relatively mild climates.
- *BAMBU-SEL* is an economic shock scenario, caused by a permanent increase in the BAMBU oil price of 400%, which is likely to occur once the global maximum of oil production is past.
- *BAMBU-CANE* is a societal shock scenario, associated with an out-of-control pandemic, not unlike recent concerns over bird flu in China and swine flu in Mexico.

Summary: Scenarios

- Scenarios are ways of exploring the future of ecosystems and their services.
- They should be plausible, but they are not predictions.
- Three different groups of scenarios have been developed by the IPCC, the MEA and the ALARM project. Each group has a different logic, but overall there are some similarities with regard to their projections.

7.7 So what do scenarios say about the possible futures for ecosystem services?

There are many scenarios and many outputs that use them. We have described three different sets in this chapter. This was intended to give an overview and some general information on their history, structure and potential within the global change debate. Now, rather than describe the effects on each ecosystem service of each of our selected scenario groups, we aim to provide some general views of the future for ecosystem services that have come from these scenarios; this is not, however, intended to be comprehensive.

7.7.1 MEA scenarios

The MEA assessed the changes in ecosystem services and their drivers across their scenarios. Some general messages are that populations will continue to grow, although the rate of growth will continue to decline. Humans will increase their use of ecosystem services. There will be continued and increasing economic growth and different patterns of consumption, and as a result the demand for provisioning services such as food, fibre and water will strongly increase no matter the scenario, although food security will still be unattainable for many people. The demand for these types of service is likely to lead to loss or degradation of other services. For example, increasing the land under agriculture would lead to loss of natural forest and grasslands and their associated services. In reality, many current services are already degraded due to the increased demand for provisioning services. Typical negative effects include declining fish stocks, deteriorating water quality and the emergence of diseases in plants, animals and humans. Some regions, such as the dryland regions of sub-Saharan Africa, are threatened by a number of different drivers of change, including intensive land use, diminishing water supply and climate change. The degree of change in land use does vary among scenarios. Order from Strength has the greatest land use change, as more land is converted to crops and grazing. More technological scenarios such as Technogarden and Adapting Mosaic, on the other hand, have relatively little land use change due to the development of more efficient agriculture, decreased meat production and lower increase in population levels.

In all MEA scenarios, the changes suggested affect biodiversity and the services from biodiversity negatively to some degree, including the complete loss of some native populations of species and their associated services. This applies often to those that provide cultural services, but also to those that provide supporting services such as pollination. The MEA also argues that ecosystems and their services are more resilient if biodiversity in all its aspects (e.g. genetic, species, functional and taxonomic) is maintained to a high level. Habitat loss is projected to occur in all four scenarios, with a subsequent loss of associated native species, which might also mean global extinctions. However, future change in habitats across the globe varies. Some habitats have already been so extensively converted to agriculture and other forms of human management that future changes are likely to be minimal, for example in the Mediterranean region. Those habitats that have experienced only moderate changes in the past, on the other hand, are more likely to be heavily impacted in the future; these include tropical and subtropical forests and grasslands.

7.7.2   SRES scenarios

More specific examples of possible future responses to global change have been developed using the emission scenarios and a combination of ecosystem models, climate models and in some cases assessment or 'valuation' of the services. Shaw *et al.* (2011) explored ecosystem service change by using two emission scenarios (A2 and B1) and three atmospheric–oceanic general circulation models, along with economic valuation modelling, to assess the impacts of global change on some selected services in California. They were particularly interested in two important services: carbon sequestration and natural non-irrigated vegetation used for cattle grazing. They used the MC1 DGVM (Chapter 2) for the vegetation, carbon, nutrient and water fluxes and a modification of the CENTURY biogeochemistry model for plant productivity, decomposition and water and nutrient cycling. Their results indicated that the use of a warmer and wetter climate model increases aboveground carbon sequestration under both emission scenarios. However, with the hotter and drier climate models and the high-emission scenario (A2), carbon storage declines steeply, especially towards the end of the century (Figure 7.18).

The economic and social benefits of the increased carbon storage capacity under the warmer and wetter scenarios are estimated at US$22–38 billion annually. However, under the hotter and drier scenarios, the costs are modelled to be between US$646 million and US$62 billion annually, depending on the model and scenario. Forage production declines over the century under all models and scenarios: 14% in annual mean production under warmer and wetter scenarios and 58% under hotter and drier scenarios. Economic losses in forage production are US$14–570 million.

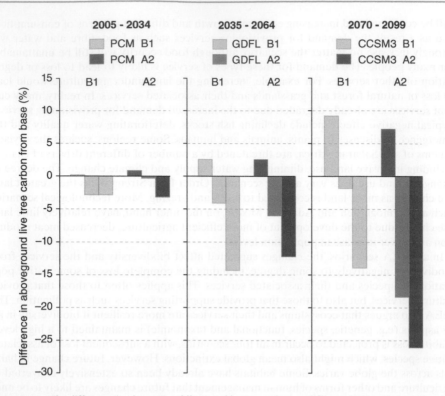

Figure 7.18   The difference in aboveground live tree biomass under two different SRES emission scenarios (A2 and B2) using three different climate models (PCM, GFDL, CCSM3) in three different time periods. Source: Shaw *et al*, 2011. Reproduced by permission of Springer

In another example, using the SRES B2 scenario with a regional climate model and a dynamic terrestrial biosphere model (AVIM2), Wu *et al*. (2010) projected that the average ecosystem net primary production (NPP) for China is likely to decrease. With an increase of 1 °C there are regional differences, with NPP responding favourably in some regions but declining in others. Under an increase of 2 °C, adverse effects become significant and key changes can occur. Increases beyond this lead to significantly greater adverse impacts. Northwestern China, which is arid today, is likely most vulnerable.

Rounsevell *et al*. (2006) used a downscaled interpretation of the global SRES scenarios for Europe to develop a number of spatially explicit land use change scenarios. They were interested in the coupling of climate change to related socio-economic factors. They used the AIFI, A2, B1 and B2 storylines for their scenario development, employing qualitative descriptions of the ranges and roles of different land use change drivers, quantitative assessments of the total area requirement of each land use type relevant to the changes in these drivers and spatial allocation rules to assign land use quantities in geographic space. The results showed most change in the use of agricultural land. As the development of crops leads to better yields, less land is needed and land abandonment occurs. Urban area increases in all scenarios, although there are differences in spatial patterns, reflecting different assumptions about urban development. Forest area increases in all scenarios; this is due in part to policy objectives, although abandoned land may also undergo succession towards natural forests. Bioenergy crops are likely to increase in some scenarios, and this likelihood has increased since this research was completed.

7.7.3   ALARM scenarios

Socio-economic drivers of biodiversity change and analyses of political and economic dynamics with regard to the storylines described in the ALARM project have been addressed by Stocker *et al.* (2012). Their results are scenario-dependent; for example, the lowest unemployment is under the SEDG scenario, while the highest occurs in the GRAS scenario. Resource use in general is highest under GRAS but decreases in SEDG. Energy usage does not decline in any of the scenarios. CO_2 emissions decline in SEDG and decline slightly in BAMBU but increase in GRAS. This is partly explained by an increase in renewable energy usage, especially in SEDG.

In a later EU project, ECOCHANGE (www.ecochange-project.eu), the ALARM scenarios were used as part of an integrated sustainability assessment (ISA) in three case study regions in Switzerland, Romania and Belgium. The study asked questions about plausible scenarios for 2050, types of landscape, the decisions made by farmers, the impacts of these decisions on sustainability, trade-offs and the strategies needed to reach desired goals (Bohunovsky *et al.*, 2012). The site in Belgium was a river catchment: a lowland area of intensive agriculture producing both food and energy crops in which there was significant pressure on biodiversity. At the site in Romania the economy was in transition, with major social and economic change and migration underway. The Swiss site was in a mountainous region sensitive to global and economic change and was very important ecologically (Table 7.3). A number of sustainability indicators were used in the study, including traffic demand, renewable energy production, protected areas, regional unemployment, income, regional self-sufficiency and quality of life.

Each ISA contained a number of stages, including scoping, envisioning, experimental and learning, all of which are more fully explained in Bohunovsky *et al.* (2012). Briefly, scoping included case study and stakeholder selection, interviews with stakeholders, research of existing literature and stakeholder workshops. The envisioning phase was based on the existing ALARM scenarios, as described earlier. In the experimental phase, visions and proposed policies were explored for adequacy and feasibility using agent-based modelling (ABM; see Chapter 2). The learning stage included an internal evaluation and assessment of lessons learnt. Various indicators expressing criteria for sustainability were chosen for each case study, and the three ALARM scenarios BAMBU, GRAS and SEDG were used. Each site had its own table and trends. The one for the Belgian site is given here by way of example (Table 7.4; Bohunovsky *et al.*, 2012).

Table 7.3   Main characteristics of the three ECOCHANGE case study sites (Bohunovsky *et al.*, 2012)

Characteristic	Belgium	Romania	Switzerland
Economy	Developed	In transition	Developed
Political	EU	Joining EU	Non-EU
Agriculture	Intensive	Extensive	Intensive
Urban pressure	High	Low	High
Governance	Federal and regional	Centralised	Regionalised
Topography	Gently undulating	Strong altitudinal gradient	Strong altitudinal gradient
Farm size	Average >50 ha	Average <5 ha	Average >10 ha
Population	195 000, dense, increasing	1600, loose, decreasing	60 000, dense, increasing
Study area	195 km²	120 km²	99 km²
Municipalities	10	2	13
Urban/rural	Peri-urban	Rather rural, medium-sized cities	Gradient from urban to rural
Agricultural sector	1%	80%	2%

Table 7.4 Evaluation of indicators in the Belgian case study. The black arrow indicates the target direction with regard to sustainability, while the coloured arrows show the direction of development of each indicator. If a coloured arrow is in the same direction as the black arrow then development is positive. As an arrow is more opposite in direction, so the development is more negative. This is also expressed in the colours: green is very positive, yellow is positive, grey is neutral, orange is negative and brown is very negative (Bohunovsky *et al.*, 2012)

Criterion Indicator	Country Scenario	Sustainable Development	Goal	Trend		
				BAMBU	GRAS	SEDG
Transport Impact	Traffic demand (distance that needs to be travelled per day)	Reduce distances	↘	↑	↑	↓
	Traffic by fuel- and non-fuel-based transport means	Higher share of renewables in the traffic sector	↗	↓	↓	↑
Energy sustainability	Renewable energy production	Higher share of renewable energy	↗	↗	↓	↗
	Renewable energy production from biomass	Higher share of biomass production without negative influence on food production	↘	↗	↘	↑
Biodiversity conservation	Protected areas	Higher share of areas for biodiversity conservation, without negative influence on the region's productivity	↗	↗	↘	↘
Employment	Regional unemployment	Lower level of unemployment and efficient use of human resources	↘	→	→	↘
Economic performance	Regional income	Higher regional income	↗	→	→	↑
	Income distribution	Higher distributive justice regarding income	↗	→	↓	↗
Regional self-sufficiency	Diversity	Strengthening regional production to meet regional demand (for high quality of life) and to make the economy resilient to external shocks	↗	→	↘	↘
	Independency		↗	↗	↓	↑
Quality of life	Work–life balance	Better balance between paid work and spare time, voluntary work etc.	↗	→	↓	↑
	Social capital	Stronger ties on the meso-scale (community, organisations, unions)	↗	→	↓	↑

Source: With thanks to Ines Omann

In the Belgian case study, the most sustainable scenario (SEDG) is also the most positive with regard to sustainable development, while the liberal growth scenario is the most negative; in fact, all indicators are negative. The more mixed scenario (BAMBU) is relatively stable by comparison. In the Romanian case study the sustainable scenario (SEDG) is strongly positive throughout, while the GRAS scenario is relatively stable. Bohunovsky *et al.* (2012) relate this to the depopulation that is happening in the study region. BAMBU trends show that traffic-related issues become important as tourism expands. In the Swiss example, once again the indicators under SEDG are positive for sustainability. On the other hand, the GRAS scenario leads to a situation where most indicators are substantially different from today: negatively so for many, including sustainability, work–life balance and protected areas.

What does this mean for ecosystem services in these case studies? In Belgium, using the ABM approach, production in terms of grassland and crops is highest under the GRAS scenario. Although organic production is highest under SEDG, it remains marginal with regard to productivity under all scenarios. In Romania, development under SEDG is sustainable. BAMBU outcomes are less sustainable than those under the GRAS scenario, because population decreases under GRAS and therefore the area for crops is reduced. Under BAMBU, recreation, such as skiing, leads to negative effects on food production and biodiversity. In Switzerland, food production initially increases in all scenarios, and under SEDG it subsequently remains constant. Meat production increases under the GRAS scenario. Land use that promotes higher biodiversity is only possible in small areas under all three scenarios. Under the SEDG scenario, land use reduces biodiversity because more intensive agriculture is still used and is in fact promoted by subsidies. However, under the liberal GRAS scenario, biodiversity is higher because subsidies no longer exist and organic crop prices are high, stimulating organic management regimes.

Summary: Scenarios and ecosystem services

- *MEA scenarios*: Populations will continue to grow, although the rate of growth will continue to decline. Humans will increase their use of ecosystem services, especially provisioning services such as food, fibre and water. Food security will still be unattainable for many people. Some ecosystem services are already being degraded by the increased demand for provisioning services. The degree of change in land use does vary among scenarios. Biodiversity and the services it provides are affected negatively to some degree under all scenarios.

- *IPCC scenarios*: Examples using a combination of ecosystem models and climate models include the impacts of global change on carbon sequestration and natural non-irrigated vegetation used for cattle grazing in California, calculation of the average ecosystem NPP for various regions in China and the use of downscaled European SRES scenarios to quantify requirements for different land use.

- *ALARM scenarios*: With regard to sustainability, examples from three different regional studies show that the more liberal scenario is likely to lead to a severe degradation of many selected indicators. However, there are regional differences in response among all three regions and between selected scenarios.

- Scenarios should be treated in a similar way to models and users should be aware that they are full of assumptions and uncertainties. They are a way of exploring possible futures, so long as they are not assumed to be 'the future'. They are one way of exploring possible changes to the services ecosystems provide to humanity. They also show clearly that the problems and relationships among ecosystem services and providers are highly complex, leading to both predictable and (more worryingly) unpredictable outcomes for some life-sustaining ecosystem services.

- Ecosystem services are vital for the survival of humanity. However, the relationships between them and the various drivers and pressures for change are highly complex, and while a number of scenarios have been developed to explore these relationships, there are likely to be many unexpected outcomes over the coming decades.

8

Cultural Ecosystem Services

'The entire cosmos is a cooperative . The sun, the moon, and the stars live together as a cooperative. The same is true for humans and animals, trees and the earth. When we realize that the world is a mutual, interdependent, cooperative enterprise … then we can build a noble environment. If our lives are not based on truth, then we shall perish.'

Buddhadasa Bhikkhu (in Swearer, 2001)

8.1 Introduction

In Chapter 7 we focused on the provisioning, regulating and supporting services provided by ecosystems to human society in the past and at present. In this chapter we concentrate on the past, present and future of cultural ecosystem services that include the conservation of species and ecosystems. Conservation has a strong biological context but has deep cultural roots that lie behind contemporary spiritual and recreational services. First, we should perhaps present our view of cultural ecosystem services, showing how they differ from the services that are of clear economic benefit to society. These services are among the most difficult to assess in any standardised comparative way with regard to societal benefit. Many cannot and *should not* be valued in an economic sense, but the current obsession with economics, the rule of the auditor and knowing the cost of everything but the value of nothing plays against this viewpoint. Cultural services can be described as the nonmaterial benefits people obtain from ecosystems. This might be through rather obvious activities such as recreation and fitness or through other, less easily defined activities such as aesthetic appreciation, cultural heritage and a person's spiritual and religious experiences within and as a response to an ecosystem, be it an old-growth natural forest or a high mountain area. Most European communities, for example, are linked in some way to the land, as exemplified by the Swedish *allemansrätt* (everyone's right) to access the greater part of the countryside, apart from private gardens. Forests in many parts of the world have strong religious associations for local populations.

Cultural services can form an important bridge to policy, as their appeal has a broad base in many communities. They can be grouped into three general categories. First, the recreation–fitness–health complex, which is of growing importance in increasingly

Ecosystem Dynamics: From the Past to the Future, First Edition.
Richard H.W. Bradshaw and Martin T. Sykes.
© 2014 John Wiley & Sons, Ltd. Published 2014 by John Wiley & Sons, Ltd.
Companion Website: www.wiley.com/go/bradshaw/sykes/ecosystem

urbanised societies. The hunting, shooting and fishing as 'sport' lobby can be included in this category, and in addition these activities have very tangible cash values: one day of grouse shooting in the Peak District National Park, UK, can be sold for well over £15 000 per person and moose hunting in Sweden can give higher cash returns than timber. Second, restoration of traditional systems of agricultural production in order to increase quality of foodstuffs is a second 'cultural' service that is of increasing interest. Food grown in traditional ways and the use of traditional animal breeds are thriving businesses in several countries. Third, conservation in the form of the protection and management of 'biodiversity' yields a widely appreciated nonmaterial benefit, and in regions with long histories of land use the target ecosystem states are often of 'cultural' origin. All these cultural services have an in-built time perspective, which is discussed later in this chapter.

As with the overall framework of this book, we need to go back in time to explore the background and interactions between society and the natural world. This should provide a better understanding of the present situation with regard to cultural values, biodiversity and conservation before we go on to look into possible futures. In the past, humans related to the physical and biological world through the need to acquire basic services such as food, water and shelter. But there were and are other needs that make us human. One is a strong need to relate to the universe on deeper levels of understanding, often expressed through some religious or spiritual aspect. In pre-literate societies this was usually linked to nature and the surroundings within which the people lived and consequently ensured the protection of special species and habitats. There are numerous examples throughout the world from many societies. We explore this spirituality and its links to conservation through sacred sites and sacred species in order to help understand where we have come from and how these values might develop in the future.

8.2   Sacred sites and species

> Tibetan Buddhists are interested in the essential being of all nature – past, present, future – including not just biodiversity, but all rocks, water, stars and the whole universe. A Tibetan sacred site does not have a function, such as conserving biodiversity; it is a connection with the essential being of a plant, of rocks, of water, of a mountain, of the sky, of the universe. To interpret a sacred site in the limited sense of Western Conservation is to misinterpret Tibetan Buddhism.
>
> Tashi Duojie, 2004 (in Salick *et al.*, 2007)

Mountains, groves of trees and other sacred sites have held a special place in the culture and belief systems of human societies throughout time. It is easy to trace this reverence back to Artemis and Diana in the Greek and Roman mythologies, but even these cultures probably modified and adopted earlier customs. Diana lived in a 'sacred grove' and was 'the goddess of woodlands and of wild creatures, probably also of domestic cattle and of the fruits of the earth' (Frazer, 1922). Her association with forests and nature was carried over into human fertility and successful childbirth, which has been a central concern of many cultures. On the North Island of New Zealand, the Tuhoe tribe of Maoris ascribed 'the power of making women fruitful' to trees (Frazer, 1922). Frazer illustrates the links between these classical goddesses and tree worship in general, with several compelling examples from around the world: the Celtic Druids and their oak worship; the Hidatsa Indians of North America and the cottonwood (*Populus deltoides*); the Wanika of eastern Africa and the coconut tree (*Cocos nucifera*); worship of clove trees (*Syzygium aromaticum*) in the Moluccas; and the Australian Dieri tribe, who believe their fathers' spirits are transformed into trees, which must therefore be protected. The list in Frazer's work is long, but surely not comprehensive. All these beliefs are forms of animism, in which plants, animals and even geographic features like mountains and rivers are endowed with spirits or souls, giving the general public in animistic societies a deep-rooted respect for nature. Aspects of these traditional beliefs have become incorporated into more recent religious frameworks, such as Shintoism, Hinduism and Buddhism. This inherent respect and regard for nature, particularly forests and trees, is a cultural relationship and should be regarded as a type of cultural service. This relationship contributes to public support for nature conservation and even to the development of environmentalism as a quasi-religious movement (Orr, 2003).

Sacred sites have frequently proven to be hotspots of biodiversity and natural targets for conservation. Metcalfe *et al.* (2010) describe a sacred site called the Three Sisters Cave complex in a fragment of Kenyan forest. Here the village elders have banned tree felling and vegetation destruction so as to preserve traditional values, and many threatened plant and animal species have survived the general exploitation and destruction of East African coastal forest that has occurred at similar sites. In another sacred site of 28 ha in Ghana is found the only remaining population of the Mona monkey subspecies (*Cercopithecus mona mona*) (Wild and McLeod, 2008). These types of preservation are probably among the oldest known methods of habitat protection (Dudley *et al.*, 2005). Sacredness at these sites can be associated with the ancestors of the local people, the presence of burial grounds or deity and nature worship. In most cases they are assumed to have originated in hunter-gatherer societies, based on a belief in the presence of nonhuman spirits in animals, plants, rocks and landscape features such as mountains, springs and rivers (Bhagwat and Rutte, 2006). Subsequent beliefs or religions, such as Christianity, can adopt the sites as their own but continue to conserve them, often justified by some new motivation relevant to their beliefs. In other cases, the sites may be destroyed along with the old religion, and this is likely to affect biodiversity conservation negatively. In modern times, with major changes from rural to urban cultures, many sacred sites are losing their sacred importance to local communities and consequently their conservation value. This whole subject is worthy of a book in its own right (see Pungetti *et al.*, 2012), but we only have space to cover a relatively few examples here. It is perhaps important to emphasise again that the sacred, religious and even medicinal aspects of these sites – in other words the ecosystem services – are what is usually important for the local people and that biodiversity conservation as we understand it is peripheral.

Summary: Sacred sites and species

- Cultural ecosystem services are among the most difficult to assess in any standardised comparative way with regard to benefits for society. Many of the services cannot and *should not* be valued in an economic sense.
- Humanity often needs to relate to the universe through deeper levels of understanding than the usual mundane everyday. This is often expressed through some religious or spiritual aspect in people's lives.
- In pre-literate societies this was usually linked to nature and the surroundings within which the society lived, and as a result ensured the protection of special species and habitats.
- Sacred sites and sacred species are found on all continents and it is important to note that for the local people it is not about conservation per se but rather about connecting to the essential 'beingness' of the plant, animal, hill or rock.
- Spiritual associations with nature contribute to public support for nature conservation and to the development of environmentalism as a quasi-religious movement.

8.2.1   Some examples from around the globe

North America   Sacred mountains are the focal point of the culture and traditions of the Mescalero Apache Indians of south-central New Mexico (Ball, 2000). In their traditions, all mountains are sacred, but some ('medicine mountains') have higher biological diversity than other mountains and thus provide more opportunities to find medicinal plants. White Mountain, Otero County (Figure 8.1) is particularly important as a medicine mountain, having greater biodiversity than others; it is sacred for ceremonial reasons but also because of its medicinal value. Sacred mountains are divided into three zones – base, middle and top – related to where medicines can be found. Power in mountains is concentrated near the top and medicines found near the top are stronger as a result. In the Mescalero creation stories, mountain tops were the areas that protruded above the waters which once covered the land and thus were the only places where the spirits floating in the air could go. Mountains are not powerful because of their medicinal value; rather, medicines are powerful because they are found on mountains, especially the mountain tops. According to Ball (2000), the mountain spirit tradition is not particularly old, being historical – that is, some hundreds of years old; maybe as few as 300 – in comparison to some Mescalero ceremonies, such as the 'Big Tipi' female initiation ceremony, which goes back to 'the beginning of time'. However, from a conservation point of view the preservation of sacred mountains and their biodiversity is important and even a few hundred years of conservation is valuable.

Asia   Mountains as sacred sites are also important in Tibetan culture, where the elements of the whole environment, both living and nonliving, are intimately linked and are therefore all viewed as sacred (Swearer, 2001). There is a hierarchy of sacredness, as in a mandala, with sacred trees in a sacred grove on a sacred mountain. Anderson *et al.* (2005) compared

Figure 8.1   White Mountain (Sierra Blanca), Otero County, New Mexico. Source: http://en.wikipedia.org/wiki/File:Sierra_Blanca_and_electricity_pole.jpg. Courtesy Dusty Matthews

sacred and nonsacred sites, using maps and geographic information systems (GISs), on one of the mountains in the eastern Himalayas, looking at their species richness and habitat diversity. They found that sacred sites were both richer and more diverse than randomly selected nonsacred sites. In addition, above a contour line called the *ri-vgag* (door of a mountain) the mountain is entirely sacred, and endemic species are more frequent above than below it. In a later paper the same group explored sacredness within habitats and the effects on conservation (Salick *et al.*, 2007). They found from field studies at this scale that sacredness preserves old-growth trees and forest cover.

Africa Ancestors are often involved in traditional religions and the conservation of sacred land. Byers *et al.* (2001) investigated Zimbabwean traditional spiritual values associated with the preservation of forest sites for both tribal ancestors and future generations. The Shona people, who live along the Musengezi River, believe that some of the forest patches are protected by ancestral spirits who have returned to live among the current inhabitants. These spirits can take physical form, such as animals, and the most powerful – former tribal chiefs – guard the sites, often taking the form of lions. The area of dry forests in the Musengezi region decreased substantially between 1960 and 1993 as a result of either the complete felling of the forest or the removal of forest patches. Much of the clearance was for agriculture and new villages. Once cleared, dry forests do not regenerate in the medium term and if they are not used for crops then the typical dry forest species are replaced by common early successional species. However, the reduction of forest was substantially less in those areas that included sacred forests than in those unconnected to sacred areas. Traditional religious beliefs have influenced behaviour in these areas but have not prevented loss and fragmentation from happening to some degree. Reasons for this may include a lack of local knowledge by local people today about where sacred sites were located and their extent, the reduction of the influence of 'spirit mediums' (who identify the sites) and the disempowerment of traditional leaders. Overall there is a clear link between the conservation of the culture and the conservation of native forests.

Oceania Australia is one of the most biologically diverse countries in the world, with a wide range of climates and ecosystems. It has been inhabited for at least 60 000 years by the Aboriginal people, comprising 250 language groups, each of which has developed close relationships with their local regions and landscapes. Sacred sites in Australia are associated with this Aboriginal culture and are usually based on land. Nature and land conservation are central at all levels – individually, societally and culturally – and cannot be separated from the survival of the Aboriginal people (Gray, 1999).

> 'When I discuss biodiversity with my people they want to know what it is. When I start explaining the concept in terms of species, animals, plants, the whole of existence, they start to realise what I am describing. It's part of their land, of their very existence as Iwingi people.'
>
> Henriette Fourmile, Polidingi tribe (in Senanayake, 1999)

Aboriginal culture is closely linked to the stories of the Dreamtime and how spirit ancestors created the stones, the plants and the animals. Once they had created the world they themselves changed into trees, rocks and stars, and thus they are themselves the sacred sites. As a result the Dreamtime continues and is never-ending: past, present and future. The ancestors are the land, and the Aborigines, who are their descendants, are directly connected to it.

The Queensland World Heritage Area comprises forest stretching over 900 000 ha and contains many rare and endangered plant and animal species, including diverse assemblages of primitive angiosperm families, as well as biota representing eight major stages in the evolutionary history of the Earth. The local Aborigines are culturally diverse in language, technology and organisation (Slikkerveer, 1999). In the north of the region lies the traditional land of the Kuku-Yalanji people. While much of the area is rainforest, there are also more open areas of sclerophyll vegetation, which provide the local people with important cultural services, including the provision of grass species, which are used to make dilly bags. These were traditionally used for ceremonial and totemic purposes linked to specific sites in the landscape, as well as for the carrying and storing of food. Nowadays they are often made for tourists. In the past these open areas were maintained by the use of fire (Chapter 5), but this is now banned by the government. As a result, wet tropical forest species are invading these open areas, leading to a loss of both diversity and cultural services.

Europe In Europe, sacred sites are probably no less common than elsewhere in the world, although they may seem fewer due to the extensive land-use change that has occurred there. In many cases the original reason for a site being sacred may no longer be clear. Other sites around the world are often clearly associated with pre-literate communities. In Europe, this link can be less obvious, as many sites have been destroyed by industrialisation and agricultural intensification or absorbed by later religions such as Christianity. Evidence of sacred land and structures can be found throughout the Holocene. Neolithic people established sacred sites such as burial sites, earthworks or henges, culminating in the great megalithic monuments we described in Chapter 6, such as the ancient temple of Newgrange in Ireland. Newgrange was not just an elaborate tomb but rather a centre of astrological, spiritual and ceremonial importance and part of a complex of temples and other related structures covering a large area. The landscape setting of Newgrange and the even larger temple at nearby Knowth were of crucial importance, and the cultural services provided must have been a dominant concern for a large community. These temples were built 500 years earlier than the Egyptian pyramids and over 1000 years before the English Stonehenge. Druids, who first appeared around 2400 years ago in some parts of Europe, are not associated with many of these famous prehistoric sites, as is sometimes believed in popular culture (Palmer and Palmer, 1997). The Druids did have sacred sites, but little evidence of these remains. The Celts swept into the British Isles from mainland Europe at the onset of the Iron Age and, while they had some similar beliefs to the Druids, they concentrated on the sacred nature of forests, groves and lakes. Rather little evidence of their sacred groves now remains, but Tim Robinson has painstakingly researched and mapped numerous landscape features associated with ancient folklore and traditions from the Burren (Figure 8.2), the Aran Islands and Connemara, western Ireland, which reach far back in time.

In European societies, the spring and summer festivals in particular accentuate the importance that even urbanised populations still attach to natural values. There are connections between contemporary traditions such as the Swedish midsummer pole, the European maypole and the more international Christmas tree and the widespread earlier forms of tree worship. There are also many examples of specific plants, animals and birds that have some spiritual links. Trees in particular, probably because of their longevity, are often very important as sacred objects and were regarded as such well before the arrival of Christianity. Oak (*Quercus robur*) is an important species in Celtic, Germanic and Viking belief systems. Yew (*Taxus baccata*) is another, and is also one of the oldest trees in Europe. A yew tree growing in the churchyard at Fortingall, Perthshire, Scotland is estimated to be between 2000 and 5000 years old (Figure 8.3), while one in the cloisters of Muckross Abbey, Ireland is at least as old as the abbey itself, which was founded in AD 1448. The yew was considered sacred to the Druids and was seen in early Celtic belief as a symbol of immortality. It was associated with fairies and devas of the forest. The site of such a tree may have been used for Beltane fires in May each year. The longevity of these trees was likely used as a

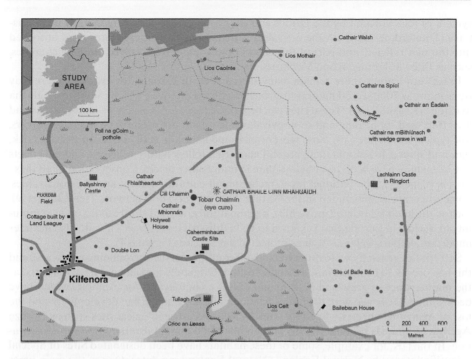

Figure 8.2   Part of the Burren, Ireland, showing the high density of ancient human settlement on the limestone (pale blue) and a traditional cultural service: Tobar Chaimín, a well whose waters are a cure for eye problems (www.foldinglandscapes.com). Source: From Tim Robinson's map of the Burren, published by Folding Landscapes, Roundstone, Co. Galway

Figure 8.3   The yew in Fortingall churchyard, Perthshire, Scotland. This image was taken from the Geograph project collection (http://commons.wikimedia.org/wiki/File:Fortingall_Yew.jpg, last accessed 22 November 2013). The copyright on this image is owned by Snaik and is licensed for reuse under the Creative Commons Attribution-ShareAlike 2.0 licence ©Snaik

symbol of reincarnation by the Druids and later as a symbol of resurrection by the Christians (Dunford, 2004). With the arrival of Christianity, many sacred sites were absorbed into the new religion, and part of this absorption would have involved creating churches, such as those at Fortingall and Muckross.

In Eastern Europe there remain many sacred sites. In Estonia, for example, there may be as many as 7000 sacred natural sites, including sacred groves, which can be up to 100 ha in extent (Kaasik, 2010). These sites can be stones, trees, lakes, rivers or marine areas and are often used for healing and soul revival, while groves are used for community prayers. The typical sacred trees are those most often found naturally in the landscape, such as oak, pine and spruce. In Estonia there are still many that follow nature religions, and the effects of Christianity have been less obvious than in many other places. Conservation of all these sites is planned at the national level, including all cultural, natural and religious values. Some areas are also included in the Natura 2000 conservation network (see Chapter 9), such as the hiiemägi (sacred grove hill), a seminatural grass and shrubland: a plant community in decline in Europe that is known to be highly species diverse. This site also includes important orchid species, as well as protected ant species (Figure 8.4).

Sacred sites generally were often treated differently from the surrounding landscape, and this has proven to be important for conservation. Originally it would not have mattered as the maintenance of grassland both in and out of a churchyard, for example, would have been similar. However, with the intensification of farming and urban development and the loss of many high-diversity habitats, churchyards and archaeological sites such as Bronze Age burial mounds became islands of high plant and animal diversity ripe for conservation. In the UK, for example, some of these habitats have been designated Sites of Special Scientific Interest by Natural England (Natural England, no date). One of the main types of habitat that is found in churchyards is seminatural grassland: a highly diverse but threatened habitat in many places in Europe. To maintain its diversity, this grassland must either be grazed by animals only moderately or be cut just two or three times a year. It is also important that the biomass is taken away in order to create or maintain a low nutrient

Figure 8.4   Paluküla hiiemägi, Kehtna Parish, Rapla County, Estonia: a sacred site conserved within the Natura 2000 network. Source: http://www.panoramio.com/user/4933208?with_photo_id=41586026. Courtesy of urmas286

status, which promotes high diversity. Unfortunately, this does not always happen and there can be either frequent grass cutting or virtual abandonment, both of which lead to lower biodiversity. Other 'special habitats' can also be found in churchyards, including complex lichen communities on old gravestones and bat habitats (Manning, no date).

In a similar way, Mediterranean olive groves develop high biodiversity if maintained by traditional methods. The olive (*Olea europaea*) was being cultivated in Syria and Crete 5000 years ago (Standish, 1960; Dudley *et al.*, 2005) and was an important sacred tree for Greek and Roman cultures. It is also central to Judaism for a number of reasons, including as a symbol of permanence during the change from a nomadic existence to permanent settlement, and is an important symbol in the Christian faith. It is a slow-growing tree and over time its groves become areas of high biodiversity, with a variety of species and habitats. However, traditional groves are under threat from two opposing trends in land use: abandonment, where rural populations are in decline, and the intensification of olive production, where pesticides, fertilisers and excessive cultivation lead to the loss of habitat and diversity (Moussouris and Regato, 1999).

8.3   Cultural landscapes: biodiverse relicts of former land use systems

The abandonment of Mediterranean olive groves is just one example of the altered pace of European landscape change, which has accelerated considerably during the last 2 centuries (Chapter 6), with many ecosystems with which people had been familiar for generations becoming rare or even disappearing. This process is primarily the replacement of one type of 'cultural' landscape by another, reflecting socio-economic developments, but the loss of traditional cultural landscapes, which can have histories of several thousand years, stimulates debate (Agnoletti, 2006). There has been a continuous record of intensive land use and cereal cultivation in the Po Valley of Italy for the last 6300 years (Kaltenrieder *et al.*, 2010), but abandonment of traditional small farms with formerly extensive grazing on less fertile soils in the nearby Tuscan uplands during the last century has led to regrowth of forest over large areas that were managed open landscapes in the past. Species composition has changed as a consequence, and there has also been a loss of local suppliers of traditionally produced foods. The increasing interest among developed countries in farm shops, eco-foods and the 'slow food' movement (www.slowfood.com) is a reaction to rapid changes in food production techniques and shows a tacit appreciation of the cultural values associated with former, traditional farming systems. This type of food production illustrates a shift in ecosystem services from provisioning to cultural: when a society has enough to eat, the cultural values of eating rituals often appear to be more significant than their provisioning value.

Sweden is one of the most forested countries in Europe and potentially among the least affected by recent changes in agricultural practice, but not only does cultural impact in Sweden have a longer and geographically more extensive history than was once believed but also many of the prized biodiversity values within Sweden owe their existence to a complex interrelationship between cultural and natural processes (Bradshaw and Hannon, 2006). In the province of Halland in southwestern Sweden, almost half of the land protected in nature reserves is valued for having communities of cultural origin that require active management for their maintenance. This demonstrates the willingness of a public authority to pay for the maintenance of cultural ecosystem services. The long-term history of Swedish forest meadows or wood pastures (Swedish: *löväng*) and hay meadows (Swedish: *slätteräng*) illustrates the importance of former cultural activity

Figure 8.5   *Scorzonera humilis* (Viper's grass) growing in a Swedish forest meadow

to current biodiversity values. Forest meadows or wood pastures are among the most cherished and visually attractive cultural landscapes that exist in Southern Scandinavia (Figure 8.5). They were developed from forested landscapes close to farms and formed part of the so-called 'in-fields'; they have counterparts in Central and Southern Europe. They are maintained by selective thinning of trees, together with mowing and controlled burning to create a mosaic of vegetation types, which include sparse deciduous trees and shrubs among herb-rich grassy meadows. Many of the trees are pollarded, and in the past they were shredded for leaf fodder. Most forest meadows have been abandoned in the last century. The parish of Mara in the province of Blekinge comprised just five farms within an area of about 10 km² in 1769, yet included 130 separate forest meadows. There is a restored forest meadow on the Råshult estate, southwest Sweden, which was the birthplace of Carl Linnaeus. It was declared one of only ten Swedish cultural reserves in 2003. The meadow system has its origins in about AD 1100, when mixed deciduous forest was cleared. There is evidence of small-scale cultivation of rye from this time, when there was a marked increase in the diversity of herbaceous plants and grasses (Lindbladh and Bradshaw, 1995). Forest made a temporary comeback for 150 years following the Black Death and local depopulation around AD 1350. Between 1700 and 1900 there was increasing pressure on the land to supply food to the rapidly expanding local population. The fields were probably overgrazed during a period of poor growing seasons and they became unproductive heathland and were subsequently abandoned, with many inhabitants emigrating to the USA during the late 1800s. The restoration of forest meadows in Sweden today is chiefly motivated by the biological value of the flora and historical interest, but also by a certain nostalgia for traditional systems of dairy cattle management. The grazing of cattle in the herb-rich meadows improves the quality of their milk.

8.4 Hunting as a cultural ecosystem service

'Laives of the Forest – a certen territorie of wooddy grounds and fruitfull pastures, priviledged for wild beasts and foules of forrest, chase, and warren, to rest and abide in, in the safe protection of the king, for his princely delight and pleasure.'

Manwood (1598), quoted in Cox (1905)

In the past, the conservation of species was associated with hunting, whether by accident or by design. Hunting can be viewed as both a provisioning and a cultural ecosystem service. Australian Aborigines and several other cultures used fire as a tool to promote hunting, providing favourable habitats for herbivores, although according to Bird *et al.* (2005) it was the hunting of burrowing animals such as lizards (by women) that benefited, rather than the hunting of larger animals (by men). Much of this type of hunting is a provisioning service. Burning also helped increase the production of local food plants (Bowman, 1998).

In Europe, the biodiversity of the forests today has been heavily influenced by different types of management over many centuries, both for provisioning and for cultural reasons. Many forests have been intensively exploited for resources such as firewood, construction materials and animal feed, often becoming open parklands as a result (Welzholz and Johann, 2007). The rich and the powerful often took charge of forest areas and controlled how they were used; by the eleventh and twelfth centuries, forests in a number of countries had become exclusively royal hunting grounds, and while the hunt clearly provided some food, the main service was the enjoyment of the activity by the privileged hunters. Forests were conserved and lower classes were excluded from hunting, fishing, grazing or the felling of trees. What precisely counted as a forest is not always clear and the definition was not the same as that we use today (Rackham, 2003). Forest was described in Norman, Plantagenet and early Tudor times as a 'portion of territory consisting of extensive waste lands, and including a certain amount of both woodland and pasture, circumscribed by defined metes and bounds, within which the right of hunting was reserved exclusively to the king, and which was subject to a special code of laws administered by local as well as central ministers' (Cox, 1905). Thus a 'forest' may in fact have been a very open landscape with few trees (Figure 8.6). Nevertheless, clear conservation measures were undertaken, if only for the king and his friends. Sometimes peasants were allowed to use selected paths through a forest, and they used this access to collect wood from the path 'by hook or by crook'.

These types of restriction were common throughout Europe during this period. Białowieża Forest, stretching though Poland and Belarus, is well known as one of the few remaining old-growth forests in Europe and was a royal reserve from the fifteenth century. Similarly, around the same time in Romania there were extensive hunting areas that conserved many wild animals, including the aurochs and European bison.

The palace and park of Fontainebleau near Paris is now a UNESCO World Heritage Site. It was originally built as a hunting lodge for royalty in the twelfth century and was later extended into a palace in the sixteenth. The associated forest of Fontainebleau, set in the Île-de-France, is a huge mixed deciduous and conifer forest covering 25 000 ha, mainly comprising oak, beech and pine species, along with a generally rich biodiversity of fungi, plants and animals, including thousands of insects (Figures 8.7 and 8.8). From the seventeenth century, some parts were protected by royal command. La Tillaie is a beech ecosystem within the forest that has not been significantly managed or cut since AD 1372 (ANVL, 2007), although there is pollen evidence for prehistoric forest disturbance by people (Lemée, 1981). However, conservation only really began in the forest in the middle of the nineteenth century, when artists of the Barbizon school, part of the movement towards realism in art based in the nearby village of Barbizon and renowned for their pioneering paintings of nature, fought to save the old trees used in their compositions. In 1861, 1091 hectares were officially declared by imperial decree to be 'artistic reserves'. This was the first such

THE KING HUNTING (1)
(FIFTEENTH CENTURY)

Figure 8.6   Hunting in the royal forests of England in the fifteenth century (Cox, 1905)

protection in the world, predating the designation of the first National Park in the USA (see Chapter 9). In 1965 the whole of the forest at Fontainebleau was protected as a 'natural site'.

Hunting, whether by the aristocracy in royal forests or by Swedish moose hunters in boreal forest, can be regarded as the development of a previous provisioning service into a luxury cultural ecosystem service. Bobwhite quail (*Colinus virginianus*) is probably the most important game bird hunted in the southeastern USA, even though it typically weighs a paltry 200 g. Native Americans are known to have used it for food alongside the more rewarding wild turkey (*Meleagris gallopavo*) and it became an important subsistence food for early European settlers. A thriving market in quail developed and the birds became rare in the northeastern USA. 500 000 individuals were exported from Alabama to the northeast during the winter of 1905–06. Gradually, quail hunting developed into a luxury pastime and today it is very big business for an exclusive clientele, with 5 million birds introduced into large hunting estates in Georgia each year (Dozier *et al.*, 2010). Quail hunting is also a key factor in ecosystem management in the southeastern USA and has contributed to the intensive prescribed burning programme practised there, based on research initiated at the Tall Timbers Research Station (www.talltimbers.org). The quail hunting business

Figure 8.7   The authors in Fontainebleau Forest, December 2012. Photo by Gina Hannon

Figure 8.8   Old-growth Fontainebleau Forest

has much in common with grouse shooting and deer stalking in the British uplands. These are cultural ecosystem services with such great economic value that they can dominate the ecosystem management focus of large regions. Gamekeepers in the UK manage 1.3 million hectares of land for hunting, exceeding the combined area of all nature reserves owned by the government and conservation organisations (Northumberland, 2011).

The interests of the hunting lobby can combine with other conservation interests to raise the profile of the management of cultural ecosystem services, but there can be disagreements, particularly in relation to the management of predators of game animals and the use of fire. The potentially conflicting motivations for nature conservation are further discussed in Chapter 9, but hunting and conservation interests account between them for the considerable increase in importance of cultural ecosystem services since industrialisation.

Summary: Hunting as a cultural ecosystem service

- The European landscape has changed rapidly during the last 2 centuries, as the traditionally managed or 'cultural' landscapes of extensive agriculture have been replaced by highly mechanised intensive agriculture.
- The process is primarily the replacement of one type of 'cultural' landscape by another, reflecting socio-economic developments.
- Hunting, although partly a provisioning service, became a cultural service for the exclusive enjoyment of the rich, as the 'common man' was excluded from royal estates.
- It has remained a cultural service in many countries and can still be maintained as the preserve of the wealthy.

9
Conservation

'Earth provides enough to satisfy every man's needs, but not every man's greed.'

Mahatma Gandhi

9.1 Conservation as we know it

Nature conservation is a topic in which biological theory rubs shoulders with practical management, and there are many diverse opinions about appropriate targets, processes and timescales. In scientific terms, biological conservation or conservation biology has been defined in a number of ways. According to Soulé (1985), conservation biology 'addresses the biology of species, communities and ecosystems that are perturbed, either directly or indirectly, by human activities or other agents'. The term was first used in 1978 at a conference on conservation biology at the University of California in San Diego. However, somewhat earlier in 1968, an Elsevier journal, *Biological Conservation*, began publishing papers covering a wide range of subject areas closely associated with conservation. In its first editorial statement in 1968, the editors described 30 subject areas, including history, man and the environment, biosphere maintenance, ecosystem disturbance, population dynamics, threatened species, reserves, parks and botanical gardens, exotic species, economics, legislation, education and conservation in various forms. So conservation can be many things, but naturally it involves humans in some way, which may be self-evident but is perhaps worth emphasising.

So what does the future hold for biodiversity and for the attempts to conserve it? The current answer is probably that the situation is dire. The Millennium Ecosystem Assessment (MEA, 2005) projects the future extinction rate of species to be 10× higher than the current rate, which is itself 1000× higher than the rate estimated from the fossil record (Figure 9.1).

Conservation as it is understood by society today was probably an issue as soon as the global population became sufficiently large to impose noticeable impacts on natural resources, and many societies must have begun to discuss the relationships between exploitation and conservation. As is often the case, it was the ancient Greeks who first (at least in writing) raised the issue of conservation. Plato expressed concern about the state of the land and the conservation of soil in his *Critias*, c. 380 BC: 'By comparison with the original territory, what is left now is, so to say, the skeleton of a body wasted

Ecosystem Dynamics: From the Past to the Future, First Edition.
Richard H.W. Bradshaw and Martin T. Sykes.
© 2014 John Wiley & Sons, Ltd. Published 2014 by John Wiley & Sons, Ltd.
Companion Website: www.wiley.com/go/bradshaw/sykes/ecosystem

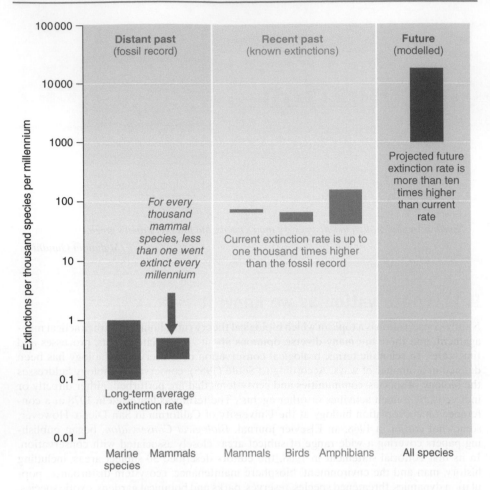

Figure 9.1 Rates of species extinctions (MEA, 2005)

by disease; the rich, soft soil has been carried off and only the bare framework of the district left ... what now we call the plains of Phelleus were covered with rich soil, and there was abundant timber on the mountains ... not so very long ago trees fit for the roofs of vast buildings were felled ... there were also many lofty cultivated trees which provided unlimited fodder for beasts ... the soil got the benefit of the yearly "water from Zeus" which was not lost, as it is today, by running off a barren ground to the sea; a plentiful supply of it was received into the soil and stored up in the layers of nonporous potter's clay. Thus the moisture absorbed in the higher regions percolated into the hollows and so all quarters were lavishly provided with springs and rivers' (Hamilton and Huntington, 1961). This reads much like the many problems associated with soil erosion caused by over-exploitation and lack of conservation in today's world.

Later, the Chinese were also concerned with conservation, and the *Book of Rites* (c. 70 BC) describes the organisations and government officials appointed to restrict hunting, fishing and tree felling to certain seasons. Conservation measures were from an early time regarded as a natural part of the management of game, fish and forest resources (Zhu, 1989), and this approach continues today in many parts of the world, with Swedish moose hunters for example considering themselves wildlife managers and conservationists (although some environmental activists might disagree). My (R.B.) grandfather, Birger Ahlgren, was a Swedish jägmästare (literally 'master of the hunt', but in reality a qualified forester), who ended a relatively undistinguished career working in a hunting and fish-tackle shop in Stockholm. He certainly regarded his earlier, field-based forest company job in northern Sweden as protecting biological values in Swedish forests during the 1920s, and not purely maximising commercial timber production, although his employers did not fully share his ideals and his career as a forester was short. For Birger, and many old-timers like him, conservation was 'wise use' of resources, although many would argue that important aspects of this approach were lost when the ruthless, focused efficiency of commercial forestry methods transformed much of boreal Sweden into spruce–moose plantations.

Examples of conservation can be found throughout history. St Cuthbert is recognised as one of the earliest conservationists in England for introducing bird protection laws to save the eider duck population on one of the Farne Islands in Northumbria in AD 676, even if his motivation was to secure the supply of comfortable bedding filled with soft eider down. As we saw in Chapter 6, human impact on ecosystems had become so severe and widespread by the 1800s that the need for conservation measures began to be widely debated, at least in Western countries. Formal conservation measures started to influence land management, even in remote, sparsely populated regions of the Western world. Conservation until this time had been a by-product of other activities, such as the protection of areas for the enjoyment (usually hunting) of the rich and noble. But change was afoot (as mentioned in Chapter 8), and in 1864 the beauty of the Yosemite Valley in the USA was formally preserved for public recreation when an area of 60 square miles came under the jurisdiction of the State of California. However, it was the Yellowstone area in northwest Wyoming, covering 2 million acres, with its 'boiling sulfur springs', that in 1874 became the first National Park in the USA and in the world. By 1890 a further three National Parks were created – Yosemite and two others: the Sequoia National Park and the General Grant National Park – to protect the giant sequoia trees of the western Sierra Nevada. There are now around 400 National Parks in the USA, which allow accessibility for all and preserve wilderness areas for future generations. 'Preserve' is usually the appropriate term as active management is kept to a minimum in the hope that these wilderness areas will retain many features that are 'natural' and relatively unmodified by human activities. National Parks of this type have now been established in many parts of the world, including Australia (1879), Canada (1885), New Zealand (1887), Sweden (1909) and Switzerland (1914). In Europe, most countries have established them in the last 100 years or so, and the International Union for Conservation of Nature (IUCN; www.iucn.org) now lists around 6600 National Parks globally.

Conservation practitioners are constrained by factors that barely affect modern research scientists, who often live within a virtual world viewed through the window of their computer screen. Practical conservation managers have to work with the ecosystems and species that currently occur within their country, in full view of the general public and politicians and within volatile budgets. Consequently, much nature conservation policy to date has been national in character. It is often static, as a common public perception is that change is unnatural. The ecologist Sir Arthur Tansley once said that many view conservation as 'keeping the landscape the way it was when I was young'. The focus of conservation has all too often been on the maintenance of stable populations of cuddly, furry animals or birds that are not small, brown or common on rubbish tips. Costly interventions with 'nature' are limited.

Summary: Conservation as we know it

- The MEA projects the future extinction rate to be 10x higher than the current one, which is itself 1000x higher than the rate estimated from the fossil record.
- The problems are not new and the ancient Greeks and Chinese raised the issue of conservation.
- Examples of conservation measures are found throughout history.
- The first National Park was created in 1874 in the USA and there now around 6600 globally.

9.2   Knowledge of the past: relevance for conservation

Practical conservation has been supported by academic conservation biologists, chiefly based in biological departments, where teaching and research on longer timescales has traditionally been weak. In the rather few cases where the palaeoecology of protected areas has been investigated, there is often evidence for prehistoric anthropogenic influence on ecosystems that has relevance for the choice of management goals today.

Rutger Sernander was an influential Professor of Botany at Linneaus' base in Uppsala, Sweden between 1908 and 1931 and was the founder of the Swedish Society for Nature Protection, but his static view of vegetation laid a dead hand on conservation biology in Sweden for many years. He was convinced that farmers were destroying the natural dynamics of Swedish forest meadows (see Chapter 7) and that if the meadows were left undisturbed, their natural values would be enhanced. In 1911 he organised the fencing of a 5 ha nature reserve at Vårdsätra, near Uppsala, with a management plan of free development, leaving all dead plant material in place. One hundred years later, no trace of the meadow remains, as a woody succession has restored the forest canopy. It took an amateur botanist, Mårten Sjöbeck, to point out that even in sparsely populated Sweden, many of the biological properties valued by scientists and the general public were a result of several thousand years of human impact with fire, axes and domestic grazing animals (Lindberg, 2009). Sjöbeck's comments are probably appropriate to a greater or lesser extent to protected sites in many parts of the world.

Tansley, the eminent British ecologist and advocate of conservation, recognised the dynamics of ecosystems early in his career through his studies of rapidly changing coastal communities in England and France. Consequently, he understood that static preservation, as advocated by Sernander, would not adequately protect many types of vegetation: active intervention and management would be necessary to conserve the biological value of dynamic ecosystems. Tansley also had the key insight that most British vegetation was largely the outcome of past land use. He was a pragmatist and realised that ecologists must collaborate with farmers and foresters in managing the current landscape, acknowledging the need for increased timber and food production in the lean years following the Second World War (Ayres, 2012). He argued for the ecological study of intensively managed ecosystems as well as seminatural ones, but was prepared to complain that the Forestry Commission was 'destroying the beauty of the still wooded Highland glens by substituting close plantations of alien conifers for the natural oak, birch and pine' (Tansley, 1939). He has rightly been dubbed the Father of Ecology, and many of his ideas still guide conservation policies today, at least in the British Isles.

9.2.1   Fire history, conservation and ecosystem restoration

Fire presents some classical nature conservation quandaries that have been resolved in different ways in different parts of the world. In many countries the use of fire by people has been regarded as one of the threats from which natural communities need protection. Fire suppression became the widely accepted management plan for public land in the USA for at least three compelling reasons. First, there were many deaths and loss of property to fires as the West was colonised by new settlers during the late 1800s to mid-1900s. Second, the loss of timber resources to wildfire was judged to be an unacceptable waste during the high demand of the Second World War. Third, scientists generally believed that fire suppression would promote desirable ecological stability. In 1944 the US Forest Service began the very effective Smokey Bear publicity campaign to engage the public in fire prevention, which became the longest running public-service advertising campaign in North American history (see Figure 5.9). Firefighting efforts were highly successful, with the area burned by wildfires reduced from an annual average of 120 000 km^2 during the 1930s to 8000–20 000 km^2 by the 1960s (Figure 9.2). Spending on fire suppression had become unacceptably high by the 1960s, however, and scientific opinion on the ecological value of fire had changed. The policy of fire suppression had caused fuel supplies to increase so that when fires did break out, they were larger, hotter and more difficult to control. Several ecosystems that had adapted to frequent burning were losing some of their natural characteristics because of fire suppression and studies of fire-scars on old trees were indicating the surprisingly short length of former fire cycles.

Heinselman (1973) worked in northeastern Minnesota in the wild Boundary Waters Canoe Area (BWCA) bordering Canada. While he was not the first to utilise fire-scar data in North America, his thorough approach set the standards for subsequent work. His

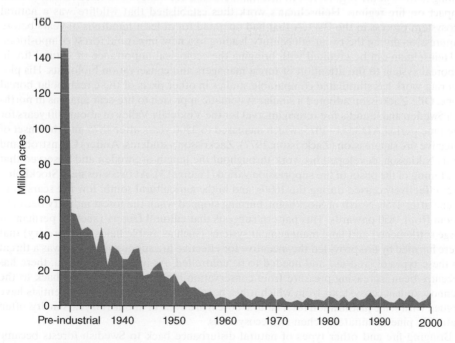

Figure 9.2   Annual area burned in the USA from all ignition sources. 1 acre = 0.4 ha. The National Interagency Fire Center

study also had significant impact because he was employed by the US Forest Service and he explicitly emphasised the management implications of his work. He concluded that 'fire should soon be reintroduced through a programme of prescribed fires and monitored lightning fires. Failing this, major unnatural, perhaps unpredictable, changes in the ecosystem will occur' (Heinselman, 1973). He reconstructed the last 400 years of fire history based on fire-scar analyses of primarily *Pinus rubra* (red pine), *Pinus banksiana* (jack pine) and *Thuja occidentalis* (northern white cedar), historical documents, aerial photos and forest maps. He showed that prior to European settlement in the late 1800s, the natural average fire return interval was approximately 120 years, while during the European settlement period (AD 1868–1911) the fire return interval was reduced to about 90 years. The early settlers were extremely careless with fire because there was little logging in the BWCA, which was the usual reason for anthropogenic burning in the forests at that time. State laws required burning after logging to reduce the wildfire risk from discarded tree remains, or 'slash'. BWCA was incorporated into the national forest administration in 1909 and the first priority was fire suppression, which was highly effective, lengthening the average fire return interval to 2000 years. The suppression policy was based on government reports that described the ravages of fire and concluded that the forests had been 'destroyed' by the fires of the previous settlement period. Such fires were always ascribed to human carelessness, and there was no recognition at that time that lightning ignitions were also common and occurred naturally (Heinselman, 1973). Fire-scar studies can only extend back as far as trees exist, and the oldest living *Pinus rubra* in Heinselman's study dated from AD 1595. Too few old trees can lead to an overestimation of older fire return intervals, but the widespread occurrence of charcoal in soils and humus in the area, taken together with long, local Holocene records of charcoal in peatlands and lake sediments, suggested that frequent fires were indeed a natural feature of this system. Native Americans have a long history in the region, but in low numbers, and are believed to have had little local impact on fire regime. Heinselman's work thus established that wildfire was a natural ecosystem process in the BWCA that had operated for at least hundreds of years before suppression during the twentieth century, leading to a new, unnatural forest composition.

Heinselman can be credited with bringing the ecological importance of natural fire in a boreal system to the attention of forest managers and conservation biologists. His pioneering work has stimulated comparable studies in other parts of the circumpolar boreal zone. Olle Zackrisson adopted a similar systematic approach to fire-scar analysis in northern Sweden and found a fire return interval for the Vindelälv Valley of about 100 years for the time period AD 1551–1875, which increased to 3500 years after 1875 and the onset of effective fire suppression (Zackrisson, 1977). Zackrisson's students, Anders Granström and Mats Niklasson, developed his work throughout the length of Sweden and observed that the timing of the onset of fire suppression varied (Figure 9.3). At two sites near Stockholm, fires effectively ceased during the 1600s, and in the agricultural south, few fire scars were found after 1750. North of Stockholm, burning stopped when the forest industry began to boom from 1850 onwards. This pattern suggests that cultural factors (such as permanent large settlements) and land management systems (such as arable fields and forestry) that were harmed by fire provided the incentive for effective fire suppression. Fire was a threat to these types of land use and needed to be controlled. As in North America, there has recently been increasing pressure from conservation biologists to bring fire back to the Fennoscandian ecosystems from which it has been excluded. The fire-scar scientists have argued with some justification for a natural boreal fire cycle of about 100 years in dry, often naturally pine-dominated, lichen-rich ecosystems.

Bringing fire and other types of natural disturbance back to Swedish forests became a hot topic following a remarkable change to the national law dictating forest management in 1993. Production and environmental goals were given equal weight, leading to a surge of interest in forest ecology aimed at identifying appropriate and practically feasible

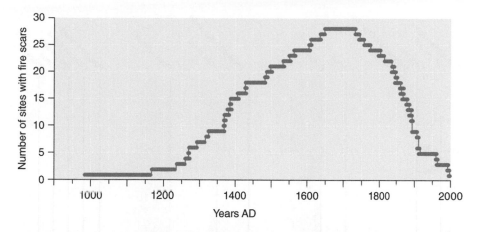

Figure 9.3   Number of Swedish sites with fire-scar evidence through time. While the increase in site number during AD 1200–1600 is an effect of time on the preservation of fire-scarred trees, the decrease in recent centuries is an effect of fire suppression (after Drobyshev *et al.*, 2010). Modified from Mats Niklasson and Anders Granstrom, (2007). With permission

conservation and ecosystem restoration measures for adoption throughout the country (Bradshaw *et al.*, 1994). The ASIO model conveniently divided the forested landscape into four zones where fire in the past had either been presumed Absent or had occurred Seldom, Infrequently or Often (Angelstam, 1998). These divisions were of course simplifications, and fire has rather different ecological consequences compared with felling, but nevertheless they have been widely used to plan where forests could most appropriately be heavily exploited, namely in areas that have burnt often in the past and are perhaps adapted to frequent disturbance. Forest certification procedures are also designed to reduce the damaging impact of forestry on forest ecosystems, and in Sweden prescribed burning on 5% of the annual clear-cut area is a requirement for certification of large forest owners by the Forest Stewardship Council, although this is hardly ever achieved.

The increasing interest in natural fire regimes for conservation and restoration management builds on information about past fire regimes. Fire-scar analyses reach back a few centuries, but do not extend beyond time periods with recognised human impact, so perhaps the 'reference' or baseline states in the Heinselman and Zackrisson studies also reflect some human impact? This could well be the case given the interesting charcoal and forest vegetation record that Jennifer Clear obtained from a small forest hollow in Vesijako Forest, in the boreal zone of southern Finland (Figure 9.4). She found regular peaks of charcoal in the sediment recurring on average every 430 years between 5000 and 2000 years ago, which is a longer fire cycle than either Heinselman or Zackrisson found in boreal forest, although one would expect some regional variation in fire return interval relating to rates of fuel accumulation and climatic conditions. Between 2000 and 700 years ago, the charcoal record is quite different. The charcoal peaks are much larger and the average fire return interval is reduced to just 180 years. While it is always difficult to link forest fire evidence directly with human activities, even today, several other studies from the region have concluded that humans began to take control of the fire regime about 2000 years ago, exploiting fire for slash-and-burn agriculture and for the improvement of forest grazing (Bradshaw *et al.*, 2010d). Charcoal is completely absent from the recent part of the Vesijako record, which covers the now familiar period of fire suppression (Figure 9.4). It has become a common theme in this book that disentangling human from 'natural' influences on ecosystem processes is difficult enough using direct observation in the present, never

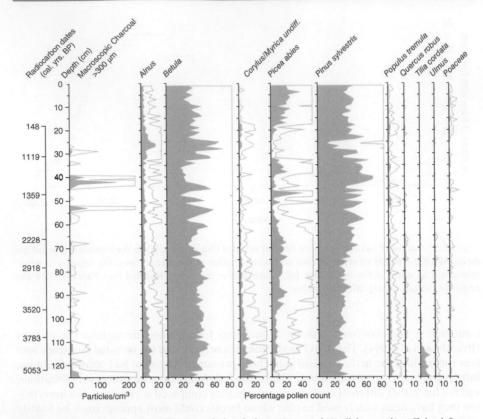

Figure 9.4 Charcoal and major pollen types during the last 6000 years in Vesijako, southern Finland. Source: Clear et al, 2013. Reproduced by permission of Elsevier

mind the past. Consequently, it is unrealistic to set up a purely 'natural' target as a goal for ecosystem restoration. The appropriate choice of fire frequency provides a good example of this problem.

Yellowstone National Park is well monitored for fires. It currently experiences about 20 lightning-initiated fires each year, plus an additional 6–10 fires that are usually accidental and of human origin. Natural fires have been allowed to burn under controlled conditions since 1972, and this management appeared to be a satisfactory way of reintroducing the ecosystem benefits of fire while controlling fuel supply to avoid catastrophic megafires. However, a major drought in 1988 resulted in more than 72 000 fires in the western USA, and about 55% (500 000 ha) of Yellowstone was affected by a series of large, hot fires. Considerable critical media comment followed this event, but despite major efforts at control there was little that could be done until autumnal rain and snow eventually extinguished the flames. The damage might well have been worse if there had not been the earlier limited tolerance of wildfire to reduce fuel loads. The Park has recovered quickly from the 1988 fires and palaeoecological research has shown that fires of this magnitude have been recorded from the past, with three periods of widespread burning during the last 750 years, suggesting that the 1988 fire was not a unique occurrence (Higuera *et al.*, 2011).

What are the benefits of restoring earlier, hopefully more natural fire regimes to modern ecosystems? The Yellowstone example showed that a single fire season could affect much of a National Park, and such sudden change is always hard to justify to the general public. However, in southeastern USA fire suppression caused large-scale ecosystem

Figure 9.5 Seedling of a longleaf pine (*Pinus palustris*), Florida, USA. Young, fast-growing seedlings tolerate ground fires that kill competitors

change and made the fire regime more dangerous for property owners, so restoration of fire has brought several benefits to biologists, hunters and the general public. Pine forests and savannahs dominated by *Pinus palustris* (longleaf pine) are estimated to have covered 37 million hectares prior to European settlement, of which 97% has now been either converted to other uses or degraded as a result of fire suppression (Varner *et al.*, 2005). Pre-settlement fire regimes comprised frequent, low-intensity surface fires, which killed off the competition for longleaf pine, whose seedlings show special adaptations to survive such fires (Figure 9.5). Florida has had the highest frequency of lightning strikes in the USA for a long time, which suits longleaf pine well. Successful fire suppression had changed the open pine ecosystems into closed forests, with reduced species diversity, increased fuel loads and less suitable habitat for the lucrative hunting of bobwhite quail. The Tall Timbers Research Station (www.talltimbers.org) has played a key role in changing public attitudes to fire and now coordinates a major prescribed fire programme in the region, which draws on knowledge of the ecosystems prior to European settlement and brings both biological and commercial benefits. This type of large-scale management of fuels is also undertaken on state lands in southwestern Australia, where programmes combine conservation, restoration and socio-economic benefits. These are useful models for other regions of the world, where fire risk is forecast to increase considerably during coming decades (see Chapter 5).

The bringing back of fire to ecosystems exemplifies a new approach to conservation, in which processes are of equal or greater importance than the more traditional focus on species preservation. The Florida example shows the benefits of gaining the confidence and support of the public and including a consideration of ecosystem services within ecosystem management. There are of course always potential conflicts and compromises. Fire releases CO_2 into the atmosphere and works against the value of forests as carbon stores and barriers to soil erosion. These conflicts are illustrated by the trees that President Kekkonen of Finland planted in the 1950s in order to increase national forest stocks and restore a heavily exploited ecosystem. His trees were burnt down several years later as part of a prescribed fire programme carried out by Metsähallitus, the Finnish Nature Protection Agency, practising the latest approach to ecosystem restoration. This stimulated an interesting public discussion and illustrates the dangers of changing policies too rapidly in ecosystem management.

9.2.2   Ecosystem restoration

Another 'close to nature' way of managing ecosystems is the use of grazing animals. As with burning, there are difficulties in defining any clear baseline or set of reference conditions that could represent a mythical 'virgin' state prior to human impact. As we saw in Chapter 3, people have been implicated in the alteration of large herbivore communities on several continents for thousands of years (Barnosky *et al.*, 2004). Palaeoecology records a long history of anthropogenic modification of large mammal populations from boreal and temperate forests, through hunting, herding and later domestication of some species (Bradshaw *et al.*, 2003). By 9000 years ago, the forests of northwestern Europe had been recolonised after the last glaciation by large mammals, but several species of both herbivores and carnivores were 'missing'. The 12 or so species of large herbivorous mammals or ungulates recorded from previous interglacials in the British Isles had become reduced to about five – moose, red deer, roe deer, aurochs and wild boar, plus possibly wild horse (Sommer *et al.*, 2011) – as a result of both natural and anthropogenic factors (Table 9.1). Some species left the region due to natural changes in Holocene climate and vegetation, such as the hippopotamus (*Hippopotamus amphibius*), while others became totally extinct, such as the straight-tusked elephant (*Palaeoloxodon antiquus*). Moose, aurochs and wild boar disappeared from the British Isles during the Holocene as a result of habitat clearance and hunting. Without human intervention, these species would still 'belong' to the region, and they can be regarded as potential targets for reintroduction (Figure 9.6).

  Studies of the past reconstruct the course of events that created the present condition and indicate how people have interfered through hunting and forest clearance. However, even prior to significant human impact there is no single set of equilibrium 'baseline' conditions that can be recognised in the recent geological past, so there is no secure reference to use as a model for future management. The much higher diversity and population size of large ungulates from past interglacials strongly suggests a diverse vegetational environment that varied between interglacials, with elephants, rhinos and many species of deer existing in their own niche. It is hard to imagine how the ungulate communities could have caused the apparent differences in habitat diversity between past interglacials themselves. Overall ungulate diversity was similar in each interglacial (Table 9.1), yet the balance between forest and grassland specialists was variable. Some interglacials (e.g. oxygen isotope stage 7 in Britain) were characterised by a dominance of grazing ungulates such as horse and mammoth, but this is likely to have been a result of the more open vegetational structure and not its cause. Grazing-adapted species are unlikely to have been responsible for the clearing of forest in the first place. One might also ask why they did not appear and open out the forest during other interglacials (e.g. oxygen isotope stage 5e). The conclusion is that other, probably climatic, factors were responsible for the primary habitat structure, which leads us into the debate initiated by Frans Vera on the 'natural' structure of temperate European forests.

Table 9.1   Ungulate faunas from British interglacials. (●), not recorded in Britain but abundant on the continent

Oxygen isotope stage	17	15	13	11	9	7	5e	1
Approximate age of stage start in thousands of years before present (BP)	700	620	520	410	330	240	130	11.5
Cervidae								
Megaloceros verticomis (Giant deer)	●	●	●					
Megaloceros savini (Giant deer)	●	●						
Megaloceros dawkinsi (Giant deer)		●	●					
Megaloceros giganteus ('Irish' giant deer)				●	●	●	●	●
Alces latifrons (Broad-fronted moose)	●		(●)					
Alces alces (Moose)								●
Cervus elaphus (Red deer)	●	●	●	●	●	●	●	●
Dama dama (Fallow deer)	●	●	●	●	●	●	●	
Capreolus capreolus (Roe deer)	●		●	●	●	●	●	●
Bovidae								
Bison schoetensacki / B. priscus (Bison)	●	●	●	●	●	●	●	
Bos pritnigenius (Aurochs)				●	●	●	●	●
Suina								
Hippopotamus amphibius (Hippopotamus)		●					●	
Sus scrofa (Wild boar)	●	●		●	●	●	●	●
Rhinocerotidae								
Stephanorhinus hundsheimensis (Rhinoceros)	●	●	●					
Stephanorhinus kirchbergensis (Merck's rhinoceros)				●	●	●		
Stephanorhinus hemitoechus (Narrow-nosed rhinoceros)				●	●	●	●	
Equidae								
Equus ferus (Wild horse)	●		●	●	●	●		
Equus altidens / hydruntinus (Small ass-like horse)	●	●		●	●			
Probascidea								
Palaeoloxodon antiquus (Straight-tusked elephant)		●	(●)	●	●	●	●	
Mammuthus trogontherii (Steppe mammoth)	●	●	(●)			●		
Total	12	12	11	12	11	12	10	5

Source: Bradshaw et al, 2003. Reproduced by permission of Elsevier

9.2.3   The wood pasture debate

Curiously, the 'natural' proportions of forested land and grassland and the structure of the forest itself have been of particular interest for recent conservation policy in much of northern and central Europe. Patches of old-growth, mixed deciduous European forests, which are often protected in nature reserves, tend to have dense, closed structures, with little light reaching the forest floor. Light-demanding species, including trees and shrubs such as *Quercus* (oak) and *Corylus* (hazel), cannot easily regenerate under these conditions and are declining in abundance. At the same time, much current valuable biodiversity, particularly among insects and herbaceous plants, is characteristic of rather open habitats, even though at the present time these conditions can often only be created and maintained by human intervention. Vera (2000) suggested that the dark contemporary forests significantly differ in structure and dynamics from the more 'natural' forests that existed earlier in the Holocene, prior to intensive anthropogenic influence. He proposed that large herbivores were sufficiently abundant at that time to help maintain open forests, permitting the survival of light-demanding species. We can call this hypothesis the 'wood pasture' hypothesis, where 'wood pasture' is used to describe a range of forest structures from

Figure 9.6   Moose: a natural British resident? (photo Richard Bradshaw)

open parkland with isolated trees to a mosaic of wooded groves and meadows, which Vera proposed were the best modern proxies for ancient, natural Holocene woodlands. A contrasting 'high forest' hypothesis, describing early–mid-Holocene forests, is widely advocated by palaeoecologists. Pollen records of early–mid-Holocene age from the temperate and boreal zones are usually so dominated by tree pollen that primaeval European forests have been interpreted as dense, dark habitats with little evidence of an extensive herb layer. The mental image that palaeoecologists have of these forests is rather like many contemporary old-growth forests in both Europe and North America (Peterken, 1996). Large herbivores are not thought to have played a major ecological role in these systems and population densities are judged to have been low.

The current balance of scientific evidence does seem to favour the high forest hypothesis and its associated small populations of large herbivores, particularly aurochs, bison and wild horse. Ancient DNA analyses can now provide palaeopopulation estimates, which indicate a European breeding population for wild horse of between 35 000 and 50 000 individuals 6000 years ago (Lorenzen *et al.*, 2011) This population was dispersed over a very large area (c. 9 million km^2), so population densities were indeed low. Past European bison populations have perhaps been underestimated (Kuemmerle *et al.*, 2011), but unfortunately there are not yet any estimates of past population size based on genetic data for bison or aurochs. The absence of preserved aurochs, bison and wild horse bones from Ireland shows that these species were almost certainly absent from this island throughout the Holocene (Woodman *et al.*, 1997), even though *Quercus* and *Corylus* were regionally very abundant in Irish forests. *Corylus* pollen was far more abundant in central and eastern Ireland during the early–mid-Holocene than on most of the European mainland, from where there are frequent fossil remains of these large mammals (Bradshaw *et al.*, 2003). This strongly suggests that large ungulates were not the key factor in the Holocene persistence of *Quercus* and *Corylus*, at least in Ireland.

A final observation that weakens the wood pasture hypothesis is that the only modern analogues for wood pasture are in managed systems with large numbers of domestic animals. In some cultural landscapes, such systems have operated for hundreds to thousands of years and are now under threat, but they are most definitely anthropogenic systems and not good analogues for how the landscape would function in the absence of human intervention.

A partial resolution of the problem posed by these competing hypotheses can be reached by a fresh appraisal of former disturbance regimes in which browsing animals play a small but integral part. The high forest hypothesis assumes that much of the landscape is under mature forest and that regeneration takes place in gaps created by the deaths of single trees. Recent ecological and palaeoecological studies have stressed the role of disturbance in creating more open forest structures. Fire, represented by charcoal in sediments, is characteristic of most areas of northwestern Europe prior to the spread of *Fagus* (beech) and *Picea* (spruce) (Bradshaw *et al.*, 1997). Both natural and cultural fire, in conjunction with animal browsing, may have created appropriate conditions for regeneration of *Quercus*, *Corylus* and *Pinus* (Scots pine) without the development of a persistent or shifting mosaic of nonforest habitat. Certainly the pollen record has underestimated the importance of herbaceous vegetation in the past, but the wood pasture hypothesis may prove to overestimate open conditions. Past natural forests incorporated a great deal of variation. They were more open in structure on sandy soils of low fertility. Seasonal flooding was more widespread than in our present heavily drained landscapes. Browsing animals and storms were also factors that contributed to open conditions, but it would be wrong to emphasise one to the exclusion of the others. Multiple disturbance factors acted together in the past to create a varied landscape that housed a great diversity of species. The message for modern conservation biology is a stress on the significance of varied disturbance regimes associated with fire, animals, disease, flooding and storms, among other agencies. Fire suppression, wildlife management and regional drainage operations have disrupted several of these disturbance agencies, with consequent changes in species composition. Rewilding should involve restoration of natural disturbance regimes and not just reintroduction of key species.

9.2.4 Reference states or baselines?

Rewilding has become a major part of conservation effort in recent years (Jackson and Hobbs, 2009). It is of course based on ecological history, and it implies a target state. In North America, Australia and other regions of the 'New World', a simple target to choose is the landscape prior to European settlement, before the major forest clearance, large-scale agriculture and fire suppression that have characterised recent centuries. The flora and fauna from this time, of which many components still survive, can be reconstructed in general terms using historical records and palaeoecology. However, several people have pointed out theoretical and practical problems associated with this approach. As discussed in Chapter 6, pre-European cultures may have altered ecosystems in important ways through modification of the fire regime or extinction of key species. Climatic conditions will often have changed from pre-settlement times, leading to new species combinations becoming suited to the area of interest, so any pre-disturbance 'baseline' is itself going to drift through time. Nevertheless, restoration of individual species or entire ecosystems is a popular conservation activity and it is useful to assess the viability of any planned restoration programme by estimating how far the present state has deviated from the target conditions. Jackson and Hobbs (2009) present different ecosystem trajectories through time and argue that where ecosystems have deviated far from reference states it is impractical to carry out major restoration programmes (Figure 9.7).

Examples of severely altered type 3 ecosystems are only too frequent, particularly where previously important species such as large herbivores are now extinct (Figure 9.8). The vertebrate fauna of Madagascar (such as the giant lemur, *Archaeoindris fontoynontii*) has been decimated since initial human contact 2500 years ago, through a combination of increasing human impact and drought (Crowley, 2010). Restoration of former communities is impossible with the extinction of so many important endemic species. In other situations,

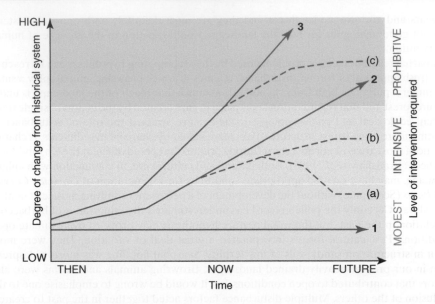

Figure 9.7 Contrasting ecosystem trajectories from historic through present to future configurations, indicating degree of change from the historic ecosystem (e.g. physical environment and species pool). Dashed lines (a–c) represent realistic management interventions of varying ambition and cost. Source: Jackson and Hobbs, 2009. Reproduced by permission of the American Association for the Advancement of Science

native species of animals and plants have been so comprehensively altered by the introduction of invasive aliens from other continents that restoration is impractical. Compared with more densely populated European countries, Sweden has many ecosystems that could be restored to near natural conditions with just moderate intervention. However, as I (R.B.) write this, I hear the raucous calls of vast flocks of Canada geese (*Branta canadensis*), which have become a dominant feature of coastal habitats in southern Sweden since their introduction in 1927. These aggressive birds have surely elbowed out a formerly greater diversity of other species to secure such a dominant position in northern Europe, and their eradication would be a major undertaking. The New Zealand government has declared the introduced *Pinus contorta* (lodgepole pine) to be a noxious weed, particularly as it can build new treelines above the native forest limit. However, despite substantial efforts at eradication, the species is so well established and successful that total removal is no longer a practical proposition (Ledgard, 2001).

In parts of the world with long records of human modification of ecosystems, including Europe, any potential baseline or pre-settlement reference state is so far in the past that this type of ecosystem restoration is even less feasible: less elements of the 'undisturbed' reference state remain, while the biota will have been influenced by climate history in a great many ways since it existed. Instead, traditional cultural landscapes, now threatened by industrial-scale land exploitation, have themselves become the focus of conservation interest, as they contain many species with rapidly declining populations. Heathlands, for example, are highly regarded by nature conservation authorities and the general public in several European countries. Some heathland patches developed in western Europe independently of human activity, but most were produced by anthropogenic burning designed to increase heathland area for winter grazing of domestic stock. Many European heathlands are several thousand years old, but their area has become reduced since industrialisation. In the county of Halland, southwest Sweden there were 150 000 ha of heathland in AD 1850, of which only 4000 ha survived in 2010. Consequently, several endangered

Figure 9.8 The extinct giant lemur (*Archaeoindris fontoynontii*), which was the largest primate recorded from Madagascar, weighing about 160 kg. Source http://en.wikipedia.org/wiki/File:Archaeoindris _fontoynonti.jpg. Courtesy Smokeybjb

species on the Swedish red list are heathland specialists, such as *Pulsatilla vernalis* (pasque flower; Figure 9.9) and *Coronella austriaca* (smooth snake), which have benefitted from the reintroduction of traditional spring burning of heathland (Figure 9.10).

These heathlands are another example of how cultural landscapes of great antiquity now hold important value (see Chapter 8). The conservation of these habitats illustrates the important principle of working with the species in hand, without worrying so much about how they arrived here. This pragmatic approach is particularly justified where cultural habitats have been in place for thousands of years, but it is also advocated by George Peterken for the 'wilding' of recent alien conifer plantations in the UK (Peterken, 1999). In southern Fennoscandia, these former cultural landscapes, which have left significant legacies in the modern landscape, retain important features of earlier, close-to-natural systems, to which they have often added their own diversity (Figure 9.11). In the rich forest meadows of southern Fennoscandia, a 'natural' disturbance regime involving fire and grazing developed into a more managed regime around AD 1100. The landscape that developed was not stable, but reflected contemporary economic developments and social change. There was heavy exploitation during the seventeenth to nineteenth centuries, followed by a major change in land use that led to abandonment of parts of the landscape and intensification of others. This major change has led to the current crisis in management of biodiversity (Figure 9.11).

Former cultural diversity, as represented in Europe by forest meadows, coppiced woodland, farmland and heathland, is currently managed within systems of nature reserves.

Figure 9.9 *Pulsatilla vernalis* (pasque flower) has benefited from reintroduction of the burning of heathlands in Halland, Sweden. Courtesy of Krister Larsson

Figure 9.10 Reintroduction of the spring burning of heathland in Halland, Sweden. Courtesy of Krister Larsson

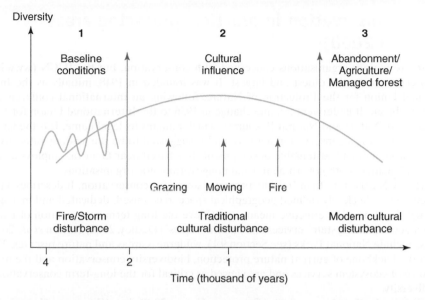

Figure 9.11   Model of changes in vascular plant diversity associated with the development and abandonment of traditional agricultural methods. Source: Bradshaw and Hannon, 2006. Reproduced by permission of CAB International

We saw in Chapter 7 how almost half of the land protected in nature reserves in southwest Sweden is valued for communities of cultural origin that require active management for their maintenance. Increasingly, small reserves are being merged into larger complexes where 'natural' and cultural values are combined. This formal recognition of the biodiversity value of cultural landscapes in southern Fennoscandia, with its relatively modest scale of anthropogenic impact compared with the Mediterranean region, indicates the importance of the cultural heritage to European nature conservation. An appropriate focus for nature conservation based on knowledge of the past places the emphasis on species and dynamic disturbance processes, with less consideration for how or when these species arrived, or for the existence of a mythical, stable reference state prior to human impact – such as the Garden of Eden before apple-picking time.

Summary: Conservation and knowledge from the past

- Studies of past ecosystem dynamics show that several valuable ecosystems are products of earlier human impact and require management for their continued existence.
- Fire is a natural disturbance agency in boreal forest and other ecosystems and its suppression in recent decades has driven unnatural ecosystem change. Prescribed burning is now a feature in the restoration of fire-dependent ecosystems.
- Past human activities are implicated in the disappearance of large mammals and other species from ecosystems. Rewilding programmes include the reintroductions of former native species. Reference states and baselines are difficult to define in both theory and practice, although past ecosystem trajectories indicate degrees of ecosystem modification from historic states.
- It is both realistic and practical to incorporate the effects of former cultural influence such as local forest clearance into conservation policy.

9.3   Conservation in practice: protected areas (Natura 2000)

There are various organisations concerned with conservation, but the IUCN (www.iucn .org) claims to be the oldest and largest. It was founded in 1948, initially as the International Union for the Protection of Nature, following an international conference in Fontainebleau. It underwent a name change in 1956 to the International Union for Conservation of Nature and Natural Resources, which remains its legal name. It is the largest professional global conservation network and is suggested to be the leading authority on the environment and sustainable development. It is based near Geneva, employs a large staff and includes both governmental and nongovernmental organisations.

The IUCN lists more than 160 000 'protected areas' for conservation. It describes a protected area as 'a clearly defined geographical space, recognised, dedicated and managed, through legal or other effective means, to achieve the long term conservation of nature with associated ecosystem services and cultural values' (Dudley, 2008; Stolton *et al.*, 2013). These include National Parks (see Section 9.1), wilderness areas and nature reserves. They form the backbone of current nature protection, biodiversity conservation and the maintenance of ecosystem services and are viewed as critical for the long-term conservation of biodiversity.

An example of such areas is the European Natura 2000 sites (European Commission, 2013a), which form a network of EU-wide nature protection areas and are the practical effect of Europe's nature conservation and biodiversity policies. These policies are based on two directives from the European Commission – the habitats directive, agreed in 1992 (European Commission, 2013c), and the protected birds directive, from 1979 (European Commission, 2013b) – and they cover the current 27 countries of the Union. They are also said to fulfil the EU's responsibilities under the United Nations Convention of Biological Diversity. More than 1000 species, both plant and animal, and 200 different types of habitat are protected, distributed among nine biogeographical regions: Alpine, Atlantic, Black Sea, Boreal, Continental, Macaronesian, Mediterranean, Pannonian and Steppic. For example:

- *The Alpine region* includes the major mountain areas of Europe. There are almost 1000 Alpine Natura 2000 sites. The distribution of species and habitats can change quickly on mountains because of the rapid increase in elevation. An increase of 100 m is equivalent to moving approximately 100 km in latitude in the lowlands (Sundseth, 2005a). In addition, because of their complex topography, involving slope, aspect and exposure, they are rich in diversity of habitats and species. Around 60% of the plants of Europe can be found on mountains, including many endemic species, especially on the upper slopes. As European mountains are relatively less disturbed than lowland areas, many animal and bird species, including wolves, bears, eagles and vultures, have retreated to these areas. Land use has until recently been restricted to traditional pasture management in seminatural grasslands, leading to high levels of biodiversity. However, much of this is changing, driven by two opposing forces: intensification of agriculture and land abandonment through rural depopulation. Climate change is a significant driver for change, but is not the only one; there is also increasing urbanisation in mountain valleys, tourism, afforestation and deforestation, alpine river dams and melting glaciers.
- *The Atlantic region* describes the northwestern and western fringes of Europe. The land is often flat and low-lying, influenced by an oceanic climate, with westerly winds, moderate rain, mild winters and cool summers. The landscape is mainly agricultural, with large urban and industrial areas. Apart from a rich bryophyte flora, there are not high levels of biodiversity, and this is a reflection of a long history of human activity

and the long-term effects of the last ice age. There is, however, abundant marine life in the coastal waters, and millions of migratory birds are seen in these coastal areas (Sundseth, 2005b). Among the 117 Atlantic Natura 2000 habitats listed are coastal heaths, calcareous grasslands, blanket and raised bogs, sand dunes, wooded dunes, machair, acidophilous beech forests, alluvial forests, salt meadows, estuarine and mudflats. Much of the original mainly deciduous coastal forest has been cleared and replaced by commercial plantations. Many problems regarding future management need to be solved, such as fragmentation due to urbanisation and the connecting of isolated fragments of useful habitats through the encouragement of corridors such as hedgerows.

- *The Continental region* covers more than a quarter of the EU, stretching from central France to eastern Poland and beyond to the Urals. It has a continental climate of cold winters and hot summers, and continentality increases eastwards. Many important European rivers flow through the region and have had a significant role to play in the economic developments within the region, leading to extensive river regulation and subsequent losses of habitats and species found on floodplains. Biodiversity is relatively high, as the area shares species with many of the surrounding regions. There are around 5000 sites, although the area of many is quite small. The habitats directive lists 149 animals and 83 rare plants in this region, along with more than a third of the birds that are listed in the birds directive. Typical habitats are beech, oak and hornbeam (*Carpinus*) forests, bog woodlands, alluvial woods and riparian forests. These are important as corridors, although many now exist only in small pockets (Sundseth, 2005d). Once again, with the intensification of agriculture biodiversity-rich habitats such as seminatural grasslands have declined substantially. One of the most famous areas is the ancient primaeval Białowieża Forest on the border between Poland and Belarus, a remnant of a much larger forest (Figure 9.12). This was originally a royal hunting park (Chapter 8) and became a nature reserve early in the twentieth century. It has a very rich flora, with many endemic species, as well as many bird species and populations of wolves and lynxes. European bison were reintroduced there in 1929.

- *The Boreal region*, with its mixture of forest and wetlands, circles the northern hemisphere and in Europe includes Fennoscandia and the Baltic states. Forest covers 60% of the region, much of which is for commercial use. The dominant species are spruce (*Picea abies*) and Scots pine (*Pinus sylvestris*) and there is a relatively poor herb and shrub layer (Sundseth, 2005c). Wetlands, comprising lakes, bogs and peatlands, are very common. Coastal areas around the Baltic have been used for grazing and haymaking, leading to the development of species-rich seminatural grasslands. Around a third of all habitats in the directive are found in the boreal zone, including forests that are now very rare, such as old-growth taiga and deciduous swamp woods. More than 5000 sites of forest and mire habitat have been identified. Although the region as a whole is relatively poor in vascular plant species, there are a number of areas (such as the islands of Öland and Gotland) in which the combination of calcareous soils and long-term historical management has led to both high biodiversity and a rich endemic flora (Purschke *et al.*, 2012). More than 50% of European bird species breed in the boreal area, including a substantial number of migratory species. Natural habitats and old-growth forests have been significantly reduced by forestry, which is a major industry in most countries of the region. The area of forest is increasing due to commercial management and through natural succession on abandoned land. The loss or severe reduction of the moderate grazing of species-rich seminatural grasslands found on the island of Öland has helped promote natural forest invasion into this species-rich ecosystem.

Figure 9.12   Richard Bradshaw in Białowieża Forest in 1992

Summary: Protected areas

- Natura 2000 sites form a network of EU-wide nature protection areas and are the practical effect of Europe's nature conservation and biodiversity policies.
- More than 1000 species, both plant and animals, and 200 different types of habitat are protected, distributed among nine biogeographical regions: Alpine, Atlantic, Black Sea, Boreal, Continental, Macaronesian, Mediterranean, Pannonian and Steppic.

9.4   Conservation and alien or invasive species

Society tends to adopt a schizophrenic approach to non-native species, which has posed practical problems for conservation. Alien species have always appealed to foresters, horticulturalists and curators of botanical gardens, but are viewed as potential threats to native ecosystems by most conservation biologists. Recently introduced 'invasive' species are not generally welcome in protected nature reserves, yet paradoxically a formerly alien species such as *Kickxia elatine* (sharp-leaved fluellen), which is an introduced weed of arable agriculture in Sweden, has become 'naturalised' and is now classed as an endangered species on the Swedish red list. An alien species can be defined as one of foreign origin that is not fully 'naturalised', which the eminent botanist David Webb took to mean that the species had arrived in the region when humans had 'ceased to be in an ordinary sense a part of nature' (Webb, 1985). As we saw in Chapter 6, this elusive point in time will vary from place to place, but for Europe it can be conveniently set to the local establishment of agriculture at the onset of the Neolithic period.

An understanding of the history of ecosystems provides a framework that can put the evaluation of alien species on a more consistent foundation (Bradshaw, 1995). The list of native species for a given region is dynamic through time because the natural ranges of all organisms undergo constant change. In the past, most plant species moved without human involvement, but during recent millennia humans have become increasingly important vectors of plant material and the alien issue has arisen. It would be logical to make a distinction between cases where dispersal by humans has enhanced a previously 'natural' process and cases where human intervention is clearly artificial. For example, it has long been suggested that human populations assisted with the spread of hazel across Europe during the early Holocene (Smith, 1970) but this spread would probably have taken place without human intervention and no one seriously questions the native status of hazel in most northwestern European countries. Similar 'assisted' spread may apply to other tree species, such as *Acer pseudoplatanus* (sycamore), although this case is controversial.

Sycamore survived the last ice age in southern Europe and spread quite naturally throughout central Europe during the early Holocene. Species of *Acer* are closely associated with species of *Fagus* (beech) throughout the world and sycamore fulfils this role in Europe. People have introduced sycamore into northwest Europe, where it has become a successful – but often despised – aggressive 'invasive' species. It is clearly well suited to the current climate of the region and could well have arrived through natural seed dispersal – in the fur of bears, for example, had they not been hunted to local extinction. Swedish managers who were tempted to weed out sycamore from nature reserves became indecisive when the endangered, red-listed butterfly *Nothocasis sertata* began to be found on sycamore plants (Figure 9.13): a legally protected species was feeding on a despised invasive alien!

Rhododendron ponticum poses a similar problem in northwestern Europe. It was a widespread native species during the last interglacial but did not manage to recolonise the British Isles during the Holocene. When planted in gardens, it quickly jumped over the fence, and it is an aggressive invader of woodland, heathland and peatland habitats (Figure 9.14). It casts a dense shade and builds an acid humus layer that prevents regeneration of the previous site occupants. Considerable effort is being made to eradicate the species from Killarney National Park in southwest Ireland and elsewhere, but the species is so successful and aggressive that these costly campaigns can only achieve a temporary local removal. The problems and paradoxes posed by sycamore and rhododendron can be resolved by adopting a more flexible and dynamic view of acceptable native species, based on knowledge of the past and a long-term perspective. Both species are potential natives in northwestern Europe. Management decisions may be made to restrict their occurrence

Sk Skäralid 11.IX 1978
E von Mentzer

Figure 9.13 *Nothocasis sertata*, a red-listed butterfly that feeds on sycamore (*Acer campestre*) in Sweden. Source: http://www2.nrm.se/en/svenska_fjarilar/n/nothocasis_sertata.html. Courtesy Bert Gustafsson, Swedish Museum of Natural History

Figure 9.14 Invasion of *Rhododendron ponticum* into native oak woodland in Killarney, Ireland (photo Richard Bradshaw)

in nature reserves, but these decisions should not be justified by declaring them to be undesirable aliens.

When species are moved between continents, the case for people assisting natural processes is harder to argue. European soil scientists are convinced of the importance of earthworms in soil formation, and Charles Darwin himself studied their activities and commented on their ecological value: 'worms prepare the ground in an excellent manner for the growth of fibrous-rooted plants and for seedlings of all kinds' (Darwin, 1881). Yet there is no evidence that earthworms lived in the midwestern states of North America prior to European settlement, although native worm species do occur in the milder southeast and Pacific Northwest. At least 15 earthworm species have been introduced to this region since the 1700s, probably from dumping of soil used for ballast in ships and on the roots of introduced plants (Tiunov *et al.*, 2006). The introduced worms have proved to be very successful invasive species and have caused major ecosystem impact. Their action reduces the thickness of the humus and moves organic material deeper into the soils, changing soil structure, nutrient availability and soil biota (Frelich *et al.*, 2006). This causes changes in seedbed conditions and a consequent loss of forest herbs. Invasive trees can be pulled out but the great worm invasion of the midwestern USA cannot be reversed, and a public campaign has been launched to limit further damage (Figure 9.15).

New Zealand has many alien species, introduced under different waves of colonisation, but mostly by Europeans. A classic case is marram grass (*Ammophila arenaria*), introduced in the late nineteenth century to help stabilise mobile coastal dune systems, where it was often planted with the nitrogen fixer, tree lupin (*Lupinus arboreus*). Planting by hand (Figure 9.16) or by tractor continued until the 1980s. It was clearly aimed at providing an ecosystem service to new settlers and farmers, stabilising both the dunes and farmland eroded by overstocking. However, it had severe effects on dune biodiversity, as it is a very invasive species, reproducing not only by seed but through fragments of the stolon, which can be spread by both wind and water (Sykes, 1987). It can in certain circumstances compete with and displace native dune vegetation, including the native dune builder pingao, the golden sand sedge (*Ficinia spiralis*; Figure 9.17). This replacement usually occurs in stable dunes with moribund pingao and seems to be unrelated to dune-building processes or dune shape. As soon as marram arrives, pingao begins to die, leaving pure marram stands (Partridge, 1995). Pingao is also severely affected by grazing (e.g. by exotic rabbits), while marram, except at the seedling stage, is not.

Figure 9.15 Campaign poster encouraging the public not to spread worms in the woods. Source. http://www.nrri.umn.edu/worms/downloads/action/Contain%20those%20crawlers_poster.pdf. Courtesy of Great Lakes Worm Watch, The Natural Resources Research Institute, University of Minnesota Duluth

Figure 9.16 Hand planting of *Ammophila arenaria* (marram grass) at Te Kaeo, New Zealand in 1974. This planting stabilised the dunes, which were then planted with *Pinus radiata* (radiata pine), an invasive alien tree species in New Zealand. Reproduced by kind permission of Archives New Zealand/Te Rua te Mahara o te Kāwanatanga, Wellington Office, [AAQA 6395 M11697]

Figure 9.17 Native dune vegetation, including *Ficinia spiralis* (pingao or golden sedge) and a few patches of marram grass (Sykes, 1987)

9.4.1 Alien species, climate change and conservation

An added issue with regard to alien species is the response of both the species and the ecosystem to rapid climate change and how conservation efforts might be affected in the future. In this section, we presume that some sort of human intervention or activity promoted the invasion of the 'alien' species and that it was not 'just' a natural migratory response to climate. There is some evidence that rapid climate change might well already be driving the expansion of alien species introduced at some point in the past by humans. A rather clear example is given by Walther *et al.* (2007) and concerns the expansion of Chinese windmill palm (*Trachycarpus fortuneii*; Figure 9.18) into a local southern Swiss landscape. Palms are most common in the tropics, but *T. fortuneii* is among the most cold-hardy and is widely cultivated on the latitudinal cold margin for palms. It is a native of South East Asia and was among the many ornamental species introduced into the parks and gardens of Europe in the eighteenth and nineteenth centuries. Large and fruiting populations were common in gardens, but it was only in the 1950s that individuals were found outside protected sites. Once outside, the populations developed to be self-sustaining in seminatural forests within 50 years (Figure 9.19). Walther *et al.* (2007) used the STASH bioclimatic model (see Chapter 2) to define the physiologically important bioclimatic limits of the species native range in Asia and used these values to explore the species invasion into the local seminatural forest in southern Switzerland. They concluded that the expansion of the species was due to changing winter temperatures, increasing frost-free days and increased growing season length.

The interactions between invasive species, climate change and ecosystem functioning have also been explored. Pollination is one of the most important processes to occur in ecosystems and is the one that has been most studied. In a literature review, Schweiger *et al.* (2010) identified climate-sensitive species interactions and mechanisms and assessed the possible future effects of climate change upon them. They concluded that the combined effects of climate change and aliens might lead to new interactions among species in communities and to disruptions of old ones. This could lead to the development of novel communities; whether this will be a negative or a positive for the ecosystem and for the

Figure 9.18   Chinese windmill palm (*Trachycarpus fortunei*) Source: http://en.wikipedia.org/wiki/File: TrachycarpusFortunei.jpg. Courtesy Fanghong

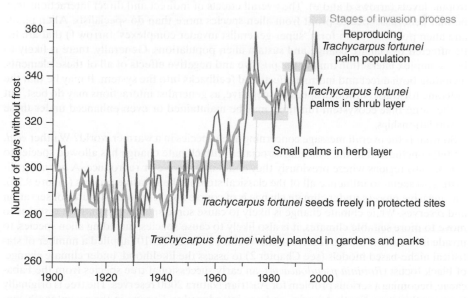

Figure 9.19   Invasion history of Chinese windmill palm (*Trachycarpus fortuneii*) into southern Switzerland. Days without frost versus year (Walther *et al.*, 2007)

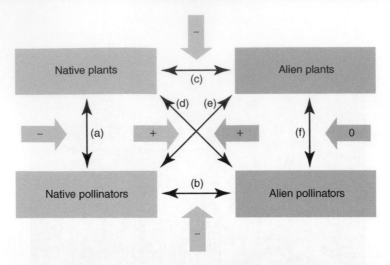

Figure 9.20   Indirect effects of climate change on the interactions among a network of native and alien plants and pollinators. Thick arrows indicate indirect effects of climate change, thin arrows (a–f) indicate interactions, blue arrows (−) indicate negative effects, green/purple arrows (+) indicate positive effects and (0) indicates no effect (Schweiger *et al.*, 2010)

conservation of biodiversity is unclear. In principle, generalist species will likely do better than specialists. Climate change is likely to affect native plant–pollinator interactions more negatively (arrow a in Figure 9.20). Alien species are likely to have negative effects on native species in the same trophic level (arrows b and c) but might have positive ones over trophic levels (arrows d and e). The overall effects of indirect and direct interactions may be that generalist species profit from alien species more than do specialists. Alien plants and alien pollinators can form 'super-generalist invader complexes' (arrow f) that are little affected by climate change and sustain alien populations. Generally, there is likely to be a complex interaction among the positive and negative effects of all of these elements, including both direct and indirect effects and feedbacks into the system. It may be that the outcome for conservation of species is negative, as generalist interactions may do best, but at the same time ecosystem functions may be maintained or even enhanced under these new relationships.

So what is the overall message concerning alien species in a warmer world? Walther *et al.* (2009) concluded that the warming associated with climate change has allowed species to extend into regions where previously they could not survive or reproduce. A warming climate also seems to influence all of the classical steps in the invasive process (Figure 9.21). The invasive process under climate change is of course also relevant to future conservation and reserves. While climate change is likely to cause some species currently in reserves to move to more suitable climates, it is also likely to cause species, including alien species, to invade into current nature reserve areas. Kleinbauer *et al.* (2010) applied a number of statistical niche-based models (see Chapter 2) to assess the likelihood, under climate change, of black locust (*Robinia pseudoacacia*), an early successional tree species from the Fabacaeae, becoming a serious problem for Austrian Natura 2000 reserves. The tree is originally from southeastern North America and was introduced to Europe in the seventeenth century and planted extensively from the eighteenth. It has become very invasive, reproducing by suckering and seeding, and once established is very difficult to eradicate. Kleinbauer *et al.*'s conclusions were that climate warming is likely to give alien species the ability to easily invade nature reserves. Montane regions are particularly susceptible, and aliens are

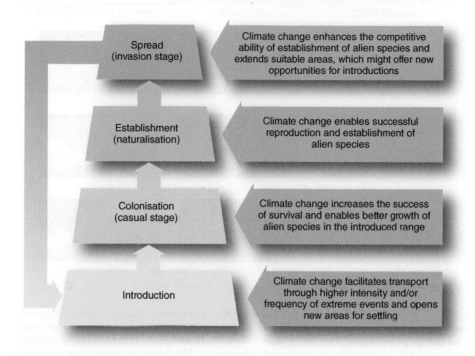

Figure 9.21 Effect of climate change on the various stages of the invasion process. Source: Walther et al. 2009. Reproduced by permission of Elsevier

currently less common there. In addition, in lowland areas increased rates of alien invasions coupled with land use change issues may further increase the pressure on remnant habitats of high biodiversity.

In a similar approach in the Mediterranean, Gritti *et al.* (2006) used the LPJ-GUESS model (see Chapter 2) to explore a number of questions related to alien invasions under climate change: Will rapid climate change drive composition shifts in Mediterranean ecosystems? Will rapid climate changes promote non-indigenous plant species to compete with native species? Are there any climatic gradients, and thus different vulnerabilities to plant invasion, across the Mediterranean Basin? What is the importance of disturbance in the invasion process in the Mediterranean? They used nine plant functional types (PFTs), including two invasive PFTs – one based on an early successional tree species found in the area (*Ailanthus altissima*) and one a herb PFT based on *Amaranthus retroflexus* – as well as two future climate scenarios from the SRES group: A1FI and B2 (see Chapter 7). As in any modelling experiment there are a range of assumptions and uncertainties that must be borne in mind, but the results showed that climate change alone had little effect on most of the simulated ecosystems in this region, although not all. The process and progress of invasion by the invasive species were very dependent on ecosystem composition and the local environment (Figure 9.22). The largest contrast was between wetter and drier areas and between the mountain and coastal areas. Overall, the rate of invasion on the shorter timescale was controlled by the amount of disturbance occurring in the ecosystem. In addition, in the longer run it is clear that eventually most Mediterranean ecosystems will come to be dominated by exotic species, irrespective of the rates of disturbance.

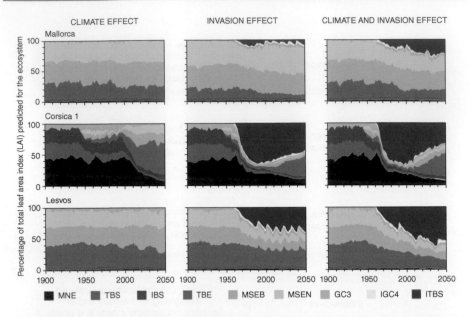

Figure 9.22 Model projections using LPJ-GUESS for three Mediterranean sites (Mallorca, Corsica and Lesvos). Three modelling experiments: left, climate change; middle, invasion in 1960 but no climate change; right, climate change plus invasion. MNE, Mediterranean needleleaved evergreen; TBS, temperate broadleaved summergreen; IBS, shade-intolerant broadleaved summergreen; TBE, temperate broadleaved evergreen; MSEB, Mediterranean shrub evergreen broadleaved; MSEN, Mediterranean shrub evergreen needleleaved; GC3, grasses C_3; IGC4, invasive grasses C_4; ITBS, invasive tree broadleaved summergreen (Gritti *et al.*, 2006)

Invasion by exotics is a worldwide phenomenon and the actual success of an invasive species is dependent not only on the species itself but also on the state of the community, the dynamics of the ecosystem and external drivers such as climate change and land use change.

> ## Summary: Alien or invasive species
>
> - Society tends to adopt an ambivalent approach to non-native species, which has posed practical problems for conservation. Exotic species can be useful for foresters, horticulturalists and curators of botanical gardens, but at the same time are potential threats to native ecosystems.
> - It can be difficult to define exotic or invasive species as there are cases where natural migrations of species were promoted accidentally or otherwise by humans. It is clearer to define alien species when they have been transferred between continents.
> - Climate change and alien invasions may lead to new interactions among species in communities, as well as to disruptions of old ones. This could lead to the development of novel communities; whether this will be a negative or a positive for the ecosystem and for the conservation of biodiversity is often unclear.
> - Invasion by aliens is a worldwide phenomenon and the actual success of an invasive species is dependent not only on the species itself but also on the state of the community, the dynamics of the ecosystem and external drivers such as climate change and land use change.

9.5 Global change, biodiversity and conservation in the future

We have seen throughout this book that history is important because the climates of the past circumscribed both taxa distributions and ecosystem dynamics not only in the past but also cast long shadows into the present and the future. There are currently a multitude of pressures on terrestrial biodiversity and while rapid climate change may seem to be one of the major forces for change, it is not the only one; others include land use and habitat change (both through intensification and abandonment of agriculture), the spread of invasive or alien species and pollution, including increasing nitrogen loads and the perhaps as yet unclear direct effects of increasing carbon dioxide on plant growth. It is also clear that these pressures affect biodiversity in different biomes differently. So, for example, in arctic biomes climate change is the main pressure and the one most likely to bring most change, while in the tropics land use change will probably have the greatest influence. In other biomes, such as the Mediterranean, most of the pressures listed have a role to play in ecosystem dynamics and biodiversity changes (Sala *et al.*, 2000).

Simulation models of the types described in Chapter 2 and scenarios of the types described in Chapter 7 have been used to project the impacts of varying pressures on biodiversity, from the point of view of habitat change and its effects on species and ecosystem services. Of course, neither the models nor the scenarios are reality, but they can be used to explore possible outcomes and to give some estimation of their different probabilities.

Many projections have been made with regard to species distributions, especially for Europe, where it is generally projected to become warmer in the north and drier in the Mediterranean, with a general shift of the types of climate currently found in the southwest towards the northeast. For example, in a study modelling range changes of 1350 European plant species, more than half were vulnerable to climate change by 2080 because of changes in the length of the growing season and in water availability, with mountain and Mediterranean species the most vulnerable (Thuiller *et al.*, 2005). However, boreal zones can expect immigration of species from the south. Mountains may well lose their more specialised mountain species, especially at higher altitudes, through both climate change and increased competition from trees and shrubs as treelines move upwards.

There is little evidence of extinctions caused by climate change, although some estimate that 20–30% of plant and animal species are at an increased risk of extinction. Habitat loss for mammals is a greater threat. Mountain plant species that are unable to disperse elsewhere, such as those in the Swedish mountains, have an increased risk of local extinction by 2080. A study shows that 10 vascular species with restricted distribution in Fennoscandia may become extinct with a doubling of CO_2 (Saetersdal *et al.*, 1998).

The rate of dispersal to new sites is likely to be a problem in many places, especially through fragmented landscapes. Some forest herbs will be required to move 2.1–3.9 km per year (depending on the climate scenario) in order to track 'their' climate space (Skov and Svenning, 2004). If there is a reliance on type of overstorey, this means that suitable forest must be available for their movement. Whether migration delays are a real problem for trees is however debatable in areas where humans heavily influence the landscape, as they may just plant a required species. The situation for other species, both plant and animal, is more problematic.

The arctic is generally expected to be severely affected by climate change involving changes in biodiversity, treelines and biome shifts. One projection suggests large ecosystem changes with a 2°C warming. Forest area between 60 and 90° N could increase by 55% (3×10^6 km^2), with a reduction in tundra of 42%. Tundra vegetation will move north, but with a significant loss in prostrate dwarf-shrub tundra. In the Barents region (northern Fennoscandia, Russia, Novaya Zemlya, Svalbard and Franz Josef Land), projections indicate a more productive and increased area of boreal needle-leaved evergreen forest northwards and up mountains (Wolf *et al.*, 2008). It remains unclear, however, whether the migration rates of trees can in fact match the rate of current and future climate change. Past migrations are discussed in Chapter 4.

Range shifts or losses could also lead to a loss of genetic diversity, which can be important for species over the long term. Alsos *et al.* (2012) assessed genetic diversity in 9581 samples from 1200 populations of 27 northern plant species from arctic and alpine bioclimatic zones near or above the treeline. They used statistical species models, two general circulation models (GCMs) and two emission scenarios to project loss of range and genetic diversity in the future. Genetic diversity loss varied among species according to dispersal attributes and genetic differentiation among populations, but all species lost some diversity under at least one scenario.

Climate change effects on freshwater species involve changes in temperature, and small streams may be more directly affected by warming temperatures and changes in precipitation than large rivers. In most of Sweden, both summer and winter precipitation is projected to increase along with an increase in heavy rain events, leading to flash floods. At the same time, there may also be general reductions in the flow of rivers in the summer due to earlier snow melt and higher evapotranspiration caused by warmer temperatures. Additionally, shorter periods of ice cover, increased temperatures and changes in the mixing regime in arctic and subarctic lakes are likely to occur. All of these changes are likely to directly influence freshwater ecosystems and the species that inhabit them. Also, as northern or boreal freshwater ecosystems are in principle less diverse than those to the south, and as species distributions at larger scales are influenced by temperature, latitudinal range shifts of species from the south to the north are likely to occur. This could lead to increasing biodiversity, which could affect native species, food webs and freshwater ecosystems in general.

One particular change in European boreal terrestrial ecosystems that is consistently projected is from mainly conifer ecosystems based on spruce to deciduous forest structured around deciduous species, including beech. While the mechanism is not entirely clear, it seems the critical phase for some species, such as Norway spruce, is the establishment phase. Seedlings or saplings are likely to be more sensitive in warmer winters, where there is less snow and thus more danger from killing frosts (Sykes and Prentice, 1996). In addition, in warmer climates spruce is possibly more sensitive to diseases. Simulations also suggest that by the end of the century Mediterranean species – even evergreen species – will be able to grow much further north. For example, the climate will become suitable for the evergreen holm oak (*Quercus ilex*), which is already planted in areas of southern England. The availability of a variety of species in both private and public gardens is an interesting issue. In such habitats they are often more protected, both from the climate and from competition, but they are also a source of propagules awaiting appropriate environmental conditions.

9.5.1   The convention on biological diversity

An attempt has been made to reduce the rate of biodiversity loss through the Convention on Biological Diversity, agreed by world governments in 2002: 'the Parties to the Convention committed themselves to achieve by 2010 a significant reduction of the current rate of biodiversity loss at the global, regional and national level as a contribution to poverty alleviation and to the benefit of all life on Earth'. It is clear that the 2010 goal was not met, but it did help to promote the development of some safeguards for biodiversity, such as protected areas, country-wide biodiversity action plans and mechanisms for monitoring and assessing biodiversity. However, according to the Global Diversity Outlook 3

(Convention on Biological Diversity, 2010) there is an ongoing and continuous decline in all three aspects of biodiversity: genes, species and ecosystems. Indications of this decline include species at risk from extinction moving closer to it (e.g. amphibians, species on coral reefs and around 25% of plant species), population sizes of vertebrate species falling by a third since 1970, natural habitats declining in many places around the globe (including wetlands, sea ice habitats, salt marshes, coral reefs), extensive fragmentation of ecosystems (particularly forest ecosystems) and declining genetic diversity.

The Secretariat of the Convention of Biodiversity issued a technical report looking at biodiversity change throughout the twenty-first century, based on models and extrapolations (Leadley et al., 2010), and reached a number of key conclusions. Current land use changes, changing river flows, pollution and the over-exploitation of marine resources have the greatest influence on current biodiversity, although climate change and ocean acidification are likely to become more important as the century progresses. The combined effects of these changing drivers could be significant for species and their ranges and for biomes. Major changes in ecosystem types are likely, such as from tropical forest to grassland (for grazing). It is likely that tipping points will be important but it is difficult to plan for any specific event. Tipping points can be identified in a number of ways, depending on the system being described. In terms of ecology, such points could be where changes in the functioning of an ecosystem have significant impacts on both biodiversity and the services provided by the ecosystem at a range of scales. Various drivers of change are involved, such as habitat change, climate change, over-exploitation, exotic species and pollution, and they can have varying impacts, driving various tipping point processes, such as positive feedbacks, irreversibility, crossing thresholds and time lags. However, it is difficult to predict when or where they might occur, because of possible complex feedback mechanisms and interactions between two drivers. If they do occur, they could lead to major biodiversity loss and to loss or degradation of important ecosystem services to humanity

In 2010, a conference in Aichi, Japan produced a revised and updated plan for biodiversity from 2011 to 2020, including the so-called 'Aichi biodiversity targets'. This framework should be translated into national biodiversity strategies and action plans within 2 years. A selection of these targets include: reducing the loss rate of natural habitats by at least 50%; establishing conservation targets of 17% of terrestrial and inland waters and 10% of marine; restoring at least 15% of degraded areas; reducing the pressures on coral reefs; promoting sustainable agriculture and forestry; reducing pollution to levels not detrimental for ecosystems; and eradicating or controlling key invasive species (Convention of Biological Diversity, no date). Whether these targets are likely to be achieved is unclear, but history does not promote confidence.

9.5.2 Atlas of biodiversity risk

2010 was also the International Year of Biodiversity. In that year an interesting and significant summary book, the *Atlas of Biodiversity Risk* (Settele et al., 2010), was published as part of the final product of a large European Commission Framework 6 project called ALARM (Assessing LArge-scale Risks for biodiversity with tested Methods; www.alarmproject.net). The project ran from 2003 to 2009 and was coordinated by Josef Settele from the Helmholtz Centre of Environmental Research (UFZ), Halle, Germany. It involved more than 250 scientists from 68 institutions in 35 countries. The atlas, which summarised the output from the project, was aimed at being a risk-communication tool for both scientists and the general public.

The loss of habitat through agricultural intensification and urban and infrastructure development is one of the major current and future threats to biodiversity. Vohland et al. (2010a) discuss priorities for nature conservation. In recent times, these priorities have

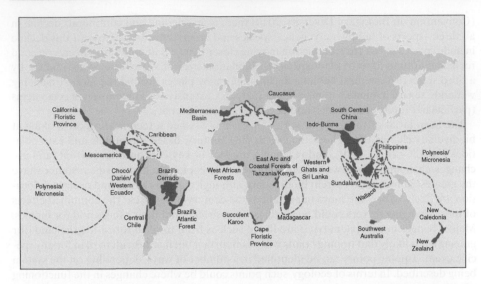

Figure 9.23 Biodiversity hotspots as conservation priorities. Source: Myers *et al.*, 2000. Reproduced by permission of Nature Publications

been dominated by the concept of biodiversity hotspots, referring to either species number or high levels of endemic species. Myers *et al.* (2000) point out that the number of species that are in danger of becoming extinct is substantially higher than the number of resources that are available to maintain them. Thus the question becomes, what species should be conserved and why? Myers *et al.* suggest the identification of hotspots of endemic species where loss of habitat is exceptional. Around 44% of all vascular plant species and 35% of all vertebrate species in four groups can be found in 25 hotspots, covering 1.4% of the Earth's surface, with tropical forests and Mediterranean habitats featuring strongly (Figure 9.23).

Hotspots are not the only way of identifying areas for conservation. Vohland *et al.* (2010a) used the Natura 2000 sites to explore other possibilities, such as representativeness of an ecoregion, provision of ecosystem services, adaptive capacity of ecosystem functions to a rapidly changing environment, retreat areas with strong climatic gradients (such as mountains) to allow for migration and the governance of biodiversity and land use for sustainability.

There has been much talk of Natura 2000 sites. Many may have originally been seen as static, supposedly unchanging entities, but under a rapidly changing climate it is likely that for at least some of the current species at these reserves, their environment and competitive relationships will change to such a degree that they can no longer exist where they are. Using the LPJ-GUESS model (see Chapter 2), Hickler *et al.* (2010) explored the changes in potential natural vegetation across Europe using 16 major tree species and 4 shrub and herbaceous PFTs. Their output was converted into vegetation classes based on the Bohn map of natural vegetation (Bohn *et al.*, 2003). Simulations were done using two GCMs, one tending hot and dry and one cool and wet, and one scenario (ALARM BAMBU). Figure 9.24 shows a typical simulation for a grid cell in southern Sweden, where there is a change from dominant Norway spruce to a beech–lime–hornbeam mixed deciduous forest (Figure 9.24). The outputs show that between 30 and 44% of the current Natura 2000 reserves are likely to see changes in their dominant vegetation, although this is dependent on the climate model used. The changes modelled are perhaps an overestimate, as such

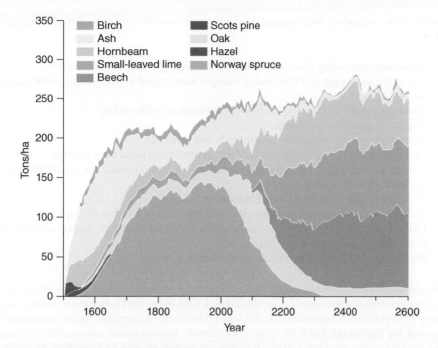

Figure 9.24 Modelled forest succession using the main tree species in a forest in central southern Sweden. Simulations begin after a disturbance in AD 1500. The climate from 1901 to 1930 is used for the period 1500–1900, but actual historical climate data are used for the twentieth century. Simulated future climates until 2100 use output from the HadCM3-A2 GCM, which is then detrended and repeated for the following 500 years (Hickler et al., 2010). Courtesy of Pensoft Publishers, Moscow and Sofia

processes as migration and some soil conditions (e.g. mycorrhizae) are not modelled and there may be substantial delays in the development of new communities and habitats. Time lags are modelled but the responses are transient, with an equilibrium with climate some way into the future. However, the results do indicate that conservation strategies for the future must take these types of transient responses into account.

Ecological networks can be seen as one answer to climate change and landscape fragmentation (Vohland et al., 2010b). These are defined as core areas, buffer zones and corridors scattered through the landscape. They are often linear, such as hedgerows or rivers, but they can be stepping stones, allowing for migration from one to the next. In Germany, forests can be seen as a network, but the potential natural forest is usually replaced by commercial plantations, often of conifers, and is thus at risk from climate change. Further research is required to see which species might benefit from such networks and how to optimise the landscape elements for such movements and provide suitable future habitats.

9.6 Conclusion

Finally, with highlights from a sister project to ALARM, MACIS (Minimisation of and Adaptation to Climate change Impacts on biodiversity) was an EU-Specific Targeted Research Project (STREP 2006–2008) that summarised, with the help of stakeholders and policy makers, what is known about climate impacts on biodiversity. Olofsson et al. (2008)

described the situation with regard to European biodiversity and climate change as follows:

1. Rapid and increasing climate change is increasing pressure on biodiversity.
2. There are increasing and continuing changes with regard to phenology, particularly in the spring.
3. Species respond differently, leading to mismatches in lifecycles.
4. Migration patterns are changing.
5. Species ranges are extending northwards and upwards.
6. Drought has imposed range contractions in southern and central Europe.
7. Range contractions in mountains and decreasing habitat space for mountain species are leading to extinctions.
8. Ecosystem types and habitats are moving northeast.
9. Habitat is transforming in southwestern Europe through forest dieback.
10. Extinctions are increasing, especially among species with low dispersal capability, small ranges and habitat specialists.
11. Mediterranean mountain species are particularly at risk.

Some conclusions were reached regarding the goals of conservation, including giving priority to ecosystems and species that are threatened. One important conclusion was that monitoring is important in assessing the rates and directions of changes. Again, corridors for dispersal are important, but how species are already moving under recent climate change must be taken into account. Ideally, the current models should be improved in order to better simulate some of the important processes that are ongoing, especially with regard to land use change, migration and competition among species. In addition, conservation planning tools for management of these changes in protected areas must be improved.

Mar Cabeza and Guy Midgely, as part of a report from the MACIS project (Berry *et al.*, 2008), summarised the major problems regarding the primary strategy of conservation, which is to establish networks and reserves that protect both species and genetic richness. Climate change is likely to produce reserves and networks with novel climates, and as a result species must migrate away or risk becoming locally extinct. Populations of plants and animals need the chance to adapt or migrate. However, this is dependent on the rate of climate change and habitat loss: if it is too fast then some of these processes do not have enough time to occur, so species face local and possibly general extinction. The issues therefore revolve around how to enhance current strategies or develop new ones.

Adaptation measures can be used, including: taking no action based on informed decision-making; relying on dynamic protected areas; developing buffer zones; controlling or preventing the invasion of alien species; understanding and managing disturbance regimes; identifying new protected areas and connecting protected areas with corridors or stepping stones; developing seed banks or captive breeding for conservation; and having the option to move species (Cabeza and Midgeley in Berry *et al.*, 2008).

Many adaptation strategies seem to be positive for biodiversity conservation, but one must be wary of unintended consequences, and some of these approaches may be riskier than others. Finally, as with most of science, further research is required. Berry *et al.* (2008) summarise such research needs, including better models at higher resolution and further basic research into the mechanisms of biotic change under changing climates (especially at range margins), the role of disturbance and of alien species, the more precise determination of what species, biomes and habitats are vulnerable and how species interact under climate change, experiments into and long-term monitoring of the risks and benefits of assisted migration, better identification of indicator species and a better understanding of refugia and their role in the changing climates throughout the Holocene.

10
Where Are We Headed?

'*Things change, Kundun.*'

Kundun, *dir. Martin Scorsese, 1997*

10.1   Introduction

In this book we have explored humans and our interactions with a changing environment from a long-term perspective. The main focus has been on the ecosystems within which we live and their goods and services, upon which our very survival depends. This has been done within the context of environmental change through the Holocene and into the current period of rapid global change. The book is all about dynamism: ecosystems are dynamic and humans have interacted dynamically with nature through them for thousands of years. What the book is saying is that we have reached a watershed in our relationships with the environment. We have created so many pressures on ecosystems and the services they provide that it is likely, without some serious and drastic action, that we will soon reach the stage at which millions of people can no longer exist. The size of the human population, our lack of awareness concerning our close reliance on nature and our desire for a better material life are dangerously linked to our belief that somehow we will fix things as we always have in the past. Consequently, we do not really grasp the seriousness of the situation.

The book addresses these issues in a number of ways, but mostly based on simulation models and data (both long- and short-term). Two chapters deal with these tools of our trade. In Chapter 2 we introduced and discussed the development of the different types of computer model that are commonly used in global change studies. There are many types of model and they can simulate a variety of factors of relevance to this book, such as climate, ecosystems, biogeochemistry, biomes, species and direct human involvement. They all have generalisations, limitations and uncertainties. These can include lack of knowledge of the actual mechanisms and lack of the experimental or field data used either to model a particular mechanism or for validation purposes. Generalisations are often made: the recognition that at broader scales only keystone species can be, or need to be, modelled is one. Also, the thousands of species that are found in nature can be summarised or generalised into a few types, based for example on traits (e.g. plant functional types, PFTs). The models discussed include broad-scale equilibrium or snapshot BIOME-type models based on relatively simple representations of a generalised PFT physiology; these can be

---

Ecosystem Dynamics: From the Past to the Future, First Edition.
Richard H.W. Bradshaw and Martin T. Sykes.
© 2014 John Wiley & Sons, Ltd. Published 2014 by John Wiley & Sons, Ltd.
Companion Website: www.wiley.com/go/bradshaw/sykes/ecosystem

linked with biogeochemistry so that the flows of carbon and water within a biome can be modelled and such measures as PFT or biome NPP can be simulated. Statistical models of species or niches have become fashionable in recent times and have both advantages and disadvantages when attempting to project responses under future climates. Models need to be dynamic, so that interactions between species and biogeochemical fluxes can be shown through time. Forest gap models were a step on the road to the development of dynamic global vegetation models (DGVMs), which integrate physiological considerations with the dynamics of a system so as to simulate carbon and water cycles and interactions among vegetation. They also form part of the next stage in global modelling of the climate system by being integrated into global circulation models (GCMs) as Earth system models (ESMs), allowing the processes and flows within an ecosystem to be modelled as feedbacks to the climate. Other types of model that are discussed include integrated assessment models (IAMs), which integrate models of the natural world such as GCMs with models that simulate some aspect of society, including for example economic, political or demographic models. Finally, agent-based models (ABMs) simulate the actions of humans as individual agents, for example making decisions about land use.

Models are generalisations and data are needed to develop these generalisations. In addition, data are required to parameterise a model and further data to validate it. In Chapter 3 we introduced the concept of these types of data and where they might originate, as well as their relevance for ecosystem dynamic studies. Data can be either direct measurements and observations or indirect proxies. Direct data can be recently gathered or can come from the past. A few data series are very long, reaching back some hundreds of years, but most are short and very recent as a result of new measuring techniques. Indirect or proxy data can take the form of historical ecology, which uses human archives such as art and historical documents to observe past ecosystem dynamics. It can also involve the production of proxy data through the analysis of such elements as pollen, macrofossils and oxygen isotopes. These data can be used to reconstruct past climates or to validate vegetation and ecosystem outputs from vegetation models simulating in the Holocene.

Chapters 2 and 3 thus described the tools and provided the background for the main ideas in the chapters that followed.

Chapter 4 explored the long-term dynamic relationships between climate and ecosystems. The idea was that if we can understand more about past climates and their interactions with species and ecosystems, we can understand more of the likely consequences of climate change into the future. Biological distribution data such as pollen data have been used extensively to reconstruct climates of the past, particularly long-term trends. On the other hand, tree-ring data can reflect short-term interannual climate variability such as a warm summer. Physical data based on oxygen isotopes tend to be used to support reconstructions based on biological data. We discussed the importance of the Milankovitch cycle with regard to vegetation dynamics and the significance and synchronicity of various events through the Holocene in terms of ecosystem responses to changing climates. We also explored, through various data–model comparisons, the roles of climate, migration biology and, importantly, humans on vegetation dynamics and ecosystem processes through the Holocene. Data–model comparison proved to be a useful tool for exploring and evaluating hypotheses of climatic control on ecosystem dynamics.

One of the important aspects in understanding the dynamics of ecosystems through both time and space is the role of episodic events or disturbances in such systems. Such events were discussed in Chapter 5. They occur through a range of spatial scales, from the fine, perhaps caused by an animal's hoof, to the broad, such as the results of a fire or insect damage, or even forest logging, covering many square kilometres. The frequency of any disturbance can affect the successional state of vegetation in an ecosystem. In some cases this can lead to the vegetation never reaching a so-called 'climax' state and thus ecosystems being maintained in some sort of dynamic equilibrium. In addition, disturbances from long ago can have direct effects on the composition of current ecosystems. Fire in particular has been an important episodic event throughout the Holocene. It can have both long- and short-term effects, depending on its intensity and spatial extent. It is often hard to distinguish between natural fires and those caused by humans, either accidentally or deliberately. Understanding the role of fire in future ecosystems is complex because modern society is so closely involved in them. Modelling tries to address fire both past and future, but this is not particularly easy, especially with regard to ignition modelling. Ignitions are often caused by humans, and their involvement should not be underestimated. Human ignition can be related to the economic activity in a region. In areas of sparse population, many ignitions are caused naturally by lightning. Attempts are currently being made to model lightning strikes, and in the future the use of weather indices such as humidity could be a useful approach in modelling. Modelling of the spread of fires is more developed and is based on empirical studies, statistics and theory. At the global scale, modelling is more generalised, and while interesting outcomes have been explored there are still many shortcomings in the approaches that have been used to date. Chapter 5 also discussed some other disturbances, including those caused by pathogens and wind damage.

In Chapter 6 we explored the impact of human exploitation on ecosystems, both in the past and in the future. The evidence suggests that such exploitation could have started at least a million years ago, although the transition from hunting and gathering to agriculture took place during the Holocene. It is this transition which has impacted most on natural ecosystems. Islands, for example, were sensitive to early human settlements and mass extinctions of native flora and fauna are not unknown as a result. Globally there has been extensive deforestation associated with changing land use over the last 5000 years, especially in Europe and North America. More recently this trend has become a feature of tropical regions as forest has been cleared for agriculture. We concluded that there are clear links between human population size and the impacts on ecosystems, and of course on climate. The last 200 years has been a period of intensification of the impacts of humans on ecosystems, but it now seems likely that there were significant and observable human impacts long prior to industrialisation.

Chapter 7 introduced one of the central concepts of the book, namely the relationship between ecosystems past and future and the services they provide. Ecosystem services have been defined and classified in a number of ways and include services that are provisioning, regulating, cultural and supporting. Human society would collapse without this wide range of service provision. Many services, such as food and water provisioning, are essential for human survival. Others, especially those classed as cultural, give people important added dimensions to their lives. Some services are easy to quantify and value in financial terms, many others are not – and probably should not be. However, services of all types are finite and the current over-exploitation of so many of them can only lead to disaster for the human race and many other life forms on the planet. In this chapter we explored the changing demand for services through the Holocene so that we could better understand how we arrived at this sorry state. The interactions between the demands of society and the provision of services, including different models and frameworks, were also discussed. We addressed the importance of maintaining biodiversity and the linkages to service provision, although we do consider the maintenance of high biodiversity to be a justifiable end in itself. Finally, we described various scenarios that are used to explore the interactions between different possible societal developments and the natural world across the next 50–100 years.

Throughout the book, a major theme has concerned interactions between humans and nature through time. These interactions go beyond simple hand-to-mouth existence and are concerned with more fundamental needs of being. The question 'How then shall we live?' perhaps sums up the dilemma. These needs were recognised by the Millennium Ecosystem Assessment (MEA) as important and were grouped under the heading 'cultural services'.

We consider cultural services and their role with regard to ecosystems and biodiversity to be so important as to warrant a separate chapter (Chapter 8). Here we explored these services by first looking at how they developed in pre-literate societies through the ideas of sacred sites and sacred species. We used examples from around the world to show the universality of such a human need. These samples also show a completely different relationship between society and nature from our current one. This naturally led into cultural landscapes and their importance with regard to biodiversity. Some cultural activities can also be classified as 'provisioning'. For example, hunting on royal estates is and was partly cultural and very exclusive, but it is also a clear provisioning service. In more recent times, protected forests have provided different cultural services, for example to the Barbizon school of painters located near Fontainebleau Forest in France.

Cultural activities can and often do contribute to the conservation of species and in Chapter 9 we discussed conservation and biodiversity. We began with a historical perspective, before moving on to current practical conservation using the example of the very extensive European Natura 2000 sites. Species move or are moved, and their migration is a constant process, as species respond to their changing environment and to other species. We discussed in earlier chapters migration through the Holocene, but in this chapter we concentrated on exotic species and their arrival into 'native' communities. We separated them from species that undergo 'natural migration', although this can be difficult as human activities can influence species spreading in many ways, both directly through inadvertent transport of seeds and indirectly through habitat modification. We also introduced climate change as an additional driver of change and showed how it can promote the invasion and establishment of some exotic species. Finally, we looked at conservation and the risks to biodiversity under various climate-change scenarios. An important conclusion was that human influence has become so embedded in several ecosystems that its removal can create something unfamiliar, and often unwanted.

10.2   Emergent themes and important underlying concepts

We have identified a number of significant emergent themes that provide at least partial answers to the questions posed in Chapter 1: (1) How have ecosystems changed in the past? (2) How much of this change is attributable to human activities? 3) How much change is anticipated for the future? 4) What are the appropriate ecosystem management measures by which to prepare for the future?

10.2.1   How have ecosystems changed in the past?

- Ecosystems are dynamic and appropriate management of this dynamism is challenging. Past societies adapted, moved or broke down in response to ecosystem dynamics, but with the current state of society and the natural world, these options are inappropriate or unavailable. There is a tendency for us to see the landscape in which we live as somehow permanent and not continually responsive to its environment. We often have a strong urge for stability and continuity, which leads to a desire either to maintain our current surroundings or even to revert to some earlier ones with an emotional attachment, such as those from our childhood. This book clearly shows that as climate changes, so too do the general environment and ecosystems, however much we might wish otherwise.
- 'Saving the planet' is often popularly proclaimed as the central environmental issue today, but from a long-term perspective the planet has already survived extreme

environmental hazards such as ice ages, enormous volcanic eruptions and meteorite impacts. Our most pressing environmental concern needs to be rephrased such that the central issue for humanity is saving some sort of acceptable society within new – and in many cases more hostile – environments. In addition, we must also improve the lot of our fellow humans. The principle could therefore be to maintain and strengthen the ecosystem services we need now and in the future and to maintain high levels of biodiversity in as many ecosystems as possible, if for no other reason than 'just in case'.

10.2.2 How much of this change is attributable to human activities?

- Ecosystems were fairly resilient to major changes in the past, at least until the start of the Industrial Revolution. There has been a tendency among ecosystem scientists to ignore the timing and extent of earlier interactions between humans and ecosystems, but these interactions (e.g. the use of fire) are proving to have occurred earlier and more extensively than was previously thought and they have left important legacies in terms of species and community structures, which carry through into the ecosystems of today.
- There is a pervasive idea that runs through most modern societies that human beings somehow have more right to exist than the other species that inhabit the world and indeed that humans have dominion over them. Our brief exploration of the world of pre-literate societies shows that this need not be so and in fact for most of human existence has not been so. Many societies had a much more holistic view of the world and all its inhabitants. The 'illusion of separateness' that humanity has adopted, together with its powerful tools for altering ecosystems, has led to widespread and in some cases catastrophic changes to ecosystems and biodiversity. These changes appear likely to continue at an ever increasing rate in many parts of the world, hastening the next great species extinction event. There have been major 'natural' extinction events in the deep past, from the first more than 400 million years ago to the fifth and most recent (the loss of the dinosaurs) 65 million years ago. The environmental consequences of the 200 years since industrialisation, coupled with the many projections about global change effects on species and ecosystems, are likely to be the causes and drivers of the sixth great species extinction on the planet Earth.
- Alien species are a hot topic within conservation. Some species that move with and without the help of human populations are probably not really alien as they would have migrated eventually given enough time and the right environment and habitats to invade. As we have shown, there is plenty of evidence of northward migrations following the last ice age, and some so-called aliens are only alien because of a lack of long-term observations. On the other hand, some species that are moved great distances, especially from continent to continent – almost always by humans – are clearly alien, and in many cases they cause serious disruption to the local fauna and flora. Countries like New Zealand have suffered serious damage to and loss of native species as a result. Many countries nowadays have strict agricultural transport and immigration rules, but alien species already in the country pose serious ecological issues that are sometimes impossible to address.
- Globalisation has improved standards of living for many, especially in the developing world, but it has also brought with it major environmental problems. Rapid economic development has had an enormous environmental impact, which poses major challenges to many ecosystems. In addition, the economic problems manifest in many countries in recent years have led to short-termism and self-interest among both politicians and policy makers and to a general reluctance to face up to environmental problems. These attitudes may only change as a response to major environmental crises arising

in the next few years. In the end, the basic issue comes down to there being too many people exploiting too few resources.

10.2.3   How much change is anticipated for the future?

- One of the problems for both humanity and ecosystems is the carrying capacity of the Earth with regard to the human population and the expanding consumption of limited resources. Matching the size of this human population to the supply of the major ecosystem services, including food supply and fresh water, is the central issue. The population today is more than 7 billion people and this is projected to rise by some estimates to more than 9 billion by 2050. Even without the effects of global warming, it could be argued that this is well beyond the carrying capacity of the Earth. The current efforts to reduce the rate of increase in atmospheric CO_2 concentration can only be described as ineffective. It appears likely that major effects of climate warming, such as increased drought and loss of agricultural land, will reduce still further the ability of the Earth to sustain an increasing human population in the near future. We have explored this question through the concept of scenarios and given some examples of different possibilities. In many ways, the greatest challenge for the survival of humanity is how to manage our resources given the fragmented and competitive nature of our many different societies. Many scenarios project that populations will continue to grow, even though the rate of growth may be less than at present. Reducing the rate of increase is important. The problems of population are many and they involve deeply held beliefs, but they do need to be faced, for until they are, the size of the human population remains a ticking timebomb. What are the prospects for successfully feeding and watering an increasing population in the future? This is not easy to forecast and many scenarios suggest that food security will remain a basic issue for large numbers of people throughout the coming century at least. Of course, this depends on many factors; for example, future food security will be affected by whether there is an interacting cooperative global society or a much more regionalised one in which countries are most concerned about their own protection and their own populations. Deterioration in food supply brings with it an increasing likelihood of political instability.
- The degree of land-use change that could occur in the future is scenario-dependent. Technological development can lead to more efficient agriculture and coupled with falling populations and reduced demand for meat could provide a better balance between agriculture and other ecosystem service provision, including biodiversity. A combination of behavioural changes and improving technology is probably the key to constructive progress.
- As far as climate change goes, most projections are currently made to 2050 or 2100. However, climate change will not suddenly stop at these points: even in the unlikely event that atmospheric CO_2 concentration is stabilised, there is a long response curve, which means that the effects on climate will continue well beyond these dates. The more likely scenario is that CO_2 concentrations will continue to increase, albeit at a slower rate, meaning even greater climate change. Among the many responses to this is an increased risk of reaching tipping points in the climate, perhaps leading to vast areas of the globe becoming just too hot for humans to live in and forcing us to abandon these regions. This would be likely to lead to migrations to areas closer to the poles, resulting in a variety of possible confrontations between residents and climate refugees.
- Continuing rapid climate change will impact ecosystems far into the future. Plant and animal species will respond to these changes by migrating into the climate space within which they can establish and reproduce. Diverse and dynamic habitats must be available to allow these processes to occur. One of the most useful ways of allowing species

movement through such habitats is to identify core areas and corridors in an interconnecting network, allowing the possibility of a local response to changing climate.

10.2.4 What are the appropriate ecosystem management measures by which to prepare for the future?

- We can identify several technical management recommendations that will contribute to mitigation of the impacts of current climate change on ecosystems. Several of these build on our knowledge of how ecosystems have changed in the past and the extent of human influence currently embedded within key ecosystem properties. Our recommendations include the following. (1) Use knowledge of the past to set management goals. This can entail acceptance of firmly embedded long-term human influence in, for example, grassland and heathland ecosystems and avoidance of the rigid adoption of 'natural' baselines. (2) Manage soils and vegetation to maximise carbon uptake and storage, buying time against a runaway greenhouse effect. Appropriate measures include restoring wetlands and blocking drainage ditches, managing soils to maximise organic content and increasing forest area, particularly old-growth forest, by increasing the length of forest rotations. (3) Manage fire regimes to minimise carbon release, both through fuel management (to limit uncontrolled wildfire) and by using grazing and browsing animals instead of fire to control unwanted vegetation successions. (4) Adopt a broader view of the conservation of cultural ecosystem services, moving away from an unemotional market control of such services towards the incorporation of elements of the innate respect for ecosystems still retained by many societies. Research-based stewardship of ecosystems, as adopted in forest certification schemes, represents a constructive way forward. (5) Retain habitat diversity, creating opportunities for dynamic successional processes. Ensure that management plans are flexible and permit broad variation within the management goals. (6) Control long-distance movement of species to limit unnatural invasive aliens and pathogens.
- With regard to the maintenance of biodiversity through the concept of ecosystem services, there is a developing tendency to value such services in economic terms, partly in order to assign them their 'true' financial value but also as a means of comparing between them, especially where selective decisions are to be made. However, as George Monbiot commented in the *Guardian* newspaper, 'The costing and sale of nature … diminishes us, it diminishes nature. By turning the natural world into a subsidiary of the corporate economy, it reasserts the biblical doctrine of dominion. It slices the biosphere into component commodities: already the government's task force is talking of "unbundling" ecosystem services, a term borrowed from previous privatisations. This might make financial sense; it makes no ecological sense. The more we learn about the natural world, the more we discover that its functions cannot be safely disaggregated' (Monbiot, 2012). There are many issues with the free market and economic valuation and we would argue that not everything is about economics: other values are worth fighting for, such as cultural services in all their many guises.
- As we have suggested throughout the book, changing climate is only one driver of change. Others, such as land use change, also have major effects on ecosystems, dynamics and biodiversity. The issue is highly complex as all drivers frequently act on ecosystems, their species and their dynamics. This complexity alters through time and the rising management challenge is how to address this dynamism in a proactive rather than a reactive way. Model projections may give guidance, but the real issue is the need for political and social will to address environmental change. The options for action also change, and management will increasingly have to be done through adaptation to environmental changes. The window of opportunity for mitigating the worst effects of

climate change is rapidly shrinking, for example. The question is how can society adapt to all these changes not once but continually into the future?

- Scientists can identify vulnerable ecosystems and species as much as they like, but persuading policymakers to actively protect the most vulnerable and promote biodiversity, especially during a period of rapid global change, is frequently unsuccessful. Significant improvements are needed with regard to the role of science and scientists in policy-making.

- A major issue for anybody concerned about ecosystems and their services in a rapidly changing world is convincing both the political elite and the public at large of the urgency of the need to preserve ecosystems, biodiversity and thus ecosystem services. There has been much failure in the recent past, and in the current economic climate it is not at all clear how the situation can be improved. In addition, the uncertainties and assumptions made within any scientific pronouncement, whether they be about climate change or biodiversity loss, lead to a lack of clarity, and thus the information loses power in the eyes of the general public. Further, there are many groups that are 'nonbelievers', actively working against any initiative to improve the situation. They either simply do not believe in the concept of rapid global change, viewing it as a scientific conspiracy mainly aimed at getting more research funds, or believe firmly that any problem can be solved by the ingenuity of humans and the power of money, or just don't care about other people or the future.

Scientists have been making technical recommendations about the environment for a long time, but with relatively little success on the ground. As our management measures suggest, the major barriers to progress lie not so much with the scientific understanding of ecosystems as with the nature of the challenges themselves and the related lack of political will to address them. There are many complex reasons for this lack of political action, but scientists must shoulder part of the blame for ineffective communication. A recent survey has shown great diversity around the world in the public perceptions of climate change and environmental problems (AXA/IPSOS, 2012). Current climate change was considered to be mostly a result of human activity by over 90% of respondents in Hong Kong, Indonesia and Mexico, while most doubters were found in the USA and Great Britain (Table 10.1).

There are several factors that influence these opinions, but one of importance for the mature economies of the world is the powerful lobbying of large companies with economic interest in fossil fuels and the maintenance of the current economic status quo. The emergence of a new, thriving 'green' economy provides opportunities for new players to emerge and creates unwelcome uncertainty for risk-shy existing industrial leaders, who are reluctant to embrace too-rapid change. As we pass the milestone of 400 ppm of CO_2 in the atmosphere, we continue to pour out over 1 million kg every second, which is a rate of increase probably never before seen in Earth history. The atmosphere last held this amount of CO_2 about 4 million years ago, when the sea level was 40 m higher than today. As United Nations (UN) climate and emission control talks are deadlocked, with unrealistic targets being continually pushed into the future, it would seem to be an appropriate time to ensure that all are properly informed of the likely consequences of inaction.

Stephen Emmott is Professor of Computational Science, University of Oxford and Head of Microsoft Research's Laboratory in Cambridge. In an attempt to reach out to new audiences, he put on a one-man show during 2012 called '10 Billion'. His Oxford University office was recreated at the Royal Court Theatre in London and he outlined to packed audiences the long-term development of our use of the Earth's resources, before showing the probable state of the world when global population reaches 10 000 000 000. This form of communication, alongside Al Gore's book/film *An Inconvenient Truth* and *The Age of Stupid*, a film narrated by the late Pete Postlethwaite, helps to fill knowledge gaps and stimulate climate change doubters. Emmott contrasted the public reaction to a large asteroid

Table 10.1 Public attitudes in different countries to the main driver of current climate change (AXA/IPSOS, 2012) http://www.axa.com/lib/en/library/axapapers/climaterisks.aspx

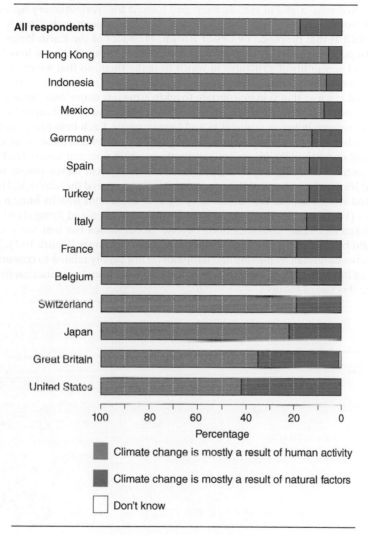

hurtling towards Earth with the problems posed by climate change. The effects of asteroid impact are immediate and easy to comprehend, while the complexity of climate–ecosystem interactions is harder to grasp, even if the eventual outcomes for society may be of comparable scale. Emmott believes that humans are optimistic and inventive when coping with conceptually simple issues but tend to ignore longer-term problems until they present unavoidable, individual crises. The current effects of climate change can be difficult to observe and 'it's immensely appealing to want to believe that this isn't a problem on this scale, or that even if it is, that we will figure out a way to stop it' (O'Callaghan, 2012).

Emmott calls for a fundamental change in human behaviour: 'that is a revolution in the way economies work, in the way governments do their job, and as a sense of collective responsibility as citizens rather than individual responsibility. Bringing about that kind of change is outside the domain of scientific expertise. That is the domain of politicians and economists and perhaps philosophers' (O'Callaghan, 2012). Emmot's call reinforces that made by the normally more cautious UN Secretary-General, who came to feel we had become locked into 'a global suicide pact' and looked for 'revolutionary action' in the developed world to replace the prevailing model of economic growth. Clearly we need a Plan B, such as that proposed by Lester Brown, President of the Earth Policy Institute (www.earth-policy.org). We need new ways of making decisions that have long-term perspectives and new political structures in which to make them, so that we can correct the mismatch between global environmental signals and a control panel that is currently wired and managed in a way that is inappropriate to addressing our major environmental issues. We urgently need to optimise our use of ecosystem services and natural capital and find an appropriate balance between availability and human needs. Each individual has to recognise better when they have a sufficient share of global resources and must break out of the vicious spiral of continually increasing consumption, which does not always lead to a proportional increase in well-being or happiness. The 2013 Human Development Report of the United Nations Environment Programme (UNEP) compared the 'ecological footprint' or estimated area of the Earth's surface used by each inhabitant with its human development index (HDI), which takes into account health, education and living standards. The report showed that many countries lived beyond global means but that fairly high HDI values could be obtained with a fair share of ecosystem resources (Figure 10.1). The relationship between human well-being and happiness is not simply related to consumption of resources or income, and it is encouraging to see an international organisation like UNEP developing this line of thought.

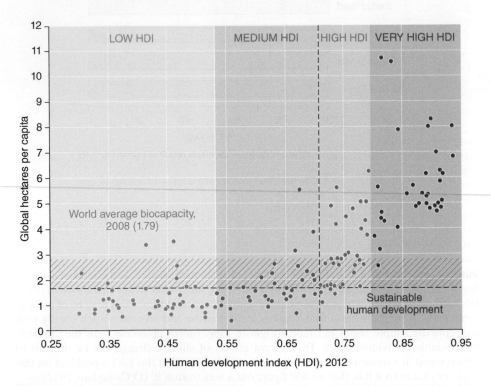

Figure 10.1 Relationship for different countries between the UN human development index (HDI) and the area of the Earth used per capita (UNDP, 2013) Source: UNDP Human Development Report 2013. Reproduced by permission of the United Nations Development Programme (UNDP)

It is way beyond the scope of this book to come up with answers to some of the most pressing political and environmental issues, so instead we finish by outlining seven major challenges threatening the long-term integrity of terrestrial ecosystems and how we have contributed to organising ideas for addressing them:

1. Land-use management to optimise the balance between agricultural production and competing ecosystem services.
2. Identification and protection of the most vulnerable and threatened ecosystems and biodiversity.
3. Management of land surfaces with regard to feedbacks to the climate system through storage and release of greenhouse gases and albedo.
4. Identification of tipping points in Earth systems and development of an early warning system.
5. Matching of the size of the human population to the supply of the major ecosystem services, including food supply and fresh water.
6. Development of an energy policy to slow the rate of atmospheric accumulation of greenhouse gases.
7. Policy and management for mediation of the ecosystem impacts of future climate change.

Our earlier chapters explored the background to challenges 1–4, concerning ecosystem dynamics, land use and tipping points, and we have built on these chapters here to identify some approaches to safeguarding a sustainable way of life as regards ecosystem manage ment. The challenges facing the Earth today and in the future demand action on several fronts, and appropriate management of ecosystems should play its part. One conclusion from this book is that the present crisis is a culmination of human activities that stretch back over millennia, rather than just decades. We can use our knowledge from the past to improve the forecasting efficiency of models, but there are few direct analogues from the past to inform modern management decisions. The current crisis is precipitated by the recent rapid increase in human population size and living standards and by improved tools for the exploitation of ecosystem services and natural capital.

Challenges 5–7 reach from science far into policy issues. The difficult trade-offs between economics and policy have temporarily paralysed the world at a time when action on all fronts is desperately needed. While the science behind appropriate policies concerning agriculture and forestry is well established, the major issue is the future use of our increasingly abundant fossil fuel reserves. Burning all of these will make the Earth uninhabitable, yet the international markets are banking on precisely that course of action, strongly encouraged by a powerful fossil-fuel lobby (Berners-Lee and Clark, 2013). We can only hope for a future of adjusted policies in which ecosystem management links with new technologies such as renewable energy sources and the exploitation of artificial photosynthesis to yield a global measure for an acceptable standard of living that enables us to pass on functioning ecosystems to future generations.

References

Aaby, B. (1983) Forest development, soil genesis and human activity illustrated by pollen and hypha analysis of two neighbouring podzols in Draved Forest, Denmark. *Danmarks Geologiske Undersøgelse II*, **114**, 1–114.

Agnoletti, M. (ed.) (2006) *The Conservation of Cultural Landscapes*. CABI, Oxford, UK.

Alcamo, J. (ed.) (1994) *IMAGE 2.0: Integrated Modeling of Global Climate Change*. Kluwer Academic Publishers, Dordrecht, The Netherlands.

Alcamo, J., Shaw, R. and Hordijk, L. (eds) (1990) *The Rains Model of Acidification: Science and Strategies in Europe*. Kluwer Academic Publishers, Dordrecht, The Netherlands.

Alcamo, J., Kreileman, G.J.J., Krol, M.S. and Zuidema, G. (1994) Modelling the global society-biosphere-climate system. Part 1: model description and testing. *Water, Air and Soil Pollution*, **76**, 1–35.

Alcamo, J., Kreileman, G.J.J., Bollen, J.C. *et al.* (1996) Baseline scenarios of global environmental change. *Global Environmental Change*, **6**, 261–303.

Alcamo, J., Leemans, R. and Kreileman, E. (eds) (1998) *Global Change Scenarios of the 21st Century: Results from the IMAGE 2.1 Model*. Pergamon, an imprint of Elsevier Science, Oxford, UK.

Aldersley, A., Murray, S.J. and Cornell, S.E. (2011) Global and regional analysis of climate and human drivers of wildfire. *Science of the Total Environment*, **409**(18), 3472–3481.

Alley, R.B. and Ágústsdóttir, A.M. (2005) The 8k event: cause and consequences of a major Holocene abrupt climate change. *Quaternary Science Reviews*, **24**, 1123–1149.

Alsos, I.G., Ehrich, D., Thuiller, W. *et al.* (2012) Genetic consequences of climate change for northern plants. *Proceedings of the Royal Society B: Biological Sciences*, **279**(1735), 2042–2051.

Amaral-Turkman, M., Turkman, K., Le Page, Y. and Pereira, J.M.C. (2011) Hierarchical space-time models for fire ignition and percentage of land burned by wildfires. *Environmental and Ecological Statistics*, **18**(4), 601–617.

Andersen, N.H. (2000) Kult og ritualer i den aeldre bondestenalder [Cult and rituals in the TRB-culture]. *KUML*, **13–57**.

Anderson, B.J., Akcakaya, H.R., Araújo, M.B. *et al.* (2009) Dynamics of range margins for metapopulations under climate change. *Proceedings of the Royal Society B: Biological Sciences*, **276**(1661), 1415–1420.

Anderson, D.M., Salick, J., Moseley, R.K. and Xiaokun, O. (2005) Conserving the sacred medicine mountains: a vegetation analysis of Tibetan sacred sites in northwest Yunnan. *Biodiversity and Conservation*, **14**(13), 3065–3091.

Anderson, L.L., Hu, F.S., Nelson, D.M. *et al*. (2006) Ice-age endurance: DNA evidence of a white spruce refugium in Alaska. *Proceedings of the National Academy of Sciences*, **103**, 12 447–12 550.

Andrews, P. L. (2007) BehavePlus fire modeling system: past, present and future. *Proceedings of the 7th Symposium on Fire and Forest Meteorology, Bar Harbor, Maine, American Meteorological Society*.

Angelstam, P.K. (1998) Maintaining and restoring biodiversity in European boreal forests by developing natural disturbance regimes. *Journal of Vegetation Science*, **9**, 593–602.

Anton, C., Young, J., Harrison, P.A. *et al*. (2010) Research needs for incorporating the ecosystem service approach into EU biodiversity conservation policy. *Biodiversity and Conservation*, **19**(10), 2979–2994.

ANVL (2007) *Massif de Fontainebleau: La longue marche vers le Parc National*. Fontainebleau, France.

Aono, Y. and Kazui, K. (2008) Phenological data series of cherry tree flowering in Kyoto, Japan, and its application to reconstruction of springtime temperatures since the 9th century. *International Journal of Climatology*, **28**, 905–914.

Applebaum, S. (1972) Crops and plants, in *The Agrarian History of England and Wales. I. AD 43–1042* (ed. H.P.R. Finberg). Cambridge University Press, Cambridge, UK, pp. 108–121.

Araújo, M.B. and Pearson, R.G. (2005) Equilibrium of species distributions with climate. *Ecography*, **28**(5), 693–695.

Araújo, M.B., Whittaker, R.J., Ladle, R.J. *et al*. (2005) Reducing uncertainty in projections of extinction risk from climate change. *Global Ecology and Biogeography*, **14**(6), 529–538.

Araújo, M.B., Thuiller, W. and Pearson, R.G. (2006) Climate warming and the decline of amphibians and reptiles in Europe. *Journal of Biogeography*, **33**, 1712–1728.

Archibald, S., Roy, D.P., Van Wilgen, B. and Scholes, R.J. (2009) What limits fire? An examination of drivers of burnt area in Southern Africa. *Global Change Biology*, **15**(3), 613–630.

Archives New Zealand (no date) Photographs – Wellington. Archives New Zealand/Te Rua Mahara o te Kāwanatanga. http://archives.govt.nz/research/guides/photographs-wellington [accessed 22 November 2013].

Attiwill, P.M. (1994) The disturbance of forest ecsosystems: the ecological basis for conservative management. *Forest Ecology and Management*, **63**, 247–300.

AXA/IPSOS (2012) *Individual Perceptions of Climate Risks*, 18.

Ayres, P. (2012) *Shaping Ecology: The Life of Arthur Tansley*. John Wiley and Sons, Ltd., Chichester, UK.

Ball, M. (2000) Sacred mountains, religious paradigms, and identity among the Mescalero Apache. *Worldviews: Global Religion, Culture and Ecology*, **4**, 264–282.

Barnosky, A.D., Koch, P.L., Ferance, R.S. *et al*. (2004) Assessing the causes of late Pleistocene extinctions on the continents. *Science*, **306**, 70–75.

Barnosky, A.D., Matzke, N., Tomiya, S. *et al*. (2011) Has the Earth's sixth mass extinction already arrived? *Nature*, **471**, 51–57.

Bartlein, P.J., Prentice, I.C. and Webb, T. III, (1986) Climatic response surfaces from pollen data for some eastern North American taxa. *Journal of Biogeography*, **13**, 35–57.

Bartlein, P.J., Harrison, S.P., Brewer, S. *et al*. (2011) Pollen-based continental climate reconstructions at 6 and 21 ka: a global synthesis. *Climate Dynamics*, **37**, 775–802.

Beerling, D.J. and Royer, D.L. (2002) Fossil plants as indicators of the Phanerozoic global carbon cycle. *Annual Review of Earth and Planetary Sciences*, **30**, 527–556.

Behre, K.-E. (1988) The role of man in European vegetation history, in *Vegetation History* (eds B. Huntley and T. Webb III,). Kluwer Academic Publishers, Dordrecht, The Netherlands, pp. 633–672.

Bennett, K.D. (1988) Holocene geographic spread and population expansion of *Fagus grandifolia* in Ontario, Canada. *Journal of Ecology*, **76**, 547–557.

Bergeron, Y., Gauthier, S., Flannigan, M. and Kafka, V. (2004) Fire regimes at the transition between mixed wood and coniferous boreal forest in northwestern Quebec. *Ecology*, **85**, 1916–1932.

Berglund, B.E. (1986) *Handbook of Holocene Palaeoecology and Palaeohydrology*. John Wiley and Sons, Ltd., Chichester, UK.

Berglund, B.E. (ed.) (1991) *The Cultural Landscape During 6000 Years in Southern Sweden – The Ystad Project*. Ecological Bulletins, Munksgaard, Copenhagen, Denmark.

Berglund, B.E. (2000) The Ystad Project – a case study for multidisciplinary research on long-term human impact. *PAGES Newsletter*, **2000–2003**, 6–7.

Berkes, F. and Folke, C. (eds) (1998) *Linking Social Ecological Systems: Management Practices and Social Mechanisms for Building Resilience*. Cambridge University Press, Cambridge, UK.

Berna, F., Goldberg, P. *et al.* (2012) Microstratigraphic evidence of in situ fire in the Acheulean strata of Wonderwerk Cave, Northern Cape province, South Africa. *Proceedings of the National Academy of Sciences*, **109**(20), E1215–E1220.

Berners-Lee, M. and Clark, D. (2013) *The Burning Question*. Profile Books, London, UK.

Berry, P., Paterson, J., Cabeza, M. *et al.* (2008) Deliverables 2.2 and 2.3: Meta-analysis of adaptation and mitigation measures across the EU25 and their impacts and recommendations how negative impacts can be avoided. http://www.macis-project.net/MACIS-deliverable -2.2-2.3.pdf [accessed 22 November 2013].

Bhagwat, S.A. and Rutte, C. (2006) Sacred groves: potential for biodiversity management. *Frontiers in Ecology and the Environment*, **4**(10), 519–524.

Bhiry, N. and Filion, L. (1996) Mid-Holocene hemlock decline in eastern North America linked with phytophagous insect activity. *Quaternary Research*, **45**, 312–320.

Bialozyt, R., Bradley, L.R. and Bradshaw, R.H.W. (2012) Modelling the spread of *Fagus sylvatica* and *Picea abies* in southern Scandinavia during the late Holocene. *Journal of Biogeography*, **39**, 655–675.

Bird, D.W., Bird, R.B. and Parker, C.H. (2005) Aboriginal burning regimes and hunting strategies in Australia's western desert. *Human Ecology*, **33**(4), 443–464.

Birdsey, R., Pregitzer, K. and Lucier, A. (2006) Forest carbon management in the United States, 1600–2100. *Journal of Environmental Quality*, **35**, 1461–1469.

Birks, H.H. and Ammann, B. (2000) Two terrestrial records of rapid climatic change during the glacial–Holocene transition (14,000–9,000 calendar years B.P.) from Europe. *Proceedings of the National Academy of Sciences*, **97**, 1390–1394.

Birks, H.J.B. (2005) Fifty years of Quaternary pollen analysis in Fennoscandia 1954–2004. *Grana*, **44**, 1–22.

Birks, H.J.B. and Birks, H.H. (2008) Biological responses to rapid climate change at the Younger Dryas–Holocene transition at Kråkenes, western Norway. *The Holocene*, **18**, 19–30.

Björse, G. and Bradshaw, R.H.W. (1998) 2000 years of forest dynamics in southern Sweden: suggestions for forest management. *Forest Ecology and Management*, **104**, 15–26.

Black, B.A. and Abrams, M.D. (2001) Influences of Native Americans and surveyor biases on metes and bounds witness-tree distribution. *Ecology*, **82**, 2574-2586.

Bogotá-A, R.G., Groot, M.H.M., Hooghiemstra, H. *et al.* (2011) Rapid climate change from north Andean Lake Fúquene pollen records driven by obliquity: implications for a basin-wide biostratigraphic zonation for the last 284 ka. *Quaternary Science Reviews*, **30**, 3321–3337.

Bohn, U., Gollop, G. and Hettwer, C. (2003) *Karte der natürlichen Vegetation Europas* [Map of the Natural Vegetation of Europe]. Landwirtschaftsverlag, Muenster, Germany.

Bohunovsky, L., Omann, I.: the EcoChange team (2012) Final report: EcoChange – Challenges in assessing and forecasting biodiversity and ecosystem change. Results of the integrated sustainability assessment/agent-based modelling for the case studies. Integrated Project EU Funded Project FP6 2006 GOCE 036866.

Bolmgren, K., Miller-Rushing, A. and Vanhoenacker, D. (2012) One man, 73 years, and 25 species. Evaluating phenological responses using a lifelong study of first flowering dates. *International Journal of Biometeorology*, 1–9.

Bond, W.J., Woodward, F.I. and Midgley, G.F. (2005) The global distribution of ecosystems in a world without fire. *New Phytologist*, **165**, 525–538.

Boose, E.R., Chamberlin, K.E. and Foster, D.R. (2001) Landscape and regional impacts of hurricanes in New England. *Ecological Monographs*, **71**(1), 27–48.

Booth, R.K. (2012) Testing the climate sensitivity of peat-based paleoclimate reconstructions in mid-continental North America. *Quaternary Science Reviews*, **29**, 720–731.

Botkin, D.B., Janak, J.F. and Wallis, J.R. (1972) Some ecological consequences of a computer model of forest growth. *Journal of Ecology*, **60**, 849–872.

Bowman, D.M.J.S. (1998) The impact of Aboriginal landscape burning on the Australian biota. *New Phytologist*, **140**(3), 385–410.

Bowman, D.M.J.S., Balch, J.K., Artaxo, P. *et al.* (2009) Fire in the earth system. *Science*, **324**, 481–484.

Bowman, D.M.J.S., Balch, J., Artaxo, P. *et al.* (2011) The human dimension of fire regimes on Earth. *Journal of Biogeography*, **38**, 2223–2236.

Box, E.O. (1981) Predicting physiognomic vegetation types with climate variables. *Vegetatio*, **45**, 127–139.

Boyle, J.F. (2007) Loss of apatite caused irreversible early-Holocene lake acidification. *The Holocene* **17**, 543–547.

Boyle, J.F., Gaillard, M.-J., Kaplan, J.O. and Dearing, J.A. (2011) Modelling prehistoric land use and carbon budgets: a critical review. *The Holocene*, **21**, 715–722.

Braconnot, P., Harrison, S.P., Kageyama, M. *et al.* (2012) Evaluation of climate models using palaeoclimatic data. *Nature Climate Change*, **2**, 417–424.

Bradley, R.S. (2011) *Global Warming and Political Intimidation*. University of Massachusetts Press, Amherst, MA, USA.

Bradshaw, C.J.A., Giam, X. and Sodhi, N.S. (2010a) Evaluating the relative environmental impact of countries. *PLoS ONE*, **5**(5), 1–16.

Bradshaw, E.G., Rasmussen, P. and Odgaard, B.V. (2005) Mid- to late-Holocene land-use change and lake development at Dallund Sø, Denmark: synthesis of multiproxy data, linking land and lake. *The Holocene*, **15**, 1152–1162.

Bradshaw, R.H.W. (1995) The origins and dynamics of native forest ecosystems: background to the use of exotic species in forestry. *Icelandic Agricultural Sciences*, **9**, 7–15.

Bradshaw, R.H.W. (2004) Past anthropogenic influence on European forests and some possible genetic consequences. *Forest Ecology and Management*, **197**, 203–212.

Bradshaw, R.H.W. (2013) Stand-scale palynology, in *The Encyclopedia of Quaternary Science* (ed. S.A. Elias). Elsevier, Amsterdam, The Netherlands, pp. 846–853.

Bradshaw, R.H.W. and Hannon, G.E. (1992) Climatic change, human influence and disturbance regime in the control of vegetation dynamics within Fiby Forest, Sweden. *Journal of Ecology*, **80**, 625–632.

Bradshaw, R.H.W. and Hannon, G.E. (2006) Long-term vegetation dynamics in southern Scandinavia and their use in managing landscapes for biodiversity, in *The Conservation of Cultural Landscapes* (ed. M. Agnoletti). CABI, Wallingford, UK, pp. 94–107.

Bradshaw, R.H.W. and Lindbladh, M. (2005) Regional spread and stand-scale establishment of *Fagus sylvatica* and *Picea abies* in Scandinavia. *Ecology*, **86**(7), 1679–1686.

Bradshaw, R.H.W. and Mitchell, F.J.G. (1999) The palaeoecological approach to reconstructing former grazing-vegetation interactions. *Forest Ecology and Management*, **120**, 3–12.

Bradshaw, R.H.W., Gemmel, P. and Björkman, L. (1994) Development of nature-based silvicultural models in southern Sweden, the scientific background. *Forest and Landscape Research*, **1**, 95–110.

Bradshaw, R.H.W., Tolonen, K. and Tolonen, M. (1997) Holocene records of fire from the boreal and temperate zones of Europe, in *Sediment Records of Biomass Burning and Global Change* (eds J.S. Clark, H. Cachier, J.G. Goldammer and B.J. Stocks). Springer-Verlag, Germany, pp. 347–365.

Bradshaw, R.H.W., Holmqvist, B.H., Cowling, S.A. and Sykes, M.T. (2000) The effects of climate change on the distribution and management of *Picea abies* in southern Scandinavia. *Canadian Journal of Forest Research*, **30**, 1992–1998.

Bradshaw, R.H.W., Hannon, G.E. and Lister, A.M. (2003) A long-term perspective on ungulate-vegetation interactions. *Forest Ecology and Management*, **181**, 267–280.

Bradshaw, R.H.W., Kito, N. and Giesecke, T. (2010b) Factors influencing the Holocene history of *Fagus. Forest Ecology and Management*, **259**, 2204–2212.

Bradshaw, R.H.W., Hannon, G.E., Rundgren, M. and Giesecke, T. (2010c) Which trees grew on the Faroe Islands before people arrived? in *Dorete: Her Book* (eds S.-A. Bengtsson, P. Buckland, P. H. Enckell and A.M. Fosaa). Faroe University Press, Torshavn, Faroe Islands, pp. 36–49.

Bradshaw, R.H.W., Lindbladh, M. and Hannon, G.E. (2010d) The role of fire in southern Scandinavian forests during the late Holocene. *International Journal of Wildland Fire*, **19**, 1040–1049.

Brewer, S., Guiot, J. and Barboni, D. (2013) Use of pollen as climate proxies, in *Encyclopedia of Quaternary Science* (ed. S.A. Elias). Elsevier, Amsterdam, The Netherlands.

BRIDGE (no date) Earth systems modelling results. http://www.paleo.bris.ac.uk/ummodel/list _of_simulations.html [accessed 22 November 2013].

Briffa, K.R., Jones, P.D., Bartholin, T.S. *et al.* (1992) Fennoscandian summers from AD 500: temperature changes on short and long time scales. *Climate Dynamics*, **7**, 111–119.

Brown, K.J., Clark, J.S., Grimm, E.C. *et al.* (2005) Fire cycles in North American interior grasslands and their relation to prairie drought. *Proceedings of the National Academy of Sciences*, **102**, 8865–8870.

Bush, M.B. and Flenley, J.R. (1987) The age of British chalk grassland. *Nature*, **329**, 434–436.

Byers, B.A., Cunliffe, R.N. and Hudak, A.T. (2001) Linking the conservation of culture and nature: a case study of sacred forests in Zimbabwe. *Human Ecology*, **29**(2), 187–218.

Cappelen, J. (2011) *Storm og ekstrem vind i Danmark. Teknisk rapport 11–12.* Danmarks Meteorologiske Institut, Copenhagen, Denmark (in Danish).

Carson, R. (1962) *Silent Spring*. Houghton Mifflin, Boston, MA, USA.

Cheddadi, R., Yu, G., Guiot, J. *et al.* (1997) The climate of Europe 6000 years ago. *Climate Dynamics*, **13**, 1–9.

Chuine, I. and Beaubien, E.G. (2001) Phenology is a major determinant of tree species range. *Ecology Letters*, **4**(5), 500–510.

Claussen, M., Kubatzki, C., Brovkin, V. and Ganopolski, A. (1999) Simulation of an abrupt change in Saharan vegetation in the Mid-Holocene. *Geophysical Research Letters*, **26**, 2037–2040.

Claussen, M., Mysak, L.A., Weaver, A.J. *et al.* (2002) Earth system models of intermediate complexity: closing the gap in the spectrum of climate system models. *Climate Dynamics*, **18**(7), 579–586.

Clear, J.L., Seppä, H., Kuosmanen, N. and Bradshaw, R.H.W. (2013) Holocene fire frequency variability in Vesijako, Strict Nature Reserve, Finland, and its application to conservation and management. *Biological Conservation*, **166**, 90–97.

Cohen, D.J. (2011) The beginnings of agriculture in China. *Current Anthropology*, **52**, S273–S293.

Colombaroli, D., Tinner, W., van Leeuwen, J. *et al.* (2009) Response of broadleaved evergreen Mediterranean forest vegetation to fire disturbance during the Holocene: insights from the peri-Adriatic region. *Journal of Biogeography*, **36**, 314–326.

Colombaroli, D., Henne, P.D., Kaltenrieder, P. *et al.* (2010) Species responses to fire, climate and human impact at tree line in the Alps as evidenced by palaeo-environmental records and a dynamic simulation model. *Journal of Ecology*, **98**, 1346–1357.

Conedera, M., Torriani, D., Neff, C. *et al.* (2011) Using Monte Carlo simulations to estimate relative fire ignition danger in a low-to-medium fire-prone region. *Forest Ecology and Management*, **261**(12), 2179–2187.

Convention on Biological Diversity (2010) *Global Biodiversity Outlook 3*. Montreal: 94.

Convention on Biological Diversity (no date) Aichi biodiversity targets. http://www.cbd.int /sp/targets/ [accessed 22 November 2013].

Costanza, R., d'Arge, R., de Groot, R. *et al.* (1997) The value of the world's ecosystem services and natural capital. *Nature*, **387**, 253–260.

Cowling, S.A., Sykes, M.T. and Bradshaw, R.H.W. (2001) Palaeovegetation-model comparisons, climate change and tree succession in Scandinavia over the past 1500 years. *Journal of Ecology*, **89**(2), 227–236.

Cox, J.C. (1905) *The Royal Forests of England*. Methuen & Co, London, UK.

Cox, P.M., Betts, R.A., Jones, C.D. *et al.* (2000) Acceleration of global warming due to carbon-cycle feedbacks in a coupled climate model. *Nature*, **408**(6809), 184–187.

Cox, P.M., Betts, R.A., Collins, M. *et al.* (2004) Amazonian forest dieback under climate-carbon cycle projections for the 21st century. *Theoretical and Applied Climatology*, **78**(1–3), 137–156.

Coxon, P., Hannon, G. and Foss, P. (1994) Climatic deterioration and the end of the Gortian Interglacial in sediments from Derrynadivva and Burren Townland, near Castlebar, County Mayo, Ireland. *Journal of Quaternary Science*, **9**(1), 33–46.

Crowley, B.E. (2010) A refined chronology of prehistoric Madagascar and the demise of the megafauna. *Quaternary Science Reviews*, **29**, 2591–2603.

Crumley, C.L. (2003) Historical ecology: integrated thinking at multiple temporal and spatial scales. *World System History and Global Environmental Change, Lund University, Sweden*. http://profesores.usfq.edu.ec/fdelgado/Teor%C3%ADa%20y%20M%C3%A9todos%20en %20Arqueolog%C3%ADa/Teor%C3%ADa%20y%20M%C3%A9todos%20en%20 Arqueolog%C3%ADa/Archivos%20Digitales/03.crumley.pdf [accessed 22 November 2013].

Dahl, E. (1998) *The Phytogeography of Northern Europe*. Cambridge University Press, Cambridge, UK.

Dahl-Jensen, D., Mosegaard, K., Gundestrup, N. *et al.* (1998) Past temperatures directly from the Greenland ice sheet. *Science*, **282**(5387), 268–271.

Daily, G.C. (ed.) (1997) *Natures Services: Societal Dependence on Natural Ecosystems*. Island Press, Washington, DC, USA.

Dambrine, E., Dupouey, J.-L. and Laut, L. (2007) Present forest biodiversity patterns in France related to former Roman agriculture. *Ecology*, **88**(6), 1430–1439.

Darwin, C. (1881) *The Formation of Vegetable Mould through the Action of Worms, with Observations on their Habits*. John Murray, London, UK.

Davis, M.B. (1963) On the theory of pollen analysis. *American Journal of Science*, **261**(10), 897–912.

Davis, M. (1976) Pleistocene biogeography of temperate deciduous forests. *Geoscience and Man*, **13**, 13–26.

Davis, M.B. (1981) Outbreaks of forest pathogens in Quaternary history. *Proceedings of the IV International Palynological Conference (1976–77)*, **3**, 216–227.

Davis, M.B. and Botkin, D.B. (1985) Sensitivity of cool-temperate forests and their fossil pollen record to rapid temperature change. *Quaternary Research*, **23**(3), 327–340.

Davis, B.A.S., Brewer, S., Stevenson, A.C. *et al.* (2003) The temperature of Europe during the Holocene reconstructed from pollen data. *Quaternary Science Reviews*, **22**, 1701–1716.

de Groot, W.J., Cantin, A.S., Flannigan, M.D. *et al.* (2013) A comparison of Canadian and Russian boreal forest fire regimes. *Forest Ecology and Management*, **294**, 23–34.

Dearing, J.A., Braimoh, A.K., Reenberg, A. *et al.* (2010) Complex land systems: the need for long time perspectives to assess their future. *Ecology and Society*, **15**(4), 21.

del Hoyo, L.V., Isabel, M.P.M. and Vega, F.J.M. (2011) Logistic regression models for human-caused wildfire risk estimation: analysing the effect of the spatial accuracy in fire occurrence data. *European Journal of Forest Research*, **130**(6), 983–996.

deMenocal, P., Ortiz, J., Guilderson, T. *et al.* (2000) Abrupt onset and termination of the African Humid Period: rapid climate responses to gradual insolation forcing. *Quaternary Science Reviews*, **19**, 347–361.

Denham, T.P., Haberle, S.G., Lentfer, C. *et al.* (2003) Origins of agriculture at Kuk Swamp in the highlands of New Guinea. *Science*, **301**, 189–193.

Devictor, V., Julliard, R., Couvet, D. and Jiguet, F. (2008) Birds are tracking climate warming, but not fast enough. *Proceedings of the Royal Society B: Biological Sciences*, **275**(1652), 2743–2748.

Dodson, J.R. and Intoh, M. (1999) Prehistory and palaeoecology of Yap, Federated States of Micronesia. *Quaternary International*, **59**, 17–26.

Dowler, R.C., Carroll, D.S. and Edwards, C.W. (2000) Rediscovery of rodents (Genus *Nesoryzomys*) considered extinct in the Galapagos Islands. *Oryx*, **34**, 109–117.

Dozier W.A. III,, Bramwell, K., Hatkin, J. and Dunkley, C. (2010) Bobwhite quail production and management guide. *Bulletin*, University of Georgia, 8.

Drobyshev, I., Niklasson, M., Granström, A. and Linderholm, H. (2010) Fire activity in Scandinavia during 1500–1900. *Northern Primeval Forests: Ecology, Conservation and Management*. Sundsvall. Poster Abstract.

Drobyshev, I., Niklasson, M. and Linderholm, H.W. (2012) Forest fire activity in Sweden: climatic controls and geographical patterns in 20th century. *Agricultural and Forest Meteorology*, **154–155**, 174–186.

Dudley, N., Higgins-Zogib, L. and Mansourian, S. (2005) Beyond belief: linking faiths and protected areas to support biodiversity conservation. WWF, Equilibrium and The Alliance of Religions and Conservation (ARC).

Dudley, N. (ed.) (2008) *Guidelines for Applying Protected Area Management Categories*. IUCN, Gland, Switzerland.

Dunford, R. (2004) Monarch of antiquity: the sacred yew in Fortingall, central Scotland, reputedly the oldest tree in Europe. http://www.sacredconnections.co.uk/holyland/fortingallycw.htm [accessed 22 November 2013].

Edwards, P.N. (2010) *A Vast Machine: Computer Models, Climate Data, and the Politics of Global Warming*. MIT Press, Boston, MA, USA.

EEA (1999) *Environmental Indicators: Typology and Overview*. Copenhagen European Environment Agency.

Ehrlich, P.R. and Ehrlich, A.H. (1981) *Extinction: The Causes and Consequences of the Disappearance of Species*. Random House, New York, NY, USA.

Ehrlich, P.R., Kareiva, P.M. and Daily, G.C. (2012) Securing natural capital and expanding equity to rescale civilization. *Nature*, **486**, 68–73.

Elias, S.A. (ed.) (2013) *Encyclopedia of Quaternary Science*. Elsevier, Amsterdam, The Netherlands.

Elmendorf, S.C., Henry, G.H.R., Hollister, R.D. *et al.* (2012) Plot-scale evidence of tundra vegetation change and links to recent summer warming. *Nature Climate Change*, **2**, 453–457.

Emanuel, W., Shugart, H. and Stevenson, M.P. (1985) Climatic change and the broad-scale distribution of terrestrial ecosystem complexes. *Climatic Change*, **7**(1), 29–43.

Emanuelsson, U., Bergendorff, C., Billqvist, M. *et al.* (2002) *Det skånska kulturlandskapet*. Naturskyddsföreningen i Skåne, Lund, Sweden.

Endicott, P., Ho, S.Y.W., Metspalu, M. and Stringer, C. (2009) Evaluating the mitochondrial timescale of human evolution. *Trends in Ecology and Evolution*, **24**, 515–521.

European Commission (2013a) Natura 2000 network: what is Natura 2000? http://ec.europa.eu/environment/nature/natura2000/index_en.htm [accessed 22 November 2013].

European Commission (2013b) The birds directive. http://ec.europa.eu/environment/nature/legislation/birdsdirective/index_en.htm [accessed 22 November 2013].

European Commission (2013c) The habitats directive. http://ec.europa.eu/environment/nature/legislation/habitatsdirective/index_en.htm [accessed 22 November 2013].

Falk, D.A., Heyerdahl, E.K., Brown, P.M. *et al.* (2011) Multi-scale controls of historical forest-fire regimes: new insights from fire-scar networks. *Frontiers in Ecology and the Environment*, **9**, 446–454.

Fang, J. and Lechowicz, M.J. (2006) Climatic limits for the present distribution of beech (*Fagus* L.) species in the world. *Journal of Biogeography*, **33**, 1804–1819.

FAO (1948) Forest resources of the world. *Unasylva*. Washington, DC.

Farquhar, G.D., von Caemmerer, S. and Berry, J.A. (1980) A biochemical model of photosynthetic CO_2 assimilation in leaves of C_3 plants. *Planta*, **149**, 78–90.

Feld, C.K., Martins da Silva, P., Sousa, J.P. *et al.* (2009) Indicators of biodiversity and ecosystem services: a synthesis across ecosystems and spatial scales. *Oikos*, **118**, 1862–1871.

Flannigan, M., Stocks, B., Turetsky, M. and Wotton, M. (2009) Impacts of climate change on fire activity and fire management in the circumboreal forest. *Global Change Biology*, **15**(3), 549–560.

Flato, G.M. (2011) Earth system models: an overview. *Wiley Interdisciplinary Reviews: Climate Change*, **2**(6), 783–800.

Foley, J.A., DeFries, R., Asner, G.P. *et al.* (2005) Global consequences of land use. *Science*, **309**, 570–574.

Foley, J.A., Ramankutty, N., Brauman, K.A. *et al.* (2011) Solutions for a cultivated planet. *Nature*, **478**, 337–342.

Foster, J.R. (1988) The potential role of rime ice deforestation in tree mortality of wave-regenerated balsam fir forests. *Journal of Ecology*, **76**(1), 172–180.

Foster, D.R. (1992) Land-use history (1730–1990) and vegetation dynamics in central New England, USA. *Journal of Ecology*, **80**, 753–772.

Foster, D.R. and Boose, E.R. (1992) Patterns of forest damage resulting from catastrophic wind in central New England, USA. *Journal of Ecology*, **80**(1), 79–98.

Foster, D.R., Zebryk, T., Schoonmaker, P. and Lezberg, A. (1992) Post-settlement history of human land-use and vegetation dynamics of a *Tsuga canadensis* (hemlock) woodlot in central New England. *Journal of Ecology*, **80**, 773–786.

Foster, D.R., Knight, D.H. and Frankin, J.F. (1998a) Landscape patterns and legacies resulting from large, infrequent forest disturbances. *Ecosystems*, **1**, 497–510.

Foster, D.R., Motzkin, G. and Slater, B. (1998b) Land-use history as long-term broad-scale disturbance, regional forest dynamics in central New England. *Ecosystems*, **1**, 96–119.

Frazer, J.G. (1922) *The Golden Bough*. Macmillan, New York, NY, USA.

Frelich, L.E., Hale, C.M., Scheu, S. *et al.* (2006) Earthworm invasion into previously earthworm-free temperate and boreal forests. *Biological Invasions*, **8**, 1235–1245.

Frolking, S., Talbot, J., Jones, M.C. *et al.* (2011) Peatlands in the earth's 21st century climate system. *Environmental Reviews*, **19**(NA), 371–396.

Fronzek, S., Carter, T.R. and Jylhä, K. (2012) Representing two centuries of past and future climate for assessing risks to biodiversity in Europe. *Global Ecology and Biogeography*, **21**(1), 19–35.

Fuller, D.Q., van Etten, J., Manning, K. *et al.* (2011) The contribution of rice agriculture and livestock pastoralism to prehistoric methane levels: an archaeological assessment. *The Holocene*, **21**, 743–759.

Fuller, J.L., Foster, D.R., Jason, S. *et al.* (1998) Impact of human activity on regional forest composition and dynamics in central New England. *Ecosystems*, **1**(1), 76–95.

Gaillard, M.-J., Sugita, S., Bunting, J. *et al.* (2008) Human impact on terrestrial ecosystems, pollen calibration and quantitative reconstruction of past land-cover. *Vegetation History and Archaeobotany*, **17**, 415–418.

Gallai, N., Salles, J.-M., Settele, J. *et al.* (2009) Economic valuation of the vulnerability of world agriculture confronted with pollinator decline. *Ecological Economics*, **68**(3), 810–821.

Gallopin, G.C. (1991) Human dimensions of global change: linking the global and the local processes. *International Social Science Journal*, **43**, 707–718.

Gallopin, G.C. (2006) Linkages between vulnerability, resilience, and adaptive capacity. *Global Environmental Change*, **16**(3), 293–303.

Giesecke, T. and Bennett, K.D. (2004) The Holocene spread of *Picea abies* (L.) Karst. in Fennoscandia and adjacent areas. *Journal of Biogeography*, **31**, 1523–1548.

Giesecke, T., Hickler, T., Kunkel, T. *et al.* (2007) Towards an understanding of the Holocene distribution of *Fagus sylvatica* L. *Journal of Biogeography*, **34**, 118–131.

Giesecke, T., Bjune, A.E., Chiverrell, R.C. *et al.* (2008) Exploring Holocene continentality changes in Fennoscandia using present and past tree distributions. *Quaternary Science Reviews*, **27**, 1296–1308.

Giesecke, T., Miller, P.A., Sykes, M.T. *et al.* (2010) The effect of past changes in inter-annual temperature variability on tree distribution limits. *Journal of Biogeography*, **37**, 1394–1405.

Giesecke, T., Bennett, K.D., Birks, H.J.B. *et al.* (2011) The pace of Holocene vegetation change: testing for synchronous developments. *Quaternary Science Reviews*, **30**, 2805–2814.

Giesecke, T., Davis, B., Brewer, S. *et al.* (2014) Towards mapping the late Quaternary vegetation change of Europe. *Vegetation History and Archaeobotany*, **23**, 75–86.

Giglio, L., Randerson, J.T., van der Werf, G.R. *et al.* (2010) Assessing variability and long-term trends in burned area by merging multiple satellite fire products. *Biogeosciences*, **7**, 1171–1186.

Gill, J.L., Williams, J.W., Jackson, S.T. *et al.* (2009) Pleistocene megafaunal collapse, novel plant communities, and enhanced fire regimes in North America. *Science*, **326**, 1100–1103.

Girod, B., Wiek, A., Mieg, H. and Hulme, M. (2009) The evolution of the IPCC emissions scenarios. *Environmental Science & Policy*, **12**, 103–118.

Glaser, B. and Birk, J.J. (2012) State of the scientific knowledge on properties and genesis of Anthropogenic Dark Earths in Central Amazonia. *Geochimica et Cosmochimica Acta*, **82**, 39–51.

Gleason, H.A. (1926) The individualistic concept of the plant association. *Bulletin of the Torrey Botanical Club*, **53**, 7–26.

Gleick, P.H., Adams, R.M., Amasino, R.M. *et al.* (2010) Climate change and the integrity of science. *Science*, **328**, 689–690.

Global Carbon Project (2012) Carbon budget 2012: an annual update of the global carbon budget and trends. http://www.globalcarbonproject.org/carbonbudget/ [accessed 22 November 2013].

Godwin, H. (1934) Pollen analysis, an outline of the problems and potentialities of the method. *New Phytologist*, **33**, 278–305.

Grabherr, G., Gottfried, M. and Pauli, H. (1994) Climate effects on mountain plants. *Nature*, **369**, 448.

Grace, J.B. (2006) *Structural Equation Modeling and Natural Systems*. Cambridge University Press, Cambridge, UK.

Granström, A. and Niklasson, M. (2008) Potentials and limitations for human control over historic fire regimes in the boreal forest. *Philosophical Transactions of the Royal Society B*, **363**, 2353–2358.

Gray, A. (1999) Indigenous peoples, their environments and territories, in *Cultural and Spiritual Values of Biodiversity, United Nations Environment Programme* (ed. D.A. Posey). Intermediate Technology Publications, London, UK, pp. 61–119.

Grimm, E.C., Donovan, J.J. and Brown, K.J. (2011) A high-resolution record of climate variability and landscape response from Kettle Lake, northern Great Plains, North America. *Quaternary Science Reviews*, **30**, 2626–2650.

Gritti, E., Smith, B. and Sykes, M.T. (2006) Vulnerability of Mediterranean basin ecosystems to climate change and invasion by exotic plant species. *Journal of Biogeography*, **33**, 145–157.

Guisan, A. and Zimmermann, N.E. (2000) Predictive habitat distribution models in ecology. *Ecological Modelling*, **135**, 147–186.

Hamilton, E. and Huntington, C. (eds) (1961) *The Collected Dialogues of Plato*. Bollingen Series LXXI. Bollingen Foundation, New York, NY, USA.

Hannon, G.E. and Bradshaw, R.H.W. (2000) Impacts and timing of the first human settlement on vegetation of the Faroe Islands. *Quaternary Research*, **54**, 404–413.

Hannon, G.E., Wastegård, S., Bradshaw, E. and Bradshaw, R.H.W. *et al.* (2001) Human impact and landscape degradation on the Faroe Islands. *Biology and Environment Proceedings of the Royal Irish Academy*, **101B**, 129–139.

Hannon, G.E., Bradshaw, R.H.W., Bradshaw, E. *et al.* (2005) Climate change and human settlement as drivers of late-Holocene vegetational change in the Faroe Islands. *The Holocene*, **15**, 639–647.

Hannon, G.E., Arge, S.V., Fosaa, A.M. *et al.* (2009) Faroe Islands, in *Encyclopedia of Islands* (eds R.G. Gillespie and D.A. Clague). University of California Press, California, USA, pp. 291–297.

Harrison, P., Vandewalle, M., Sykes, M.T. *et al.* (2010) Identifying and prioritising services in European terrestrial and freshwater ecosystems. *Biodiversity and Conservation*, **19**(10), 2791–2821.

Haxeltine, A. and Prentice, I.C. (1996) BIOME3: an equilibrium terrestrial biosphere model based on ecophysiological constraints, resource availability, and competition among plant functional types. *Global Biochemical Cycles*, **10**(4), 693–709.

Haxeltine, A., Prentice, I.C. and Creswell, I.D. (1996) A coupled carbon and water flux model to predict vegetation structure. *Journal of Vegetation Science*, **7**, 651–666.

Head, L. (1999) Holocene human impacts in Australia and the Western Pacific. *Quaternary International*, **59**, 1–3.

Heikkilä, M., Edwards, T.W.D., Seppä, H. and Sonninen, E. (2010) Sediment isotope tracers from Lake Saarikko, Finland, and implications for Holocene hydroclimatology. *Quaternary Science Reviews*, **29**, 2146–2160.

Heikkinen, R.K.L., Araújo, M.B., Virkkala, R. *et al.* (2006) Methods and uncertainties in bioclimatic envelope modelling under climate change. *Progress in Physical Geography*, **30**(6), 751–777.

Heinselman, M.L. (1973) Fire in the virgin forests of the Boundary Waters Canoe Area, Minnesota. *Quaternary Research*, **3**, 329–382.

Helama, S., Seppä, H., Bjune, A.E. and Birks, H.J.B. (2012) Fusing pollen-stratigraphic and dendroclimatic proxy data to reconstruct summer temperature variability during the past 7.5 ka in subarctic Fennoscandia. *Journal of Paleolimnology*, **48**, 275–286.

Hickler, T., Vohland, K. *et al.* (2010) Vegetation on the move – where do conservation strategies have to be redefined? in *Atlas of Biodiversity Risk* (eds J. Settele, L. Penev, T. Georgiev *et al.*). Pensoft Publishers, Sofia-Moscow, Russia, p. 280.

Hickler, T., Vohland, K., Feehan, J. *et al.* (2012) Projecting the future distribution of European potential natural vegetation zones with a generalized, tree species-based dynamic vegetation model. *Global Ecology and Biogeography*, **21**, 50–63.

Higuera, P.E., Peters, M.E., Brubaker, L.B. and Gavin, D.G. (2007) Understanding the origin and analysis of sediment-charcoal records with a simulation model. *Quaternary Science Reviews*, **26**, 1790–1809.

Higuera, P.E., Whitlock, C. and Gage, J.A. (2011) Linking tree-ring and sediment-charcoal records to reconstruct fire occurrence and area burned in subalpine forests of Yellowstone National Park, USA. *The Holocene* **21**(2), 327–341.

Holdridge, L.R. (1947) Determination of world plant formations from simple climatic data. *Science*, **105**, 367–368.

Holzschuh, A., Dudenhoffer, J.H. and Tscharntke, T. (2012) Landscapes with wild bee habitats enhance pollination, fruit set and yield of sweet cherry. *Biological Conservation*, **153**, 101–107.

Hougner, C., Colding, J. and Söderqvist, T. (2006) Economic valuation of a seed dispersal service in the Stockholm National Urban Park, Sweden. *Ecological Economics*, **59**(3), 364–374.

Howe, S. and Webb, T. III, (1983) Calibrating pollen data in climatic terms: improving the methods. *Quaternary Science Reviews*, **2**, 17–51.

Hu, F.S., Hampe, A. and Petit, R.J. (2009) Paleoecology meets genetics: deciphering past vegetational dynamics. *Frontiers in Ecology and the Environment*, **7**(7), 371–379.

Huang, C.C. and Su, H. (2009) Climate change and Zhou relocations in early Chinese history. *Journal of Historical Geography*, **35**, 297–310.

Hudson, J.M.G., Henry, G.H.R. and Cornwell, W.K. (2011) Taller and larger: shifts in Arctic tundra leaf traits after 16 years of experimental warming. *Global Change Biology*, **17**, 1013–1021.

Huntingford, C., Fisher, R.A., Mercado, L. *et al.* (2008) Towards quantifying uncertainty in predictions of Amazon dieback. *Philosophical Transactions of the Royal Society B: Biological Sciences*, **363**(1498), 1857–1864.

Huntley, B. and Birks, H.J.B. (1983) *An Atlas of Past and Present Pollen Maps for Europe 0-13000 Years Ago*. Cambridge University Press, Cambridge, UK.

Huntley, B., Berry, P.M., Cramer, W. and McDonald, A.P. (1995) Modelling present and potential future ranges of some European higher plants using climate response surfaces. *Journal of Biogeography*, **22**, 967–1001.

Huntley, B., Barnard, P., Altwegg, R. *et al.* (2010) Beyond bioclimatic envelopes: dynamic species range and abundance modelling in the context of climatic change. *Ecography*, **33**(3), 621–626.

Huntley, B., Allen, J.R.M., Barnard, P. *et al.* (2013) Species' distribution models indicate contrasting late-Quaternary histories for Southern and Northern Hemisphere bird species. *Global Ecology and Biogeography*, **22**, 277–288.

Hutchinson, G.E. (1957) Concluding remarks. *Cold Spring Harbor Symposia on Quantitative Biology*, **22**, 415–427.

IPCC (2000) *Emissions Scenarios 2000*. Cambridge University Press, Cambridge, UK.

IPCC (2007a) *Climate Change 2007: Impacts, Adaptation and Vulnerability. Contribution of Working Group II to the Fourth Assessment Report of the Intergovernmental Panel on Climate Change*. Cambridge University Press, Cambridge, UK and New York, NY, USA.

IPCC (2007b) *The Physical Science Basis: Contribution of Working Group 1 to the Fourth Assessment Report of the Intergovernmental Panel on Climate Change* Cambridge University Press, Cambridge, UK and New York, NY, USA.

Iverson, J. (1944) *Viscum, Hedera* and *Ilex* as climatic indicators. *Geologiska Föreningen Förhandlingar*, **66**, 463–483.

Iversen, J. (1956) Forest clearance in the Stone Age. *Scientific American*, **194**, 36–41.

Jackson, T. (2009) Prosperity without growth? The transition to a sustainable economy. Sustainable Development Commission, UK.

Jackson, S.T., Futyma, R.P. and Wilcox, D.A. (1988) A paleoecological test of a classical hydrosere in the Lake Michigan dunes. *Ecology*, **69**, 928–936.

Jackson, S.T. and Hobbs, R.J. (2009) Ecological restoration in the light of ecological history. *Science*, **325**, 567–569.

Jacobsen, T. and Adams, R.M. (1958) Salt and silt in Ancient Mesopotamian agriculture. *Science*, **128**, 1251–1258.

Jalas, J. and Suominen, J. (eds) (1972–1994) *Atlas Florae Europaeae: Distribution of Vascular Plants in Europe*. The Committee for Mapping the Flora of Europe and Societas Biologica Fennica Vanamo, Helsinki, Finland.

Jensen, J. (2001) *Danmarks Oldtid. Stenalder 13.000-2.000 f.Kr.* Gyldendal, Copenhagen, Denmark.

Johnsen, S.J., Dahl-Jensen, D., Gundestrup, N. *et al.* (2001) Oxygen isotope and palaeotemperature records from six Greenland ice-core stations: Camp Century, Dye-3, GRIP, GISP2, Renland and NorthGRIP. *Journal of Quaternary Science*, **16**, 299–307.

Johnson, E.A. and Miyanishi, K. (2008) Testing the assumptions of chronosequences in succession. *Ecology Letters*, **11**, 419–431.

Jones, E.W. (1945) The structure and reproduction of the virgin forest of the north temperate zone. *New Phytologist*, **44**(2), 130–148.

Jonzén, N., Lindén, A., Ergon, T. *et al.* (2006) Rapid advance of spring arrival dates in long-distance migratory birds. *Science*, **312**, 1959–1961.

Kaasik, A. (2010) Conserving sacred natural sites in Estonia: the diversity of sacred lands in Europe. *Proceedings of the Third Workshop of the Delos Initiative Inari/Aanaar, Finland IUCN and Vantaa, Finland: Metsähallitus Natural Heritage Services*.

Kaltenrieder, P., Tinner, W. and Ammann, B. (2005) Zur Langzeitökologie des Lärchen-Arvengürtels in den südlichen Walliser Alpen. *Botanica Helvetica*, **115**, 137–154.

Kaltenrieder, P., Procacci, G., Vannière, B. and Tinner, W. (2010) Vegetation and fire history of the Euganean Hills (Colli Euganei) as recorded by Lateglacial and Holocene sedimentary series from Lago della Costa (northeastern Italy). *The Holocene*, **20**, 679–695.

Kaplan, J.O., Krumhardt, K.M. and Zimmermann, N. (2009) The prehistoric and preindustrial deforestation of Europe. *Quaternary Science Reviews*, **28**, 3016–3034.

Kaplan, J.O., Krumhardt, K.M., Ellis, E.C. *et al.* (2010) Holocene carbon emissions as a result of anthropogenic land cover change. *The Holocene*, **21**, 775–791.

Kasischke, E.S. and Turetsky, M.R. (2006) Recent changes in the fire regime across the North American boreal region: spatial and temporal patterns of burning across Canada and Alaska. *Geophysical Research Letters*, **33**(9), L09703.

Keeley, J.E. and Rundel, P.W. (2005) Fire and the Miocene expansion of C4 grasslands. *Ecology Letters*, **8**, 683–690.

Keeling, D., Molloy, K. and Bradshaw, R. (1989) Megalithic tombs in south-west Donegal. *Archaeology Ireland*, **3**(4), 152–154.

Keith, D.A., Akcakaya, H.R., Thuiller, W. *et al.* (2008) Predicting extinction risks under climate change: coupling stochastic population models with dynamic bioclimatic habitat models. *Biology Letters*, **4**(5), 560–563.

Klein, A.-M., Vaissière, B.E., Cane, J.H. *et al.* (2007) Importance of pollinators in changing landscapes for world crops. *Proceedings of the Royal Society B: Biological Sciences*, **274**(1608), 303–313.

Kleinbauer, I., Dullinger, S., Peterseil, J. and Essl, F. (2010) Climate change might drive the invasive tree *Robinia pseudacacia* into nature reserves and endangered habitats. *Biological Conservation*, **143**(2), 382–390.

Kottek, M., Grieser, J., Beck, C. *et al.* (2006) World map of the Köppen-Geiger climate classification updated. *Meteorologische Zeitschrift*, **15**(3), 259–263.

Krause, J., Fu, Q., Good, J.M. *et al.* (2010) The complete mitochondrial DNA genome of an unknown hominin from southern Siberia. *Nature*, **464**, 894–897.

Krinner, G., Lezine, A.M., Braconnot, P. *et al.* (2012) A reassessment of lake and wetland feedbacks on the North African Holocene climate. *Geophysical Research Letters*, **39**, L07701.

Kröpelin, S., Verchuren, D., Lézine, A.-M. *et al.* (2008) Climate-driven ecosystem succession in the Sahara: the past 6000 years. *Science*, **320**, 765–768.

Kuemmerle, T., Hickler, T., Olofsson, J. *et al.* (2011) Reconstructing range dynamics and range fragmentation of European bison for the last 8000 years. *Diversity and Distributions*, **18**, 47–59.

Kullman, L. (2001) 20th century climate warming and tree-limit rise in the southern Scandes of Sweden. *AMBIO*, **30**, 72–80.

Kullman, L. (2008) Early postglacial appearance of tree species in northern Scandinavia: review and perspective. *Quaternary Science Reviews*, **27**(27–28), 2467–2472.

Lamb, H.H. (1967) The early Medieval warm epoch and its sequel. *Palaeogeography, Palaeoclimatology, Palaeoecology*, **1**, 13–37.

Lamb, H.H. (1977) Climate: Present, Past and Future. *Vol. 2: Climatic History and the Future*. Methuen, London, UK.

Landis, D.A., Wratten, S.D. and Gurr, G.M. (2000) Habitat management to conserve natural enemies of arthropod pests in agriculture. *Annual Review of Entomology*, **45**(1), 175–201.

Larsson, K. and Simonsson, G. (2003) Den halländska skogen – människa och mångfald. *Meddelande*. Halmstad, Länsstyrelsen Halland, **7**, 74.

Lawson, I.T., Edwards, K.J., Church, M.J. *et al.* (2008) Human impact on an island ecosystem: pollen data from Sandoy, Faroe Islands. *Journal of Biogeography*, **35**, 1130–1152.

Leadley, P., Pereira, H.M., Alkemade, R. *et al.* (2010) *Biodiversity Scenarios: Projections of 21st Century Change in Biodiversity and Associated Ecosystem Services*. Secretariat of the Convention on Biological Diversity, Montreal, Canada. CBD Technical Series No. 50.

Leakey, A.D.B., Ainsworth, E.A., Bernacchi, C.J. *et al.* (2012) Photosynthesis in a CO_2-rich atmosphere, in *Photosynthesis: Plastid Biology, Energy Conversion and Carbon Assimilation, Advances in Photosynthesis and Respiration* (eds J.J. Eaton-Rye, B.C. Tripathy and T.D. Sharkey). Springer, New York, NY, USA, pp. 733–768.

Leathwick, J.R., Whitehead, D. and McLeod, M. (1996) Predicting changes in the composition of New Zealand's indigenous forests in response to global warming: a modelling approach. *Environmental Software*, **11**(1–3), 81–90.

Ledgard, N. (2001) The spread of lodgepole pine (*Pinus contorta*, Dougl.) in New Zealand. *Forest Ecology and Management*, **141**, 43–57.

Leemans, R. and Cramer, W. (1991) The IIASA climate database for mean monthly values of temperature, precipitation and cloudiness on a terrestrial grid. International Institute for Applied Systems Analysis, Laxenburg, Austria, RR-91-18.

Lehsten, V., Harmand, P., Palumbo, I. and Arneth, A. (2010) Modelling burned area in Africa. *Biogeosciences*, **7**, 3199–3214.

Lemée, G. (1981) Contribution à l'histoire des landes de la forêt de Fontainebleau d'après l'analyse pollinique des sols. *Bulletin de la Societé Botanique de France: Lettres Botaniques*, **128**(3), 189–200.

Lenihan, J.M. (1993) Ecological response surfaces for North American boreal tree species and their use in forest classification. *Journal of Vegetation Science*, **4**, 667–680.

Li, C. and Apps, M.J. (1996) Effects of contagious disturbance on forest temporal dynamics. *Ecological Modelling*, **87**(1–3), 143–151.

Li, F., Zeng, X.D. and Levis, S. (2012) A process-based fire parameterization of intermediate complexity in a dynamic global vegetation model. *Biogeosciences*, **9**(7), 2761–2780.

Lieth, H. (1975) Modeling primary productivity of the world, in *Primary Productivity of the Biosphere* (ed. H. Lieth and R.H. Whittaker). Springer, New York, NY, USA, pp. 237–263.

Lindberg, T. (2009) Hundra års sökande. *Skogen*, Stockholm.

Lindbladh, M. and Bradshaw, R. (1995) The development and demise of a Medieval forest-meadow system at Linnaeus' birthplace in southern Sweden: implications for conservation and forest history. *Vegetation History and Archaeobotany*, **4**(3), 153.

Lindbladh, M., Bradshaw, R.H.W. and Holmqvist, B.H. (2000) Pattern and process in south Swedish forests during the last 3000 years, sensed at stand and regional scales. *Journal of Ecology*, **88**, 113–128.

Linder, P. and Östlund, L. (1998) Structural changes in three mid-boreal Swedish forest landscapes, 1885–1996. *Biological Conservation*, **85**, 9–19.

Lischke, H., Lotter, A.F. and Fischlin, A. (2002) Untangling a Holocene pollen record with forest model simulations and independent climate data. *Ecological Modelling*, **150**, 1–21.

Lisiecki, L.E. and Raymo, M.E. (2005) A Pliocene-Pleistocene stack of 57 globally distributed benthic $\delta^{18}O$ records. *Paleoceanography*, **20**, 1–17.

Lisitsyna, O.V., Giesecke, T. and Hicks, S. (2011) Exploring pollen percentage threshold values as an indication for the regional presence of major European trees. *Review of Palaeobotany and Palynology*, **166**, 311–324.

Liu, Z., Wang, Y., Gallimore, R. *et al.* (2007) Simulating the transient evolution and the abrupt change of the Northern Africa atmosphere-ocean-terrestrial ecosystem in the Holocene. *Quaternary Science Reviews*, **26**, 1818–1837.

Liu, Y.Q., Stanturf, J. and Goodrick, S. (2010) Trends in global wildfire potential in a changing climate. *Forest Ecology and Management*, **259**(4), 685–697.

Lorenzen, E.D., Nogués-Bravo, D., Orlando, L. *et al.* (2011) Species-specific responses of Late Quaternary megafauna to climate and humans. *Nature*, **479**, 359–365.

Lotter, A.F., Heiri, O., Brooks, S. *et al.* (2012) Rapid summer temperature changes during Termination 1a: high-resolution multi-proxy climate reconstructions from Gerzensee (Switzerland). *Quaternary Science Reviews*, **36**, 103–113.

Lowe, J.J. and Walker, M.J.C. (1997) *Reconstructing Quaternary Environments*. Prentice Hall, Harlow, UK.

Lu, H., Zhang, H., Wang, S. *et al.* (2011) Multiphase timing of hominin occupations and the paleoenvironment in Luonan Basin, Central China. *Quaternary Research*, **76**(1), 142–147.

Lyell, C. (1830–1833) *Principles of Geology: Being an Attempt to Explain the Former Changes of the Earth's Surface, by Reference to Causes Now in Operation.* John Murray, London, UK.

Lynch, A.H., Beringer, J., Kershaw, P. *et al.* (2007) Using the paleorecord to evaluate climate and fire interactions in Australia. *Annual Review of Earth and Planetary Sciences*, **35**, 215–239.

MacArthur, R.H. and Wilson, E.O. (1967) *The Theory of Island Biogeography.* Princeton University Press, Princeton, NJ, USA.

MacDonald, G.M., Bennett, K.D., Jackson, S.T. *et al.* (2008) Impacts of climate change on species, populations and communities: palaeobiogeographical insights and frontiers. *Progress in Physical Geography*, **32**, 139–172.

Mack, M.C., Bret-Harte, M.S., Hollingsworth, T.N. *et al.* (2011) Carbon loss from an unprecedented Arctic tundra wildfire. *Nature*, **475**(7357), 489–492.

Madsen, T. (1982) Settlement systems of early agricultural societies in East Jutland, Denmark: a regional study of change. *Journal of Anthropological Archaeology*, **1**, 197–236.

Magnússon, B., Magnússon, S.H. and Fridriksson, S. (2009) Developments in plant colonization and succession on Surtsey during 1999–2008. *Surtsey Research*, **12**, 57–76.

Magri, D. and Palombo, M.R. (2012) Early to Middle Pleistocene dynamics of plant and mammal communities in South West Europe. *Quaternary International*, **288**, 63–72.

Manabe, S. and Stouffer, R.J. (1980) Sensitivity of a global climate model to an increase of CO_2 concentration in the atmosphere. *Journal of Geophysical Research: Oceans and Atmospheres*, **85**(NC10), 5529–5554.

Mann, M.E., Bradley, R.S. and Hughes, M.K. (1998) Global-scale temperature patterns and climate forcing over the past six centuries. *Nature*, **392**, 779–787.

Manning, D. (no date) The living churchyard. http://www.buildingconservation.com/articles/living/living.htm [accessed 22 November 2013].

Manwood, J. (1598) *A Treatise of the Laws of the Forest.* Company of Stationers, London, UK.

Marlon, J.R., Bartlein, P.J. and Whitlock, C. (2006) Fire-fuel-climate linkages in the northwestern USA during the Holocene. *The Holocene*, **16**, 1059–1071.

Marlon, J.R., Bartlein, P.J., Daniau, A.L. *et al.* (2013) Global biomass burning, a synthesis and review of Holocene paleofire records and their controls. *Quaternary Science Reviews*, **65**, 5–25.

Marsh, G.P. (1864) *Man and Nature: Or, Physical Geography as Modified by Human Action.* Scribner, New York, NY, USA.

Martin, T.J. and Ogden, J. (2006) Wind damage and response in New Zealand forests: a review. *New Zealand Journal of Ecology*, **30**(3), 295–310.

Martínez, J., Vega-Garcia, C. and Chuvieco, E. (2009) Human-caused wildfire risk rating for prevention planning in Spain. *Journal of Environmental Management*, **90**(2), 1241–1252.

Matthews, R.B., Gilbert, N.G. Roach, A. *et al.* (2007) Agent-based land-use models: a review of applications. *Landscape Ecology*, **22**, 1447–1459.

Mbida, C.M., van Neer, W., Doutrelepont, H. and Vrydaghs, L. (2000) Evidence for banana cultivation and animal husbandry during the first millennium BC in the forest of southern Cameroon. *Journal of Archaeological Science*, **27**, 151–162.

McGuffie, K. and Henderson-Sellers, A. (2005) *A Climate Modelling Primer.* John Wiley and Sons, Ltd., Chichester, UK.

McKinley, D.C., Ryan, M.G., Birdsey, R.A. *et al.* (2011) A synthesis of current knowledge on forests and carbon storage in the United States. *Ecological Applications*, **21**, 1902–1924.

McWethy, D.B., Whitlock, C., Wilmshurst, J.M. *et al.* (2010) Rapid landscape transformation in South Island, New Zealand, following initial Polynesian settlement. *Proceedings of the National Academy of Sciences*, **14**, 21343–21348.

MEA (2005) *Millennium Ecosystem Assessment: Ecosystems and Human Well-Being Biodiversity Synthesis.* World Resources Institute, Washington, DC, USA.

MEA (no date) Guide to the Millennium Assessment Reports. http://www.unep.org /maweb/en/index.aspx [accessed 22 November 2013].

Medina-Elizalde, M., Burns, S.J., Lea, D.W. *et al.* (2010) High resolution stalagmite climate record from the Yucatán Peninsula spanning the Maya terminal classic period. *Earth and Planetary Science Letters*, **298**, 255–262.

Melillo, J.M., McGuire, A.D., Kicklighter, D.W. *et al.* (1993) Global climate change and terrestrial net primary production. *Nature*, **363**, 234–240.

Mesarovic, M. and Pestel, E. (1974) *Mankind at the Turning Point: The Second Report to the Club of Rome*. Reader's Digest Press, New York, NY, USA.

Metcalfe, K., Ffrench-Constant, R. and Gordon, I. (2010) Sacred sites as hotspots for biodiversity: the Three Sisters Cave complex in coastal Kenya. *Oryx*, **44**(01), 118–123.

Miles, J. (1987) Vegetation succession: past and present perceptions, in *Colonization, Succession and Stability* (eds A.J. Gray, M.J. Crawley and P.J. Edwards). Blackwell, Oxford, UK, pp. 1–29.

Miller, G.H., Fogel, M.L., Magee, J.W. *et al.* (2005) Ecosystem collapse in Pleistocene Australia and a human role in megafaunal extinction. *Science*, **309**, 287–290.

Miller, P.A., Giesecke, T., Hickler, T. *et al.* (2008) Exploring climatic and biotic controls on Holocene vegetation change in Fennoscandia. *Journal of Ecology*, **96**, 247–259.

Mindzie, C.M., Doutrelepont, H., Vrydaghs, L. *et al.* (2001) First archaeological evidence of banana cultivation in central Africa during the third millennium before present. *Vegetation History and Archaeobotany*, **10**, 1–6.

Mitchell, F.J.G. (1993) The biogeographical implications of the distribution and history of the strawberry tree, *Arbutus unedo* in Ireland, in *Biogeography of Ireland: Past, Present and Future* (eds M.J. Costello and K.S. Kelly). Irish Biogeographical Society, Dublin, Ireland, 35–44.

Mitchell, F.J.G. and Cole, E. (1998) Reconstruction of long-term successional dynamics of temperate woodland in Białowieża Forest, Poland. *Journal of Ecology*, **86**(6), 1042–1059.

Molinari, C., Lehsten, V., Bradshaw, R.H.W. *et al.* (2013) Exploring potential drivers of European biomass burning over the Holocene: a data-model analysis. *Global Ecology and Biogeography*, **22**(12), 1248–1260.

Møller, P.F. (1997) Biodiversity in Danish natural forests. A comparison between unmanaged and managed woodlands in East Denmark. Danmarks Geologiske Undersøgelse Rapport 1997/41.

Molloy, K. and O'Connell, M. (2004) Holocene vegetation and land-use dynamics in the karstic environment of Inis Oírr, Aran Islands, western Ireland: pollen analytical evidence evaluated in light of the archaeological record. *Quaternary International*, **113**, 41–64.

Monbiot, G. (2012) Putting a price on the rivers and rain diminishes us all. *The Guardian*, 6 August. http://www.theguardian.com/commentisfree/2012/aug/06/price-rivers-rain-greatest -privatisation [accessed 22 November 2013].

Mooney, S.D., Harrison, S.P., Bartlein, P.J. *et al.* (2011) Late Quaternary fire regimes of Australasia. *Quaternary Science Reviews*, **30**, 28–46.

Moorcroft, P.R., Hurtt, G.C. and Pacala, S.W. (2001) A method for scaling vegetation dynamics: the ecosystem demography model (ED). *Ecological Monographs*, **71**(4), 557–586.

Morin, X. and Thuiller, W. (2009) Comparing niche- and process-based models to reduce prediction uncertainty in species range shifts under climate change. *Ecology*, **90**(5), 1301–1313.

Moussouris, Y. and Regato, P. (1999) Forest harvest: Mediterranean woodlands and the importance of non-timber forest products to forest conversation. *Arborvitae Supplement*. WWF and IUCN, Gland, Switzerland.

Munro, N. (2001) The value of history. http://www.philosophypathways.com/essays/munro3.html [accessed 22 November 2013].

Murray, M.B., Cannell, M.G.R. and Smith, R.I. (1989) Date of budburst of 15 tree species in Britain following climatic warming. *Journal of Applied Ecology*, **26**(2), 693–700.

Myers, N., Mittermeier, R.A., Mittermeier, C.G. *et al.* (2000) Biodiversity hotspots for conservation priorities. *Nature*, **403**(6772), 853–858.

NASA (no date) Global fire maps. http://lance-modis.eosdis.nasa.gov/cgi-bin/imagery/firemaps .cgi [accessed 22 November 2013].

Natural England (no date) Sites of special scientific interest. http://www.sssi.naturalengland .org.uk/Special/sssi/index.cfm [accessed 22 November 2013].

Needham, J. (1959) *Science and Civilization in China: Volume 3, Mathematics and the Sciences of the Heavens and the Earth.* Cambridge University Press, Cambridge, UK.

Neilson, R.P. (1995) A model for predicting continental-scale vegetation distribution and water balance. *Ecological Applications*, **5**(2), 362–385.

Neilson, R.P. and Marks, D. (1994) A global perspective of regional vegetation and hydrologic sensitivities from climatic change. *Journal of Vegetation Science*, **5**(5), 715–730.

Neilson, R.P., King, G.A. and Koerper, G. (1992) Toward a rule-based biome model. *Landscape Ecology*, **7**(1), 27–43.

Nesje, A., Dahl, S.O., Thun, T. and Nordli, Ø. (2008) The Little Ice Age glacial expansion in western Scandinavia: summer temperature or winter precipitation? *Climate Dynamics*, **30**, 789–801.

Nichols, G.E. (1935) The hemlock-white pine-northern hardwood region of eastern North America. *Ecology*, **16**(3), 403–422.

Nielsen, A.B., Giesecke, T., Theuerkauf, M. *et al.* (2012) Quantitative reconstructions of changes in regional openness in north-central Europe reveal new insights into old questions. *Quaternary Science Reviews*, **47**, 131–149.

Niemi, J. (2009) *Environmental Monitoring in Finland 2009–2012.* Finnish Environment Institute, Helsinki, Finland.

Niklasson, M., Zin, E., Zielonka, T. *et al.* (2010) A 350-year tree-ring fire record from Białowieża Primeval Forest, Poland: implications for Central European lowland fire history. *Journal of Ecology*, **98**, 1319–1329.

Northumberland, H.D. (2011) Game on. *The Economist.* http://www.economist.com/blogs /blighty/2011/08/shooting-and-conservation [accessed 22 November 2013]

Notaro, M. (2008) Response of the mean global vegetation distribution to interannual climate variability. *Climate Dynamics*, **30**, 845–854.

O'Callaghan, T. (2012) Can the planet survive 10 billion people? *New Scientist*, 24 July. http:// www.newscientist.com/blogs/culturelab/2012/07/tiffany-ocallaghan-culturelab-editorbefore -packed.html [accessed 22 November 2013].

O'Connell, M. (1986) Reconstruction of local landscape development in the Post-Atlantic based on palaeoecological investigations at Carrownaglogh prehistoric field system, County Mayo, Ireland. *Review of Palaeobotany and Palynology*, **49**, 117–176.

O'Connell, M. and Molloy, K. (2001) Farming and woodland dynamics in Ireland during the Neolithic. *Biology and Environment: Proceedings of the Royal Irish Academy*, **101B**(1/2), 99–128.

Odgaard, B.V. (1994) Holocene vegetation history of northern West Jutland, Denmark. *Nordic Journal of Botany*, **14**, 402.

Odgaard, B.V. and Rasmussen, P. (2000) Origin and temporal development of macro-scale vegetation patterns in the cultural landscape of Denmark. *Journal of Ecology*, **88**, 733–748.

Ohlson, M., Brown, K.J., Birks, H.J.B. *et al.* (2011) Invasion of Norway spruce diversifies the fire regime in boreal European forests. *Journal of Ecology*, **99**, 395–403.

Oishi, R., Abe-Ouchi, A., Prentice, I.C. and Sitch, S. (2009) Vegetation dynamics and plant CO_2 responses as positive feedbacks in a greenhouse world. *Geophysical Research Letters*, **36**, 5.

Olofsson, J., Hickler, T., Sykes, M.T. *et al.* (2008) Climate change impacts on European biodiversity – observations and future projections. MACIS (Minimisation of and Adaptation to Climate Change Impacts on Biodiversity). Deliverable 1.1.

Olson, J.S., Watts, J.A. and Allison, L.J. (1983) Carbon in live vegetation of major world ecosystems. Oak Ridge National Laboratory, Oak Ridge, TN, USA, ORNL-5862.

Orr, M. (2003) Environmental decline and the rise of religion. *Zygon*, **38**, 895–910.

Oswald, W.W. and Foster, D.R. (2012) Middle-Holocene dynamics of *Tsuga canadensis* (eastern hemlock) in northern New England, USA. *The Holocene*, **22**(1), 71–78.

Overballe-Petersen, M.V., Nielsen, A.B., Halsall, K. and Bradshaw, R.H.W. (2012) Long-term forest dynamics at Gribskov, eastern Denmark with early-Holocene evidence for thermophilous broadleaved tree species. *The Holocene*, **23**(2), 243–254.

Pacala, S.W., Canham, C.D. and Silander, J.A. Jr, (1993) Forest models defined by field measurements: I. The design of a northeastern forest simulator. *Canadian Journal of Forest Research*, **23**, 1980–1988.

Palmer, M. and Palmer, N. (1997) *Sacred Britain*. Judy Piatkus, London, UK.

Parducci, L., Jørgensen, T., Tollefsrud, M.M. *et al.* (2012) Glacial survival of boreal trees in Northern Scandinavia. *Science*, **335**, 1083–1086.

Parfitt, S.A., Barendregt, R.W., Breda, M. *et al.* (2005) The earliest record of human activity in northern Europe. *Nature*, **438**, 1008–1012.

Parker, A.G., Goudie, A.S., Anderson, D.E. *et al.* (2002) A review of the mid-Holocene elm decline in the British Isles. *Progress in Physical Geography*, **26**(1), 1–45.

Parmesan, C. (2007) Influences of species, latitudes and methodologies on estimates of phenological response to global warming. *Global Change Biology*, **13**, 1860–1872.

Parton, W.J., Scurlock, J.M.O., Ojima, D.S. *et al.* (1993) Observations and modeling of biomass and soil organic matter dynamics for the grassland biome worldwide. *Global Biogeochemical Cycles*, **7**(4), 785–809.

Partridge, T.R. (1995) *Interaction between Pingao and Marram on Sand Dunes*. Department of Conservation, Wellington, New Zealand

Peglar, S.M. (1993) The mid-Holocene *Ulmus* decline at Diss Mere, Norfolk, UK: a year-by-year pollen stratigraphy from annual laminations. *The Holocene*, **3**, 1–13.

Peglar, S.M. and Birks, H.J.B. (1993) The mid-Holocene *Ulmus* fall at Diss Mere, South-East England: disease and human impact? *Vegetation History and Archaeobotany*, **2**, 61–68.

Peñuelas, J. and Boada, M. (2003) A global change-induced biome shift in the Montseny Mountains (NE Spain). *Global Change Biology*, **9**, 131–140.

Peñuelas, J., Ogaya, R., Boada, M. and Jump, A.S. (2007) Migration, invasion and decline: changes in recruitment and forest structure in a warming-linked shift of European beech forest in Catalonia (NE Spain). *Ecography*, **30**, 829–837.

Peterken, G.F. (1996) *Natural Woodlands*. Cambridge University Press, Cambridge, UK.

Peterken, G.F. (1999) Applying natural forestry concepts in an intensively managed landscape. *Global Ecology and Biogeography*, **8**(5), 321–328.

Peterken, G.F. (2000) Rebuilding networks of forest habitats in lowland England. *Landscape Research*, **25**, 291–303.

Peterken, G.F. and Mountford, P. (1998) Long-term change in an unmanaged population of wych elm subjected to Dutch elm disease. *Journal of Ecology*, **86**(2), 205–218.

Petersen, K.S., Rasmussen, K.L., Rasmussen, P. and von Platen-Hallermund, F. (2005) Main environmental changes since the Weichselian glaciation in the Danish waters between the North Sea and the Baltic Sea as reflected in the molluscan fauna. *Quaternary International*, **133–134**, 33–46.

Pickett, S.T.A. and White, P.S. (1985) *The Ecology of Natural Disturbance and Patch Dynamics*. Academic Press, Orlando, FL, USA.

Pigott, C.D. and Huntley, J.P. (1981) Factors controlling the distribution of *Tilia cordata* at the northern limits of its geographical range. III. Nature and causes of seed sterility. *New Phytologist*, **87**, 817–839.

Pigott, C.D. and Walters, S.M. (1954) On the interpretation of the discontinuous distributions shown by certain British species of open habitats. *Journal of Ecology*, **42**, 95–116.

Postma-Blaauw, M.B., Bloem, J., Faber, J.H. *et al.* (2006) Earthworm species composition affects the soil bacterial community and net nitrogen mineralization. *Pedobiologia*, **50**(3), 243–256.

Power, M.J., Marlon, J., Ortiz, N. *et al.* (2008) Changes in fire regimes since the Last Glacial Maximum: an assessment based on a global synthesis and analysis of charcoal data. *Climate Dynamics*, **30**, 887–907.

Pöyry, J., Luoto, M., Heikkinen, R.K. *et al.* (2009) Species traits explain recent range shifts of Finnish butterflies. *Global Change Biology*, **15**(3), 732–743.

Prentice, I.C. (1985) Pollen representation, source area, and basin size: toward a unified theory of pollen analysis. *Quaternary Research*, **23**(1), 76–86.

Prentice, I.C. and Leemans, R. (1990) Pattern and process and the dynamics of forest structure: a simulation approach. *Journal of Ecology*, **78**, 340–355.

Prentice, I.C. and Helmisaari, H. (1991) Silvics of north European trees: compilation, comparisons and implications for forest succession modelling. *Forest Ecology and Management*, **42**, 79–93.

Prentice, I.C., Bartlein P.J. and Webb, T. III, (1991) Vegetation and climate change in eastern North America since the last glacial maximum. *Ecology*, **72**(6), 2038–2056.

Prentice, I.C., Cramer, W., Harrison, S.P. *et al.* (1992) A global biome model based on plant physiology and dominance, soil properties and climate. *Journal of Biogeography*, **19**, 117–134.

Prentice, I.C., Sykes, M.T. and Cramer, W. (1993a) A simulation model for the transient effects of climate change on forest landscapes. *Ecological Modelling*, **65**, 51–70.

Prentice, I.C., Sykes, M.T., Lautenschlager, M. *et al.* (1993b) Modelling global vegetation patterns and terrestrial carbon storage at the last glacial maximum. *Global Ecology and Biogeography Letters*, **3**, 67–76.

Prentice, I.C., Bondeau, A., Cramer, W. *et al.* (2007) Dynamic global vegetation modelling, quantifying terrestrial ecosystem responses to large-scale environmental change, in *Terrestrial Ecosystems in a Changing World* (eds P.D. Canadell, D.E. Pataki and L.F. Pitelka). Springer-Verlag, Berlin, Germany, pp. 175–192.

Prentice, I.C., Kelley, D.I., Foster, P.N. *et al.* (2011) Modeling fire and the terrestrial carbon balance. *Global Biogeochemical Cycles*, **25**, doi:10.1029/2010GB003906.

Price, C. and Asfur, M. (2006) Inferred long term trends in lightning activity over Africa. *Earth Planets and Space*, **58**, 1197–1201.

Price, T.D. and Bar-Yosef, O. (2011) The origins of agriculture: new data, new ideas – an introduction to supplement 4. *Current Anthropology*, **52**, S163–S174.

Pungetti, G., Oviedo, G. and Hooke, D. (eds) (2012) *Sacred Species and Sites: Advances in Biocultural Conservation*. Cambridge University Press, Cambridge, UK.

Purschke, O., Sykes, M., Reitalu, T. *et al.* (2012) Linking landscape history and dispersal traits in grassland plant communities. *Oecologia*, **168**(3), 773–783.

Rackham, O. (2003) *Ancient Woodland: Its History, Vegetation and Uses in England*. Castlepoint Press, Dalbeattie, UK.

Raffa, K.F., Aukema, B.H., Bentz, B.J. *et al.* (2008) Cross-scale drivers of natural disturbances prone to anthropogenic amplification: the dynamics of bark beetle eruptions. *BioScience*, **58**(6), 501–517.

Ralska-Jasiewiczowa, M., Nalepka, D. and Goslar, T. (2003) Some problems of forest transformation at the transition to the oligocratic/*Homo sapiens* phase of the Holocene interglacial in northern lowlands of central Europe. *Vegetation History and Archaeobotany*, **12**, 233–247.

Raskin, P., Gallopin, G., Gutman, P. *et al.* (1998) *Bending the Curve: Towards Global Sustainability*. PoleStar Series Report 8. Global Scenario Group, Stockholm Environmental Institute, Stockholm, Sweden.

Rasmussen, P. (2005) Mid- to late-Holocene land-use change and lake development at Dallund Sø, Denmark: vegetation and land-use history inferred from pollen data. *The Holocene*, **15**, 1116–1129.

Rasmussen, P. and Bradshaw, E. (2005) Mid- to late-Holocene land-use change and lake development at Dallund Sø, Denmark, chronology and soil erosion history. *The Holocene*, **15**, 1105–1115.

Raunkiaer, C. (1934) *The Life-Forms of Plants and Statistical Plant Geography*. Clarendon Press, Oxford, UK.

Reale, O. and Dirmeyer, P. (2000) Modeling the effects of vegetation on Mediterranean climate during the Roman Classical Period. Part I: Climate history and model sensitivity. *Global and Planetary Change*, **25**, 163–184.

Ridgwell, A.J. (2003) Implications of coral reef buildup for the controls on atmospheric CO_2 since the Last Glacial Maximum. *Paleoceanography*, **18**(4), 7:1–7:9.

Roberts, R.G., Flannery, T.F., Ayliffe, L.K. *et al.* (2001) New ages for the last Australian megafauna: continent-wide extinction about 46,000 years ago. *Science*, **292**, 1888–1892.

Rockström, J., Steffen, W., Noone, K. *et al.* (2009) A safe operating space for humanity. *Nature*, **461**, 472–475.

Roebroeks, W. and Villa, P. (2011) On the earliest evidence for habitual use of fire in Europe. *Proceedings of the National Academy of Sciences*, **108**, 5209–5214.

Rosegrant, M.W., Tokgoz, S. and Bhandary, P. (2013) The new normal? A tighter global agricultural supply and demand relation and its implications for food security. *American Journal of Agricultural Economics*, **95**(2), 303–309.

Rotherham, I.D. (2008) Lessons from the past – a case study of how upland land-use has influenced the environmental resource. *Aspects of Applied Biology*, **85**, 85–91.

Rothermel, R.C. (1972) *A Mathematical Model for Predicting Fire Spread in Wildland Fuels*. Intermountain Forest and Range Experiment Station, Ogden, UT, USA.

Rotmans, J. (1990) *IMAGE: An Integrated Model to Assess the Greenhouse Effect*. Kluwer Academic Publishers, Dordrecht, The Netherlands.

Rounsevell, M.D.A. and Metzger, M.J. (2010) Developing qualitative scenario storylines for environmental change assessment *WIREs Climate Change*, **1**, 606–619.

Rounsevell, M.D.A., Reginster, I., Araújo, M.B. *et al.* (2006) A coherent set of future land use change scenarios for Europe. *Agriculture, Ecosystems and Environment*, **114**, 57–68.

Rounsevell, M., Dawson, T. and Harrison, P. (2010) A conceptual framework to assess the effects of environmental change on ecosystem services. *Biodiversity and Conservation*, **19**(10), 2823–2842.

Rounsevell, M.D.A., Robinson, D.T. and Murray-Rust, D. (2012) From actors to agents in socio-ecological systems models. *Philosophical Transactions of the Royal Society B: Biological Sciences*, **367**(1586), 259–269.

Ruddiman, W.F. (2003) The Anthropogenic greenhouse era began thousands of years ago. *Climatic Change*, **61**, 261–293.

Ruddiman, W.F., Cruficix, M.C. and Oldfield, F.A. (2011a) Introduction to the early-Anthropocene Special Issue. *The Holocene*, **21**, 713.

Ruddiman, W.F., Kutzbach, J.E. and Vavrus, S.J. (2011b) Can natural or anthropogenic explanations of late-Holocene CO_2 and CH_4 increases be falsified? *The Holocene*, **21**, 865–879.

Rudolf, K. (1930) Grundzüge der nacheiszeitlichen Waldgeschichte Mitteleuropas. *Beihefte Zum Botanischen Centralblatt*, **47**, 111–176.

Rulli, M.C., Saviori, A. and D'Odorico, P. (2013) Global land and water grabbing. *Proceedings of the National Academy of Sciences*, **110**(3), 892–897.

Saarnak, C.F. (2001) A shift from natural to human-driven fire regime: implications for trace-gas emissions. *The Holocene*, **11**, 373–375.

Sachs, M.M. (2009) Cereal germplasm resources. *Plant Physiology*, **149**, 148–151.

Saetersdal, M., Birks, H.J.B. and Peglar, S.M. (1998) Predicting changes in Fennoscandian vascular-plant species richness as a result of future climatic change. *Journal of Biogeography*, **25**(1), 111–122.

Sala, O.E., Chapin, F.S. III, Armesto, J.J. *et al.* (2000) Global biodiversity scenarios for the year 2100. *Science*, **287**, 1770–1774.

Salick, J., Amend, A., Anderson, D. *et al.* (2007) Tibetan sacred sites conserve old growth trees and cover in the eastern Himalayas. *Biodiversity and Conservation*, **16**(3), 693–706.

Sanderson, M. (1999) The classification of climates from Pythagoras to Koeppen. *Bulletin of the American Meteorological Society*, **80**(4), 669–673.

Scheiter, S. and Higgins, S.I. (2009) Impacts of climate change on the vegetation of Africa: an adaptive dynamic vegetation modelling approach. *Global Change Biology*, **15**(9), 2224–2246.

Schumacher, S., Bugmann, H. and Mladenoff, D.J. (2004) Improving the formulation of tree growth and succession in a spatially explicit landscape model. *Ecological Modelling*, **180**, 175–194.

Schweiger, O., Biesmeijer, J.C., Bommarco, R. *et al.* (2010) Multiple stressors on biotic interactions: how climate change and alien species interact to affect pollination. *Biological Reviews*, **85**(4), 777–795.

Scott, A.C. and Glasspool, I.J. (2006) The diversification of Paleozoic fire systems and fluctuations in atmospheric oxygen concentration. *Proceedings of the National Academy of Sciences*, **103**, 10 861–10 865.

Segerström, U., Bradshaw, R.H.W., Hornberg, G. and Bohlin, E. (1994) Disturbance history of a swamp forest refuge in northern Sweden. *Biological Conservation*, **68**, 189–196.

Segurado, P. and Araújo, M.B. (2004) An evaluation of methods for modelling species distributions. *Journal of Biogeography*, **31**(10), 1555–1568.

Seidl, R., Schelhaas, M.-J. and Lexer, M.J. (2011) Unraveling the drivers of intensifying forest disturbance regimes in Europe. *Global Change Biology*, **17**, 2842–2852.

Senanayake, R. (1999) Voices of the earth, in *Cultural and Spiritual Values of Biodiversity, United Nations Environment Programme* (ed. D.A. Posey). Intermediate Technology Publications, London, UK.

Seppä, H. and Birks, H.J.B. (2001) July mean temperature and annual precipitation trends during the Holocene in the Fennoscandian tree-line area: pollen-based climate reconstructions. *The Holocene*, **11**, 527–539.

Seppä, H. and Hicks, S. (2006) Integration of modern and past pollen accumulation rate (PAR) records across the arctic tree line: a method for more precise vegetation reconstructions. *Quaternary Science Reviews*, **25**, 1501–1516.

Seppä, H., Hannon, G.E. and Bradshaw, R.H.W. (2004) Holocene history of alpine vegetation and forestline on Pyhakero Mountain, northern Finland. *Arctic, Antarctic and Alpine Research*, **36**, 607–614.

Seppä, H., Alenius, T., Bradshaw, R.H.W. *et al.* (2009) Invasion of Norway spruce (*Picea abies*) and the rise of the boreal ecosystem in Fennoscandia. *Journal of Ecology*, **97**, 629–640.

Settele, J., Penev, L., Georgiev, T. *et al.* (eds) (2010) *Atlas of Biodiversity Risk*. Pensoft Publishers, Sofia-Moscow, Russia.

Settele, J., Carter, T.R., Kühn, I. *et al.* (2012) Scenarios as a tool for large-scale ecological research: experiences and legacy of the ALARM project. *Global Ecology and Biogeography*, **21**(1), 1–4.

Shaw, M.R., Pendleton, L., Cameron, D.R. *et al.* (2011) The impact of climate change on California's ecosystem services. *Climatic Change*, **109**, 465–484.

Shiferaw, B., Prasanna, B.M., Hellin, J. and Bänzinger, M. (2011) Crops that feed the world 6. Past successes and future challenges to the role played by maize in global food security. *Food Security*, **3**, 307–327.

Shuman, B., Newby, P., Huang, Y. and Webb, T. III, (2004) Evidence for the close climatic control of New England vegetation history. *Ecology*, **85**(5), 1297–1310.

Shuman, B.N., Newby, P. and Donnelly, J.P. (2009) Abrupt climate change as an important agent of ecological change in the Northeast U.S. throughout the past 15,000 years. *Quaternary Science Reviews*, **28**, 1693–1709.

Sitch, S., Smith, B., Prentice, I.C. *et al.* (2003) Evaluation of ecosystem dynamics, plant geography and terrestrial carbon cycling in the LPJ dynamic global vegetation model. *Global Change Biology*, **9**, 161–185.

Skoglund, P., Malmström, H., Raghavan, M. *et al.* (2012) Origins and genetic legacy of Neolithic farmers and hunter-gatherers in Europe. *Science*, **336**, 466–469.

Skov, F. and Svenning, J.-C. (2004) Potential impact of climatic change on the distribution of forest herbs in Europe. *Ecography*, **27**(3), 366–380.

Slikkerveer, L.J. (1999) Ethoscience, 'Tek' and its application to conservation, in *Cultural and Spiritual Values of Biodiversity, United Nations Environment Programme* (ed. D.A. Posey). Intermediate Technology Publications, London, UK.

Smith, A.G. (1970) The influence of Mesolithic and Neolithic man on British vegetation: a discussion, in *Studies in the Vegetational History of the British Isles* (ed. D. Walker and R.G. West). Cambridge University Press, Cambridge, UK, pp. 81–96.

Smith, B., Prentice, I.C. and Sykes, M.T. (2001) Representation of vegetation dynamics in the modelling of terrestrial ecosystems: comparing two contrasting approaches within European climate space. *Global Ecology and Biogeography*, **10**, 621–637.

Smith, M.A., Williams, A.N., Turney, C.S.M. and Cupper, M.L. (2008) Human environment interactions in Australian drylands: exploratory time-series analysis of archaeological records. *The Holocene*, **18**, 389–401.

Smol, J.P., Birks, H.J.B. and Last, W.M. (eds) (2001) *Tracking Environmental Change Using Lake Sediments*. Kluwer Academic Publishers, Dordrecht, The Netherlands.

Sommer, R.S., Benecke, N., Lõugas, L. *et al.* (2011) Holocene survival of the wild horse in Europe: a matter of open landscape? *Journal of Quaternary Science*, **26**(8), 805–812.

Soulé, M.E. (1985) What is conservation biology? *BioScience*, **35**(11), 727–734.

Spangenberg, J.H., Bondeau, A., Carter, T.R. *et al.* (2012) Scenarios for investigating risks to biodiversity. *Global Ecology and Biogeography*, **21**(1), 5–18.

Sparks, T.H. (1999) Phenology and the changing pattern of bird migration in Britain. *International Journal of Biometeorology*, **42**, 134–138.

Sprugel, D.G. (1976) Dynamic structure of wave-generated *Abies balsamea* forests in the north-eastern United States. *Journal of Ecology*, **64**(3), 889–911.

Standish, R. (1960) *The First of Trees: The Story of the Olive*. Phoenix House, London, UK.

Steadman, D.W. (1993) Biogeography of Tongan birds before and after human impact. *Proceedings of the National Academy of Sciences*, **90**, 818–822.

Steadman, D.W. (1995) Prehistoric extinctions of Pacific Island birds: biodiversity meets zooarchaeology. *Science*, **267**, 1123–1131.

Steadman, D.W. and Martin, P.S. (2003) The late Quaternary extinction and future resurrection of birds on Pacific islands. *Earth-Science Reviews*, **61**, 133–147.

Steadman, D.W., Stafford, T.W., Donahue, D.J. and Jull, A.J.T. (1991) Chronology of Holocene vertebrate extinction in the Galapagos Islands. *Quaternary Research*, **36**, 126–133.

Steffen, W., Sanderson, R.A., Tyson, P.D. *et al.* (2004) *Global Change and the Earth System: A Planet Under Pressure*. Springer, Berlin, Germany.

Steffen, W., Crutzen, P.J. and McNeill, J.R. (2007) The Anthropocene: are humans now overwhelming the great forces of nature? *Ambio*, **36**(8), 614–621.

Stocker, A., Omann, I. and Jäger, J. (2012) The socio-economic modelling of the ALARM scenarios with GINFORS: results and analysis for selected European countries. *Global Ecology and Biogeography*, **21**(1), 36–49.

Stocks, B.J., Mason J.A., Todd, J.B. *et al.* (2002) Large forest fires in Canada, 1959–1997. *Journal of Geophysical Research*, **108**, doi:10.1029/2001JD000484.

Stolton, S., Shadie, P. and Dudley, N. (2013) *IUCN WCPA Best Practice Guidance on Recognising Protected Areas and Assigning Management Categories and Governance Types*. Best Practice Protected Area Guidelines Series No. 21. IUCN, Gland, Switzerland.

Stott, P.A., Stone, D.A. and Allen, M.R. (2004) Human contribution to the European heatwave of 2003. *Nature*, **432**(7017), 610–614.

Sugita, S. (2007a) Theory of quantitative reconstruction of vegetation. I. Pollen from large sites REVEALS regional vegetation composition. *The Holocene*, **17**(2), 229–241.

Sugita, S. (2007b) Theory of quantitative reconstruction of vegetation. II. All you need is LOVE. *The Holocene*, **17**(2), 243–257.

Sundseth, K. (2005a) Natura 2000 in the alpine region. Office for the Official Publications of the European Commission, Luxembourg.

Sundseth, K. (2005b) Natura 2000 in the Atlantic region. Office for the Official Publications of the European Commission, Luxembourg.

Sundseth, K. (2005c) Natura 2000 in the boreal region. Office for the Official Publications of the European Commission, Luxembourg.

Sundseth, K. (2005d) Natura 2000 in the continental region. Office for the Official Publications of the European Commission, Luxembourg.

Svenning, J.-C. and Skov, F. (2004) Limited filling of the potential range in European tree species. *Ecology Letters*, **7**(7), 565–573.

Swearer, D.K. (2001) Principles and poetry, places and stories: the resources of Buddhist ecology. *Daedalus*, **130**(4), 225–241.

Sykes, M.T. (1987) *Sand Dune Vegetation of Southern New Zealand*. PhD monograph, University of Otago, Dunedin, New Zealand.

Sykes, M.T. and Prentice, I.C. (1995) Boreal forest futures: modelling the controls on tree species range limits and transient responses to climate change. *Water, Air, and Soil Pollution*, **82**, 415–428.

Sykes, M.T. and Prentice, I.C. (1996) Climate change, tree species distributions and forest dynamics, a case study in the mixed conifer/northern hardwoods zone of northern Europe. *Climatic Change*, **34**, 161–177.

Sykes, M.T., Prentice, I.C. and Cramer, W. (1996) A bioclimatic model for potential distributions of north European tree species under present and future climates. *Journal of Biogeography*, **23**, 203–233.

Syphard, A.D., Radeloff, V.C., Keeley, J.E. *et al.* (2007) Human influence on California fire regimes. *Ecological Applications*, **17**(5), 1388–1402.

Tainter, J.A. (2006) Archaeology of overshoot and collapse. *Annual Review of Anthropology*, **35**, 59–74.

Tallavaara, M. and Seppä, H. (2011) Did the mid-Holocene environmental changes cause the boom and bust hunter-gatherer population size in eastern Fennoscandia? *The Holocene*, **22**(2), 215–225.

Tansley, A.G. (1935) The use and abuse of vegetational concepts and terms. *Ecology*, **16**(3), 284–307.

Tansley, A.G. (1939) *The British Islands and their Vegetation*. Cambridge University Press, Cambridge, UK.

Thevenon, F., Bard, E., Williamson, D. and Beaufort, L. (2004) A biomass burning record from the West Equatorial Pacific over the last 360 ky: methodological, climatic and anthropic implications. *Palaeogeography, Palaeoclimatology, Palaeoecology*, **213**, 83–99.

Thieme, H. (1997) Lower Palaeoithic hunting spears from Germany. *Nature*, **385**, 807–810.

Thomas, J.A., Simcox, D.J. and Clarke, R.T. (2009) Successful conservation of a threatened Maculinea butterfly. *Science*, **325**(5936), 80–83.

Thonicke, K., Spessa, A., Prentice, I.C. *et al.* (2010) The influence of vegetation, fire spread and fire behaviour on biomass burning and trace gas emissions: result from a process-based model. *Biogeosciences*, **7**, 1991–2011.

Thuiller, W. (2003) BIOMOD – optimizing predictions of species distributions and projecting potential future shifts under global change. *Global Change Biology*, **9**(10), 1353–1362.

Thuiller, W. (2004) Patterns and uncertainties of species range shifts under climate change. *Global Change Biology*, **10**, 2020–2027.

Thuiller, W., Lavorel, S., Araújo, M.B. *et al.* (2005) Climate change threats to plant diversity in Europe. *Proceedings of the National Academy of Sciences*, **102**(23), 8245–8250.

Tilman, D., Socolow, R., Foley, J.A. *et al.* (2009) Beneficial biofuels – the food, energy, and environment trilemma. *Science*, **325**, 270–271.

Tinner, W., Nielsen, E.H. and Lotter, A.F. (2007) Mesolithic agriculture in Switzerland? A critical review of the evidence. *Quaternary Science Reviews*, **26**, 1416–1431.

Tiunov, A.V., Hale, C.M., Holdsworth, A.R. and Vsevolodova-Perel, T.S. (2006) Invasion patterns of Lumbricidae into the previously earthworm-free areas of northeastern Europe and the western Great Lakes region of North America. *Biological Invasions*, **8**(6), 1223–1234.

Tollefsrud, M.M., Kissling, R., Gugerli, F. *et al*. (2008) Genetic consequences of glacial survival and postglacial colonization in Norway spruce: combined analysis of mitochondrial DNA and fossil pollen. *Molecular Ecology*, **17**(18), 4134–4150.

Tonkov, S. (2003) Holocene palaeovegetation of the northwestern Pirin Mountains (Bulgaria) as reconstructed from pollen analysis. *Review of Palaeobotany and Palynology*, **124**, 51–61.

Trombold, C.D. and Israde-Alcantara, I. (2005) Paleoenvironmental and plant cultivation on terraces at La Quemada, Zacatecas, Mexico: the pollen, phytolith and diatom evidence. *Journal of Archaeological Science*, **32**, 341–353.

Tsukada, M. (1982a) Late-Quaternary development of the *Fagus* forest in the Japanese archipelago. *Japanese Journal of Ecology*, **32**, 113–118.

Tsukada, M. (1982b) Late-Quaternary shift of *Fagus* distribution. *Botanical Magazine Tokyo*, **95**, 203–217.

Turner, C. (1970) The middle Pleistocene deposits at Marks Tey, Essex. *Philisophical Transactions of the Royal Society B: Biological Sciences*, **257**, 374–437.

Turner, J. (1962) The *Tilia* decline: an anthropogenic interpretation. *New Phytologist*, **70**, 328–341.

Tzedakis, P.C., Lawson, I.T., Frogley, M.R. *et al*. (2002) Buffered tree population changes in a Quaternary refugium: evolutionary implications. *Science*, **297**, 2044–2047.

Tzedakis, P.C., Pälike, H., Roucoux, K.H. and de Abreau, L. (2009) Atmospheric methane, southern European vegetation and low-mid latitude links on orbital and millennial timescales. *Earth and Planetary Science Letters*, **277**, 307–317.

UNDP (2013) Human Development Report 2013. http://hdr.undp.org/en/mediacentre/humandevelopmentreportpresskits/2013report/ [accessed 22 November 2013].

University of East Anglia (no date) Climate Research Unit data. http://www.cru.uea.ac.uk/data/ [accessed 22 November 2013].

Valdes, P. (2011) Built for stability. *Nature Geoscience*, **4**, 414–416.

Valsecchi, V., Finsinger, W., Tinner, W. and Ammann, B. (2008) Testing the influence of climate, human impact and fire on the Holocene population expansion of *Fagus sylvatica* in the southern Prealps (Italy). *The Holocene*, **18**(4), 603–614.

van Leeuwen, J.F.N., Froyd, C.A., van der Knapp, W.O. *et al*. (2008) Fossil pollen as a guide to conservation in the Galapagos. *Science*, **322**, 1206.

van Mantgem, P.J., Stephenson, N.L., Byrne, J.C. *et al*. (2009) Widespread increase of tree mortality rates in the western United States. *Science*, **323**, 521–524.

Vandermeer, J.H. (1972) Niche theory. *Annual Review of Ecology and Systematics*, **3**, 107–132.

Vannière, B., Power, M.J., Roberts, N. *et al*. (2011) Circum-Mediterranean fire activity and climate changes during the mid-Holocene environmental transition (8500–2500 cal. BP). *The Holocene*, **21**, 53–73.

Varner, J.M., Gordon, D.R., Putz, F.E. and Hiers, J.K. (2005) Restoring fire to long-unburned *Pinus palustris* ecosystems: novel fire effects and consequences for long-unburned ecosystems. *Restoration Ecology*, **13**(3), 536–544.

Vendramin, G.G., Fady, B., Gonzáles-Martínez, S.C. *et al*. (2008) Genetically depauperate but widespread: the case of an emblematic Mediterranean pine. *Evolution*, **62**(3), 680–688.

Vera, F.W.M. (2000) *Grazing Ecology and Forest History*. CAB International, Wallingford, UK.

Vigne, J.-D., Carrère, I., Briois, F. and Guilaine, J. (2011) The early process of mammal domestication in the Near East. *Current Anthropology*, **52**, S255–S271.

Visser, M. and Rienks, F. (2003) Shifting links: Climate change disrupts food chains. *Levende Natuur* **104**, 110–113.

Vohland, K., Hickler, T., Feehan, J. *et al*. (2010a) Priority setting for nature conservation, in *Atlas of Biodiversity Risk* (eds J. Settele, L. Penev, T. Georgiev, R. Grabaum, V. Grobelnik, V. Hammen, S. Klotz, M. Kotarac and I. Kuehn). Pensoft Publishers, Sofia-Moscow, Russia.

Vohland, K., Klotz, S. and Balzer, S. (2010b) Ecological networks as one answer to climate change, in *Atlas of Biodiversity Risk* (eds J. Settele, L. Penev, T. Georgiev, R. Grabaum, V. Grobelnik, V. Hammen, S. Klotz, M. Kotarac and I. Kuehn). Pensoft Publishers, Sofia-Moscow, Russia.

von Humboldt, A. and Bonpland, A. (2008) *Essay on the Geography of Plants*. The University of Chicago Press, Chicago, IL, USA and London, UK.

von Post, L. (1916) Skogsträdpollen I sydsvenska torfmosselagerföljder. *Geologiska Föreningens I Stockholm Förhandlingar*, **38**, 384–394.

Walther, G.-R., Berger, S. and Sykes, M.T. (2005) An ecological footprint of climate change. *Proceedings of the Royal Society B: Biological Sciences*, **272**, 1427–1432.

Walther, G.-R., Gritti, E.S., Berger, S. *et al.* (2007) Palms tracking climate change. *Global Ecology and Biogeography*, **16**, 801–809.

Walther, G.-R., Roques, A., Hulme, P.E. *et al.* (2009) Alien species in a warmer world, risks and opportunities. *Trends in Ecology and Evolution*, **24**(12), 686–693.

Wang, J. and Whitlock, M. C. (2003) Estimating effective population size and migration rates from genetic samples over space and time. *Genetics*, **163**(1), 429–446.

Wanner, H., Beer, J., Butikofer, J. *et al.* (2008) Mid- to Late Holocene climate change: an overview. *Quaternary Science Reviews*, **27**, 1791–1828.

Wardle, D.A., Walker, L.R. and Bardgett, R.D. (2004) Ecosystem properties and forest decline in contrasting long-term chronosequences. *Science*, **305**, 509–513.

Waton, P.V. (1983) *A Palynological Study of the Impact of Man on the Landscape of Central Southern England, with Special Reference to the Chalklands*. PhD thesis. University of Southampton, Southampton, UK.

Watt, A.S. (1947) Pattern and process in the plant community. *Journal of Ecology*, **35**, 1–22.

Webb, D.A. (1985) What are the criteria for presuming native status? *Watsonia*, **15**, 231–236.

Welzholz, J.C. and Johann, E. (2007) History of protected forest areas in Europe, in *COST Action E27 Protected Forest Areas in Europe – Analysis and Harmonisation (PROFOR): Results, Conclusions and Recommendations* (eds G. Frank, J. Parviainen, K. Vandekerhove, J. Latham, A. Schuck and D. Little). Federal Research and Training Centre for Forests, Natural Hazards and Landscape (BFW), Vienna, Austria.

White, J.E.J. (1997) The history of introduced trees in Britain, in *Native and Non-Native in British Forestry* (ed. P.R. Ratcliffe). Institute of Chartered Foresters, Edinburgh, UK, pp. 4–8.

Whitlock, C. and Tinner, W. (2010) Fire in the earth system: a palaeoperspective. *Pages*, **18**, 53–96.

Whitney, G.G. and DeCant, J.P. (2003) Physical and historical determinants of the pre- and post-settlement forests of northwestern Pennsylvania. *Canadian Journal of Forest Research*, **33**, 1683–1697.

Wild, R. and McLeod, C. (eds) (2008) *Sacred Natural Sites: Guidelines for Protected Area Managers*. IUCN, Gland, Switzerland.

Wilkins, J., Schoville, B.J., Brown, K.S. and Chazan, M. (2012) Evidence for early hafted hunting technology. *Science*, **338**(6109), 942–946.

Wilkinson, B.H. (2005) Humans as geologic agents: a deep-time perspective. *Geology*, **33**(3), 161–164.

Willerslev, E., Cappellini, E., Boomsma, W. *et al.* (2007) Ancient biomolecules from deep ice cores reveal a forested southern Greenland. *Science*, **317**(5834), 111–114.

Williams, J.W. and Jackson, S.T. (2007) Novel climates, no-analog communities, and ecological surprises. *Frontiers in Ecology and the Environment*, **5**, 475–482.

Williams, D.F., Peck, J., Karabanov, E.B. *et al.* (1997) Lake Baikal record of continental climate response to orbital insolation during the past 5 million years. *Science*, **278**, 1114–1117.

Williams, J.W., Jackson, S.T. and Kutzbach, J.E. (2007) Projected distributions of novel and disappearing climates by 2100 AD. *Proceedings of the National Academy of Sciences*, **104**(14), 5738–5742.

Williams, J.W., Blois, J.L. and Shuman, B.N. (2011) Extrinsic and intrinsic forcing of abrupt ecological change: case studies from the late Quaternary. *Journal of Ecology*, **99**, 664–677.

Willis, K.J., Braun, M., Sumegi, P. and Toth, A. (1997) Does soil change cause vegetation change or vice versa? A temporal perspective from Hungary. *Ecology*, **78**(3), 740–750.

Willis, K.J., Gillson, L. and Brncic, T.M. (2004) How virgin is virgin rainforest? *Science*, **304**, 402–403.

Wohlfarth, B., Björck, S., Possnert, G. and Holmquist, B. (1998) An 800-year long, radiocarbon-dated varve chronology from south-eastern Sweden. *Boreas*, **27**(4), 243–257.

Wolf, A., Møller, P.F., Bradshaw, R.H.W. and Bigler, J. (2004) Storm damage and long-term mortality in a semi-natural, temperate decicuous forest. *Forest Ecology and Management*, **188**, 197–210.

Wolf, A., Callaghan, T.V. and Larson, K. (2008) Future changes in vegetation and ecosystem function of the Barents Region. *Climatic Change*, **87**(1–2), 51–73.

Woodman, P., McCarthy, M. and Monaghan, N. (1997) The Irish quaternary fauna project. *Quaternary Science Reviews*, **16**, 129–159.

Woodward, F.I. (1987) *Climate and Plant Distribution*. Cambridge University Press, Cambridge, UK.

Woodward, F.I., Smith, T.M. and Emanuel, W.R. (1995) A global land primary productivity and phytogeography model. *Global Biochemical Cycles*, **9**, 471–490.

Wotton, B.M., Nock, C.A. and Flannigan, M.D. (2010) Forest fire occurrence and climatic change in Canada. *International Journal of Wildland Fire*, **19**, 253–271.

Wramneby, A., Smith, B. and Samuelson, P. (2010) Hot spots of vegetation-climate feedbacks under future greenhouse forcing in Europe. *Journal of Geophysical Research-Atmospheres*, **115** (D21119).

Wrangham, R. (2009) *Catching Fire: How Cooking Made Us Human*. Profile Books, London, UK.

Wright, H.E.Jr., Kutzbach, J.E., Webb, T. III, *et al.* (1993) *Global Climates since the Last Glacial Maximum*. University of Minnesota Press, Minneapolis, MN, USA.

Wu, S., Yin, Y., Zhao, D. *et al.* (2010) Impact of future climate change on terrestrial ecosystems in China. *International Journal of Climatology*, **30**(6), 866–873.

Wu, X., Zhang, C., Goldberg, P. *et al.* (2012) Early pottery at 20,000 years ago in Xianrendong Cave, China. *Science*, **336**, 1696–1700.

Zackrisson, O. (1977) Influence of forest fires on the north Swedish boreal forest. *Oikos*, **29**, 22–32.

Zeder, M.A. (2011) The origins of agriculture in the Near East. *Current Anthropology*, **52**, S221–S235.

Zhu, J. (1989) Nature conservation in China. *Journal of Applied Ecology*, **26**(3), 825–833.

Zimmermann, N.E., Edwards, T.C., Graham, C.H. *et al.* (2010) New trends in species distribution modelling. *Ecography*, **33**, 985–989.

Zobler, L. (1986) *A World Soil File for Global Climate Modelling*. NASA Technical Memorandum 87802. NASA Goddard Institute for Space Studies, New York, NY, USA.

Glossary

α The Priestley–Taylor coefficient of annual moisture availability – the ratio of actual to equilibrium evapotranspiration, a measure of the total annual drought stress that limits plant growth

Abies balsamea Balsam fir

ABMs (agent-based models) In this type of modelling, several agents or individuals within a system are modelled interacting together and the outcomes and the results of these interactions on the system as a whole are observed. An agent makes decisions or exhibits behaviours; it can be an individual, a household, an organisation, a plant or an animal, and it can be described by various characteristics, behaviours or constraints. Such a model can be used to explore spatial and temporal patterns of behaviour and influence on, for example, the properties of an ecosystem

Acer pseudoplatanus Sycamore

Acer saccharum Sugar maple

Acidification Loss of nutrient basic ions (calcium, magnesium, potassium) from soil or freshwater and their replacement by acidic ions (hydrogen and aluminium), lowering pH

Adaptive capacity The capacity that an ecosystem or a human social system has to adapt to environmental change

Aerosols Fine particles or droplets that are suspended in a gas, typically particles in the atmosphere that reflect sunlight

AFE (*Atlas Florae Europeaeae*) Individual maps of the European distribution of some plant species and subspecies on a 50 km grid. Published both digitally and in books by the Finnish Museum of Natural History, University of Helsinki, http://www.luomus.fi/english/botany/afe/ (last accessed 22 November 2013)

Ageratum conyzoides Goatweed

Albedo Shortwave solar radiation is reflected back to space from the Earth's surface, and albedo is a measure of the degree of reflectivity of the surface. Snow reflects large amounts of radiation and so has high albedo; dark surfaces absorb more solar energy and so have low albedo

Ecosystem Dynamics: From the Past to the Future, First Edition.
Richard H.W. Bradshaw and Martin T. Sykes.
© 2014 John Wiley & Sons, Ltd. Published 2014 by John Wiley & Sons, Ltd.
Companion Website: www.wiley.com/go/bradshaw/sykes/ecosystem

Alien species Species that have travelled long distances (intercontinental), usually with the help (intended or otherwise) of humans

Alkenones Long-chain organic compounds produced by phytoplankton, whose chemical structure is influenced by water temperature at formation and so can be used to reconstruct past sea surface temperature

Allerød A warm, moist period between about 14 000 and 12 900 years ago, during the end of the last glaciation. It is named after a Danish village and is most easily seen in sediments from northern Europe, but it is also recorded from Greenland ice cores

Alnus glutinosa Black alder

Alnus Alder species

Angiosperms Flowering plants, as distinguished from ferns, mosses and plants without flowers

Anthropocene A proposed division of time to reflect the global effect of human activity on ecosystems, beginning around the time of the Industrial Revolution

Anthropogenic The impact of humans, usually on the environment; e.g. anthropogenic climate change – climate change that can be attributed to the actions of humans

Apatite Calcium phosphate mineral

Arbutus unedo Strawberry tree

Ard A primitive plough that cuts a furrow in the earth but does not turn the soil

Aurochs Extinct ancestor of domestic cattle, recorded from Europe, Asia and North Africa. The last recorded aurochs died in Poland in AD 1627

Beech *Fagus*

Beltane A Celtic festival held on the last day of April, midway between the spring equinox and the summer solstice

Betula papyrifera Paper birch

Betula pubescens Downy birch

Betula Birch species

Bioclimatic envelope model A model of a species distribution mapped in space based on gridded data of various climate variables. Two approaches are discussed here. (1) The first uses variables (bioclimatic) that have clear physiological constraints on species, e.g. coldest month temperature. The distribution map produced is then compared to known natural distributional maps. (2) In another, more common approach, statistics can be used to select significant bioclimatic variables that best describe the species' current mapped distribution from a wide range of environmental variables provided. Once the bioclimatic limits are known, projections of future distributions can be made under different climate scenarios

Bioclimatic variables Variables, such as temperature of coldest month or amount of summer precipitation, that can be used to describe species range limits or the environmental envelope

Biodiversity Biological diversity – the number of different species (e.g. plants, animals, microorganisms) and the variation among them. This can cover either a region or the Earth as a whole. As a concept, it can refer to ecosystem, species and genetic diversity

Biogeochemistry model A model that simulates the transfer of carbon, water and other elements between the atmosphere and ecosystems

Biogeochemistry The study of the chemical cycling between the Earth and the atmosphere. Particularly of concern here are the cycling of carbon, water, nitrogen and other elements through ecosystems

Biomass The total mass of plant material produced in a given area

Biome An area or region on the Earth in which the environmental conditions are similar enough to produce a similar sort of ecosystem, e.g. the boreal forest biome or tropical forest biome

Biota The total number of plants and animals in a region

Biotemperature Based on annual growing season length and temperature

Bioturbation Mixing of the sediments, e.g. in a lake

Black box Used to describe a model where the actual workings are not explained

Boreal forest Conifer-dominated forest found in northern Europe, Asia and North America

Boreal Circumpolar northern temperate region dominated by conifers

Bronze Age An archaeological period characterised by the use of copper and bronze. It was preceded by the Stone Age and followed by the Iron Age

Business-as-usual scenarios Used in policy analysis over the short term, where policy effects are likely to dominate

C_3 plants Plants that carry out C_3 photosynthesis (almost all) by taking up carbon dioxide into a 3-carbon compound. RUBISCO, the enzyme involved in photosynthesis, is also the enzyme involved in the uptake of CO_2. Photosynthesis takes place throughout the leaf. More efficient than C_4 plants in cool and moist climates

C_4 plants Plants that carry out C_4 photosynthesis (very few) by taking up carbon dioxide into a 4-carbon compound. Uses PEP carboxylase as the enzyme in the uptake of CO_2. It then 'delivers' the CO_2 directly to RUBISCO for photosynthesis. C_4 plants are more efficient in hot, dry climates than C_3 plants

Calcite deposit Calcite is the most stable form of the mineral calcium carbonate, which is the major constituent of limestone

Calluna vulgaris Heather

Canonical correspondence A statistical technique that can summarise complex multivariate data in two or more dimensions

Canopy conductance The transfer of water to the atmosphere and carbon dioxide to vegetation at the level of the canopy, e.g. in the forest canopy

Canopy exchange The exchange of water, carbon and energy at the level of the vegetation canopy, e.g. forest canopy

Carbon cycle The exchange or flow of carbon between the atmosphere, the biosphere and the oceans

Carbon flux The flow of carbon from one pool to another, e.g. from the atmosphere to the terrestrial biosphere or vice versa

Carbon sequestration The removal from the atmosphere of carbon for long-term storage, e.g. in the wood of forest trees

Carbon storage *see Carbon sequestration*

Carya Hickory species

Cassiope tetragona White arctic heather

Chironomids Non-biting midges that look similar to mosquitoes

Choristoneura fumiferana Spruce budworm

Chronosequences A classic research tool in studies of succession, included in many introductory textbooks. They are series of sites that vary in age since formation or latest catastrophic disturbance and they have been used to infer dynamic ecosystem processes over millennial timescales. They come with a definite health warning

Cladocera A group of crustaceans called water fleas that includes *Daphnia*

Climax forest The endpoint in a succession, e.g. from bare or disturbed ground. A typical succession in a temperate climate would be from herbaceous vegetation such as grass to shrubs, to taller light-demanding (tree) species, to shade-tolerant large tree species that overtop the earlier vegetation, often excluding it. Forests are often a mosaic of climax vegetation and patches undergoing succession after disturbance. Climax vegetation is usually the tallest vegetation that an area can support in the local climate

Coleoptera A group of insects better known as beetles

Continentality The influence on climate of distance to an open ocean from a continental interior. Measures of continentality usually include the annual amplitude in average monthly temperatures

Corylus Hazel species

Dark respiration The release of carbon dioxide by plants, with and without the presence of light

Dehesa An agroforestry system from the Iberian Peninsula that incorporates grazing and forest products such as game, fungi, honey, cork and wood

Dendroctonus ponderosae Mountain pine beetle

Deva Spiritual forces or beings that are part of nature, probably derived from supernatural entities in Buddhism and Hinduism

DGVMs (dynamic global vegetation models) Models that aim to simulate through time at broad spatial scales the dynamics of the carbon and water cycles and their interactions with vegetation. They have been used to project potential vegetation and these cycles in the past and the future using gridded climate data from general circulation model (GCM) output

Diatoms Mostly unicellular aquatic algae whose cell wall is made of silica

DPSIR Provides a structure within which different indicators can be evaluated in terms of environmental policy and policy responses. There are five elements involved: *Drivers* are the underlying factors that cause environmental changes, e.g. increasing atmospheric greenhouse gases. These lead to *Pressures*, which are the actual variables that influence the environment, such as changes in temperature and precipitation. The *State* is that element of the environment that will be influenced by these changes, e.g. a specific crop production or species presence/absence and distribution, while the *Impacts* in this case would refer to food insecurity or biodiversity loss as a result. *Responses* include how society will respond to these impacts, for example by changing relevant policies

Drivers of ecosystem change The possible causes or forcing factors of ecosystem change

Dryas integrifolia White mountain avens

Early successional species Plant species that take advantage of disturbed ground. They are light-demanding species and usually grow quickly, reproduce and spread rapidly. Slower-growing shade-tolerant species that enter the same area at the same time or later usually overtop them, and the early successional species die out until the next disturbed or open patch of ground appears

Earth system models of intermediate complexity (EMIC) Models that describe the natural Earth system by simulating some of the important processes in a way that is more simplified than in complex models but more complex than in simple conceptual models

Ecological networks Ecological networks are one answer to climate change and landscape fragmentation. They can be core areas, buffer zones or corridors, providing an interconnecting network. They are often linear, such as hedgerows or rivers, but can be stepping stones, allowing for migration from one to the next

Ecological response surfaces Describe the way in which the abundance of a species is related to two or more environmental variables that have a physiological basis

Econometric model The application of statistical analysis to economic data, permitting the development of models

Ecophysiognomic life forms Combinations of plant structures that are related to environmental conditions such as broad-leaved, evergreen trees or stem-succulent shrubs

Ecosystem services Defined by the Millennial Ecosystem Assessment (MEA). This definition was enlarged as part of the EU-funded RUBICODE project (www.rubicode.net) to 'benefits that humans obtain from ecosystems that support, directly or indirectly, their survival and quality of life. These include provisioning, regulating and cultural services that directly affect people, and supporting services needed to maintain the direct services. They are a subset of ecosystem processes, which include roles that are not easily definable in terms of human needs' (enlarged from MEA, 2005)

Ecosystem A system that includes all living organisms (biotic factors) in an area and its physical environment (abiotic factors), functioning together as a unit. An ecosystem is made up of plants, animals, microorganisms, soil, rock, minerals, water sources and the local atmosphere, all interacting with one another. Word origin coined in 1930 by Roy Clapham, to denote the physical and biological components of an environment considered in relation to each other as a unit

Ecotone A boundary zone between two different sorts of plant community, e.g. the zone between a forest and a grassland. It can be a sharp boundary or more gradual, where species from both communities can be found

Edaphic factors The characteristics of a soil, such as texture (e.g. clay or silt), moisture content, nutrient level and pH

Eltonian niche A species' actual place in the environment and its relationships to other organisms

Empetrum hermaphroditum Crowberry

Equilibrium model A model that is not dynamic. It usually gives a snapshot of a species' or a biome's response to the environment, e.g. BIOME, STASH

Ericaceae The heather family

ESMs (Earth system models) Integrate climate models (GCMs) with ecosystem models (DGVMs) so that the vegetation and the associated flows of carbon and water can feedback to the atmosphere, which may then affect the climate system

Evapotranspiration A combination of evaporation (the transfer of water to the atmosphere) from surfaces, e.g. soil and the transpiration from plants (movement of water through the plant and its release as water vapour into the atmosphere through the stomata)

Exotic species A species that is promoted by some sort of human intervention or activity (accidentally or by design) and whose distribution is not 'just' a natural migratory response to climate. The movement of species from one continent to another, for example, is usually carried out by humans, accidentally or otherwise (see alien species)

Exploratory scenarios storylines Plausible storylines about different socio-economic developments that can be compared over the next century

Fagus sylvatica European beech

Fagus Beech species

Fire regimes Frequency and extent of burning. These are important characteristics of interactions between fire and the ecosystem, but which characteristics are described, measured or modelled depends on the spatial and temporal scales under study (Whitlock
et al., 2010)

Fire scars Scar formed on a tree when a passing fire kills a section of the stem

Foraminifera Single-celled, primarily marine organisms with calcium carbonate shells that
are preserved in sediment and can be used for climatic reconstructions

Forb A herbaceous flowering plant, excluding graminoids (grasses, sedges and rushes)

Forest certification FSC/PEFC Certification that timber has been produced to agreed
environmental and socially responsible standards. The Forest Stewardship Council
(FSC) is an international nongovernment organisation, while the Programme for the
Endorsement of Forest Certification (PEFC) was largely developed by the forest
industry itself

Forest dieback Death of trees over a large area without an easily observed direct cause

FPC (foliar projective cover) The area of ground covered by foliage directly above it

Fraxinus excelsior European ash

Fraxinus Ash species

Fundamental niche The potential niche of a species – based on the abiotic or environmental variables to which the species responds, such as temperature, humidity, light, soil
water and nutrients. Excludes the effect of competition with other species

Fynbos A highly biologically diverse region of shrub and heath vegetation found in the
Mediterranean climate of the Western Cape of South Africa

GCM (general circulation model) Usually refers to a global model (though regional models are available: RCMs) that simulates aspects of the climate system, including atmosphere, oceans and land surface interactions

GDDs (growing degree days) The annual total of the number of degrees per day over a
specified limit, usually 5 °C. They can be used to express the growing season length for
plants. Ecosystem models use them to define limits for a particular species or plant functional type (PFT)

Generalist species Species that are common and are tolerant of a wide range of environmental conditions. They can usually be found in a number of different plant communities

Genus A rank used in the classification of organisms. It forms the first part of the Latin
name of a species, e.g. *Homo* in *Homo sapiens*

Glacial refugia The locations of relict populations during glaciations of species that are
more widespread during interglacial periods

Graminoid Grasses, sedges and rushes

Greenhouse gases A gas in the atmosphere that contributes to greenhouse warming, e.g.
carbon dioxide, methane

Grinnellian niche Determined by the potential habitat or area in which a species lives

Habitat model A statistical approach to the description of species ranges in terms of their
environment or habitat

Heartwood The inner, dead section of woody tissue, as distinguished from the living, outer
sapwood

Henge A prehistoric circular site with an earthen bank and ditch, often containing a circle
of stones or posts

Herbaceous A non-woody plant that usually dies down at the end of the growing season

HFTs (human functional types) An HFT has a niche that is best described by its location and its function (e.g. a farmer) as well as traits or attributes, preferences, decision-making strategies, actions and responses

Hibiscus diversifolius Swamp hibiscus

Hindcasting Use of models to look into the past

Holocene The period of time since the end of the last glaciation, comprising about 11 500 years

Hominid The group consisting of all modern and extinct great apes (i.e. the modern humans, chimpanzees, gorillas and orang-utans, plus all their immediate ancestors)

Hominin The group consisting of modern humans, extinct human species and all of our immediate ancestors (including members of the genera *Homo*, *Australopithecus*, *Paranthropus* and *Ardipithecus*)

Hotspots of biodiversity An area or region where the level of biodiversity is high

Hutchinsonian niche A multidimensional hypervolume of environmental variables in which a species could potentially exist

In-field Cultivated land and hay meadows close to a farm where grazing is limited. Distinguished from the more distant out-fields, where extensive grazing is permitted

Integrated assessment model A model composed of various coupled submodels or modules that simulate the effects of various aspects of the environment, including climate, on the ecology and the socio-economy of selected regions within a single framework

Interglacial A period lasting thousands of years with warmer temperatures and less ice cover than during glaciations

Invasive species New species in a region that show rapid population increases and local dominance. They are often exotic species

Iron Age An archaeological period characterised by the use of iron. It was preceded by the Bronze Age and followed by the Mediaeval period

Island biogeography Study of the factors that affect species richness in isolated communities, typically small islands

Isoline maps A map where equal values are connected by lines called isolines

Isopoll A line on a map that connects points that have the same pollen deposition. *see Isoline maps*

Isotherm A line on a map that connects points that have the same temperature. *see Isoline maps*

Jägmästare Literally 'master of the hunt', but in reality a qualified Swedish forest manager

Juglans Walnut species

Juniperus communis Juniper

Keystone species Species, plant or animal, that play very important roles within the local ecosystem and whose absence can significantly affect the ecosystem

LAI (leaf area index) The area of green leaf in a canopy above a unit of ground area

Lambdina fiscellaria Hemlock looper

Lava tubes Small caves in volcanic lava, formed during cooling

LGM (last glacial maximum) Time period when most ice sheets were at their greatest extent during the last glaciation, c. 20 000–25 000 years ago

LIDAR (Light Detection and Ranging) surveys Surveys generating land cover data of extremely high resolution, from which tree canopy heights, leaf area and biomass can be measured and ancient field systems can be mapped based on small variations in ground elevation

Lithology Description and classification of soft sediments and rock

Little Ice Age Cold period with the advance of glaciers in e.g. Scandinavia and Switzerland, observed in the northern hemisphere between AD 1350 and 1850

Loess Accumulations of wind-blown mineral particles about 25–50 microns in diameter

Machair Type of sand-dune pasture found in wet and windy conditions

Macrocharcoal Larger charcoal fragments with minimum size 125–250 microns

Maintenance or autotrophic respiration The energy production and release of carbon dioxide by a plant required just for it to maintain itself

Mandala Traditional sacred art form in Buddhist and Hindu religions, representing the cosmos metaphysically or symbolically

Mechanistic model A model based on understanding and simulation of the actual processes or mechanisms found within a system

Megafauna Large animal species, often defined as those weighing more than 44 kg

Megapode A group of chicken-like birds with large feet

Mesolithic Middle Stone Age period characterised by stone tools and weapons with small blades and by the first settled communities

Metapopulation A group of separated populations that have limited exchange of individuals between them

Miami model Predicts net primary production (NPP) from annual average precipitation and temperature to produce the 'Miami model' global map of terrestrial NPP. So called as it was first presented at a symposium in Miami in 1971

Midden Piles of ancient domestic waste. Kitchen middens provide insight into past diet. Pack-rat middens are samples of local vegetation

Migrational lag A hypothesis suggesting that many species are dispersal-limited and there can be long delays in filling habitats that are climatically suitable for species presence

Milankovitch cycles The effects of cyclical changes in the Earth's orbit around the sun upon climate

Monocotyledons The major division of the flowering plants. Monocotyledon seedlings produce a single first leaf, while dicotyledons have two

Montreal models Models that predicted primary productivity from evapotranspiration. *see Miami model*

Multivariate statistics Statistics that use a number of input variables to produce a number of output variables

Narrative A qualitative approach to describing alternative futures

Neanderthals An extinct division of the genus *Homo* and the closest relative to *Homo sapiens*

NEE (net ecosystem exchange) Net carbon movement between an ecosystem and the atmosphere. The term is often used by climate modellers, who have an atmospheric view of carbon flows, so that uptake by ecosystems means NEE is negative (in terms of the atmosphere), while flow to the atmosphere means it is positive. *see NEP*

Neolithic New Stone Age. Directly follows the Mesolithic and identified by the onset of agriculture, the first pottery, woven materials and polished stone tools

NEP (net ecosystem productivity) The total carbon balance of an ecosystem. It is NEE plus carbon exchanges associated with e.g. leaching and volatilisation. Ecologists with an ecosystem perspective view carbon flows into ecosystems as positive, making NEP positive, while flow from the atmosphere makes it negative. *see NEE*

Net assimilation rate Measures photosynthetic activity. The rate of increase in total dry weight in a plant per unit leaf area

Niche The Hutchinsonian niche is a multidimensional hypervolume of environmental variables in which a species could potentially exist. It includes the idea of the fundamental niche, which is the potential niche of a species based on the abiotic or environmental variables to which it responds, such as temperature, humidity, light, soil water, nutrients etc., plus the realised or actual niche. This is the part of the fundamental niche a species actually occupies after it has been excluded from other parts by competition with other species

Niche-based model An equilibrium model that simulates the niche of a particular species based on various environmental variables such as temperature of the coldest month

Normative scenarios These relate to 'desired futures', so that the storyline describes events and relationships that lead from the current situation to a wished-for future, e.g. policy scenarios like the Convention on Biological Diversity or EU goals on renewable energy

Nothofagus Southern beech species

NPP (net primary production) Gross primary production (from photosynthesis) minus maintenance or autotrophic respiration

Oak *Quercus*

Olive *Olea europaea*

Ophiostoma ulmi Fungus associated with Dutch elm disease

Ostracods A group of crustaceans typically 1 mm in size found in marine and freshwater habitats

Ostrya European hop hornbeam species

Out-field *see In-field*

Oxygen isotopes The ratio of light ^{16}O to heavy ^{18}O, which varies in water, ice and the calcium carbonate and silica dioxide found in the shells of microorganisms preserved in sediments, coral reefs and speleothems. The oxygen isotope ratio can be used to reconstruct past temperatures. There is a global record of variation in this ratio, which has been divided into numbered stages based on marine records (e.g. marine oxygen isotope stage 7)

Oxyria digyna Mountain sorrel

Palaeobotanist A botanist who studies floras of the past using fossilised and preserved material

Palaeoclimate Past climate stretching back through the entire history of the Earth

Palaeoecologist An ecologist who studies relationships between organisms and their environment in the past

Palaeoecology A broad subject area comprising elements of palaeontology and palaeobotany, which are disciplines with long traditions, drawing on fossil material covering many millions of years

Palaeolithic The Stone Age. A long period of human history beginning with the first stone tools made by hominins and developing into the Mesolithic during the early Holocene

Palynology The study of pollen and spores used for the reconstruction of vegetation dynamics

Parameterise Assign values to model components using data

Partial least squares analysis Multivariate statistical analytical technique used to investigate the relationships between predicted and observed variables via linear regression

Pathogen A small infectious agent, e.g. virus, bacterium or fungus, that causes disease in a host organism

Percolation The movement and filtering of fluid through a porous material, e.g. water percolating through soil

PET (potential evapotranspiration) The amount of evapotranspiration that would occur if there were enough water available. Air and surface temperatures and insolation influence this, and PET (the demand) can be greater than the actual evapotranspiration, e.g. in dry areas

PFTs (plant functional types) Grouping of plants based on their functional characteristics and response to the environment, e.g. needleleaved conifers

Phenology Various events that occur in the annual life cycle of a plant or animal, e.g. a tree leafing out in the spring, flowering, fruiting and then dropping leaves. Each is a phenological stage in the annual cycle

Phenotype The outward appearance and development of an organism, including its observable traits

Phenotypic adaptation The outward (nongenetic) adaptation of an organism to its physical environment

Phenotypic plasticity The flexible response of an organism to its current environment. For example, a plant's response to shading might be to grow taller

Photosynthesis The process whereby plants convert light from the sun into carbohydrates, which are then used for energy for growth and reproduction

Phytogeography (plant geography) Understanding of the distribution of species or vegetation types through space based on their response to various environmental variables

Phytolith Microscopic structures made of silica that occur in plants and are preserved in sediments and archaeological sites. Different plant groups have phytoliths with characteristic shapes and sizes, which can be used to help reconstruct the composition of past vegetation

Phytosociology A field of study that explores the associations or relationships formed among species and the identification and classification of possible communities

Picea abies Norway spruce

Picea Spruce species

Pinus albicaulis Whitebark pine

Pinus banksiana Jack pine

Pinus cembra Arolla pine

Pinus contorta Lodgepole pine

Pinus palustris Longleaf pine

Pinus rubra Red pine

Pinus strobus White pine

Pinus sylvestris Scots pine

Pinus Pine species

Plant macrofossil The preserved remains of seeds, buds, leaves and other plant parts that are macroscopic (can be seen with the naked eye), which are often identifiable to species level

Plantago lanceolata Ribwort plantain

Podzol A type of infertile soil found in cool, wet climates on nutrient-poor parent material. Organic matter, iron and aluminium are leached from the upper layers and precipitated further down the soil profile

Populus Poplar species

Priestley–Taylor coefficient *see* α

Primaeval The natural state of e.g. forest prior to human intervention. Largely a theoretical concept that is not easy to recognise in the field

Protozoa Single-celled organisms that can move

Proxy data Observations of systems made to gain indirect records of e.g. climate. For example, tree-ring widths as a proxy for summer temperature

Pteridium esculentum Austral bracken

Quaternary The last c. 2 million years of geological time, with rapid changes in climate and extent of global ice cover

Quercus ilex Holm oak

Quercus robur Penduculate oak

Quercus rubra Red oak

Quercus Oak species

Radiocarbon dating System for dating organic matter that has died during the last c. 60 000 years. Dating is based on measurement of the remaining radioactive ^{14}C in the sample

Raingreen Trees that produce their leaves at the start of the wet season and lose them at the start of the dry season, when water becomes limited

Realised or actual niche The part of the fundamental niche actually occupied by a species after it has been excluded from other parts by competition with other species

Red list List of threatened species produced by the International Union for Conservation of Nature and Natural Resources

Refugia Sites where small populations survive adverse conditions. *see Glacial refugia*

Rhizopoda Group of single-celled organisms that includes the testate amoebae. *see Testate amoebae*

Ruderal An early colonising plant of bare or disturbed ground. Can be annual or short-lived (perennial)

Runoff The flow of water over the surface of the ground, usually caused by rain or snowmelt that has not percolated down into the soil

Salix arctica Arctic willow

Sapwood The living outer part (outside the vascular cambium) of a tree trunk

Sclerophyll vegetation Shrubs with leathery leaves often found in dry Mediterranean regions

Sclerophyllous Plants with hard, thick, leathery leaves. Have resistance to both water loss and herbivory by animals

Scolytus spp. Bark beetle

Scots pine *Pinus sylvestris*

Selaginella densa Prairie club-moss

SES (socio-ecological system) Any system composed of a societal (or human) subsystem and an ecological (or biophysical) subsystem

Shredding A traditional European tree-pruning method that removes the side branches of a tree for animal fodder

Silvics The attributes (e.g. height) that are used to describe a particular forest tree species

Slash-and-burn (swidden) A system of temporary cultivation in which trees are cleared by felling and burning. A short period of crop cultivation is then usually followed by grazing of domestic animals

SNG (seminatural grassland) A traditionally managed grassland that is maintained by hay-making or by light-to-moderate grazing of cattle, horses or sheep (this type of management may have taken place for hundreds of years). Little or no fertiliser is added. These extensive grasslands are usually highly species diverse, especially in comparison to modern intensively cultivated grasslands

Specialist species A species that is only found in a restricted habitat or in restricted environmental conditions

Species area curves The relationship between the number of plant species found in a particular area and the size of the area

Speleothems Stalagmites and stalactites

Stochastic process A process that includes a random element

Stomata Pores, most often found on leaves, through which CO_2, water vapour and O_2 move in and out of a plant. Surrounded by special cells – guard cells – which open and close them depending on the environmental conditions. They must be open in order for the diffusion of CO_2 into the leaf for photosynthesis to occur. They also allow water vapour to be released into the atmosphere

Stomatal densities The number of stomata that are found in a leaf per mm^2

Subfossil record The remains of plants and animals found in sediment where the dead organic material is still preserved and not replaced by minerals as in true fossils. Most Quaternary biological records are subfossil

Summergreen trees Trees that produce their leaves in the spring and lose them in the autumn

Taphonomy The study of processes, e.g. transport, burial, decomposition, that affect biological materials as they become incorporated into sediments

Taxon A unit of classification of organisms, e.g. species, genus

Terra preta Dark earth soils of anthropogenic origin in central Amazonia

Testate amoebae Unicellular animals with shells that live on peat surfaces. The shells are preserved in the peat and can be used to reconstruct the former water table fluctuations

Thermohaline collapse The thermohaline circulation is part of the global oceanic circulation, relating to the temperature and salinity of the water. The component that involves the Gulf Stream in the Atlantic flows from the equator along the coast of North America into the high latitudes and has significant warming effects on the climate, particularly of Europe. Some general circulation model (GCM) simulations of the future have suggested that as a result of anthropogenic climate change there may be a slowing of this circulation, which could lead to cooling, particularly in northwestern Europe

Thuja occidentalis Eastern white cedar

Tilia cordata Small-leaved lime tree

Tilia Lime species

Tipping point A situation in which an ecosystem experiences a shift to a new state, with significant changes to biodiversity and the services provided. It could be regional or global in scale

Totemic An object such as a plant, animal or rock that is given special significance or reverence in some way. Often thought to relate to pre-literate societies, but also applies in modern society

Transfer function A mathematical method for describing the response of a system to an input, e.g. the change in plant abundance in response to altered temperature

Tsuga canadensis North American hemlock

Tsuga Hemlock species

Ulex Gorse species

Ulmus Elm species

Ungulates Large mammals with hooves

Uniformitarianism A theory first proposed by Charles Lyell, stating that geological processes that alter the Earth, such as erosion and mountain building, operated in the same way in the past as they do today

Water cycle The hydrological cycle – the movement of water from the ocean to the atmosphere to the land surface and back to the ocean/the atmosphere. The processes involved include evaporation, condensation, precipitation, infiltration and runoff

Water flux *see Water cycle*

Weighted averaging A transfer function method in which the average value of an environmental variable is weighted by species abundance and used to reconstruct the variable in the past. *see Transfer function*

White-eye A group of small perching birds, mostly native in tropical regions

Yew *Taxus baccata*

Younger Dryas A cold period between about 12 900 and 11 700 years ago, during the end of the last glaciation. It is named after a flower that was abundant at the time and is most easily seen in sediments from the North Atlantic region and in Greenland ice cores

Index

Ecosystem Dynamics: From the Past to the Future, First Edition.
Richard H.W. Bradshaw and Martin T. Sykes.
© 2014 John Wiley & Sons, Ltd. Published 2014 by John Wiley & Sons, Ltd.
Companion Website: www.wiley.com/go/bradshaw/sykes/ecosystem

Printed and bound by CPI Group (UK) Ltd, Croydon, CR0 4YY

27/10/2024

14580306-0001